SYSTEMS RELIABILITY ASSESSMENT

ISPRA —
— COURSES

ON RELIABILITY AND RISK ANALYSIS

A series devoted to the publication of courses and educational seminars given at the Joint Research Centre, Ispra Establishment, as part of its education and training program. Published for the Commission of the European Communities, Directorate-General Telecommunications, Information Industries and Innovation, Scientific and Technical Communications Service.

The publisher will accept continuation orders for this series which may be cancelled at any time and which provide for automatic billing and shipping of each title in the series upon publication. Please write for details.

SYSTEMS RELIABILITY ASSESSMENT

Proceedings of the Ispra Course held at the Escuela Tecnica Superior de Ingenieros Navales, Madrid, Spain, September 19–23, 1988 in collaboration with Universidad Politecnica de Madrid

Edited by

A.G. COLOMBO

Commission of the European Communities, Joint Research Centre, Ispra Site, Ispra, Italy

and

A. SAIZ DE BUSTAMANTE

Universidad Politecnica de Madrid, Escuela Tecnica Superior de Ingenieros Navales, Madrid, Spain

KLUWER ACADEMIC PUBLISHERS

DORDRECHT / BOSTON / LONDON

ISBN-13: 978-94-010-6777-5 e-ISBN-13: 978-94-009-0649-5
DOI: 10.1007/978-94-009-0649-5

Publication arrangements by
Commission of the European Communities
Directorate-General Telecommunications, Information Industries and Innovation,
Scientific and Technical Communications Unit, Luxembourg

EUR 12924

Softcover reprint of the hardcover 1st edition 1990

Published by Kluwer Academic Publishers,
P.O. Box 17, 3300 AA Dordrecht, The Netherlands.

Kluwer Academic Publishers incorporates the publishing programmes of
D. Reidel, Martinus Nijhoff, Dr W. Junk and MTP Press.

Sold and distributed in the U.S.A. and Canada
by Kluwer Academic Publishers,
101 Philip Drive, Norwell, MA 02061, U.S.A.

In all other countries, sold and distributed
by Kluwer Academic Publishers Group,
P.O. Box 322, 3300 AH Dordrecht, The Netherlands.

Printed on acid-free paper

TABLE OF CONTENTS

SECTION 3 - HUMAN RELIABILITY AND UNCERTAINTY ANALYSIS

INTRODUCTION

This book presents models and methods for systems reliability assessment, human reliability analysis and uncertainty management. It includes fourteen contributions which are grouped into three sections. Section 1 deals with basic reliability methods and applications. The papers by Saiz de Bustamante and Perlado introduce the stochastic processes and the Monte Carlo method, respectively. Sanz Fermandez de Cordoba and Gonzales discuss important practical implications of the use of reliability methods. The former refers to the aerospace industry. The latter considers nuclear power plants.

Session 2 presents some advances in systems reliability techniques. The paper by Contini and Poucet illustrates the mathematical analysis of fault trees and event trees. It includes a discussion on the logical analysis of non-coherent fault trees and considerations on the major measures of criticality and importance of a component. The paper by Babbio is devoted to Petri nets. First, the formalism of this relatively new technique is given. Then, stochastic Petri nets are introduced as a tool to describe the behaviour of systems in time. Finally, by some fully developed examples, it is shown how this approach can be used to represent and evaluate complex stochastic systems. Limnios introduces the notion of failure delay systems and gives the lifetime structure for the evaluation of reliability measures. A reservoir is studied as an example of a failure delay system. The paper by Amendola deals with the DYLAM approach, which is a powerful tool to develop stochastic models of dynamic systems. The paper summarizes the principles on which DYLAM is based and reviews the principal applications made to assess the reliability and safety of complex dynamic systems. The paper by Izquierdo and Sanchez-Perea illustrates the DYLAM-TRETA approach, which is a development of DYLAM. In particular, it includes a package for the numerical solution of ordinary non-linear differential equations. The paper discusses the safety assessment of protection systems.

Session 3 is devoted to human reliability and uncertainty analysis. The paper by Amendola reviews methods for human reliability assessment. Emphasis is given to the cognitive models that have been developed to simulate the behaviour of an operator in his decision making and actions. The paper by De Keyser and Woods is concerned with fixation errors. They discuss the identification of the main descriptive patterns of fixation errors and how to build new systems to reduce this type of error. Carbolino reviews and compares the most important theories of quantitative reasoning under uncertainty currently implemented in expert systems: probability theory, belief functions theory and possibility theory. He points out the difference between uncertainty due to partial ignorance of the information sufficient to establish the truth-value of a statement and uncertainty due to the vagueness of the meaning of a statement. The consequences of this difference are then discussed and stressed. The paper by Colombo deals with the estimation of reliability parameters by combining data from various sources. The problem is discussed in a Bayesian framework, and three typical situations are considered: combination of expert opinions, combination of expert opinions and experimental data, and combination of parameters. The last paper, by Guedez Soares, gives an overview of methods to quantify and combine uncertainty in a system reliability analysis.

The Editors

SECTION 1

BASIC RELIABILITY METHODS AND APPLICATIONS

STOCHASTIC PROCESSES, MARKOV CHAINS AND SYSTEMS AVAILABILITY

A. Saiz de Bustamante
Universidad Politécnica de Madrid
E-28040 Madrid

ABSTRACT. The principles of random processes are applied to the Poisson process, as the simplest case of a Markov chain. Later on the theory of Markov modelling is reviewed. Availability of systems is presented as result of system behaviour as a finite stationary Markov chain. It is assumed that systems satisfy the Markov and homogeniety property.

1. INTRODUCTION

The availability of a system is the probability to perform its intended function at any time during its life.

By a "system" it is meant a group of components that work together to accomplish an specific function -synergy-.

A failure is the inability of a system (component) to perform its intended function, and may occur as a result of defects in the system (component), wear, tear, or because unexpected stresses.

A failure is classified as catastrophic if the failure is complete and sudden.

A degradation failure is a partial and gradual failure.

A failure is complete when the deviation in characteristics of the system (component) is such as to cause complete lack of the required function.

The maintainability of a system is the probability of restoring it to specific condition after its failure within a given period of time when the maintenance is performed in accordance with prescribed procedures.

Maintenance is defined as all the actions necessary to restore a system (component) to specified condition.

When a system component fails, or is repaired, the system state -X- changes. This process depends on time and or hazard. Therefore system behaviour can be modelled by an stochastic process or random function as analyzed at section 2.

Most engineering system can be treated as a finite stationary Markov chain -section 4- assuming,

A. G. Colombo and A. Saiz de Bustamante (eds.), Systems Reliability Assessment, 3–22.

(i) Markov property: The future system state depends on the present state and is not affected by the state of any past time.

(ii) Homogeniety or stationary property: The state transition probabilities depend only on time differences and not on initial time.

The analysis of maintained systems is done on section 5, as a Markov chain, which combines system components states with system logic.

(iii) Notation

X	Random variable
$X(t)$	System state
$X(t), t \geq 0$	Random process (continuous time)
$X(n), n \in \mathbb{N}$	Random process (discrete time)
F_X	Probability distribution function
λ	Process intensity; Failure rate
μ	Repair rate
P_i	Probability that the system assumes the state i
$\underline{P}(P_1, \ldots, P_m)$	State probability vector
P_{ij}	Transition probability from state i to state j
$\mathbf{P} = \{P_{ij}\}$	Transition matrix
q_{ij}	Transition rate from state i to state j
$\mathbf{A} = \{q_{ij}\}$	Transition rate matrix
$\underline{\Pi}(\pi_1, \ldots, \pi_m)$	Steady state probability vector
δ_{ij}	Kronecker delta.

2. RANDOM PROCESSES

A random process $X(t,\omega)$ on a probability space (Ω, ζ, P) define a function $T * \Omega \to E$, being T the index set, E the state space, and Ω the sample space, whose elements or points are represented by ω. ζ is a σ-field of events defined on Ω, and P a probability measure.

When a r.v. or stochastic procecess is represented, we normally make abstraction of the variable ω -see notation-.

A random process is characterized by its state space E, or range of its possible values; its index set, or range of parameter values and its dependence relationship.

Thus the two dimentional function $X(t,\omega)$ represents a random variable for fixed t; and a real function for fixed ω.

Both index -t-, or states, X, can be discrete or continuous.

A stochastic process is statistically specified by its nth order probability distribution function.

$$F_X(x_1, x_2, \ldots, x_n ; t_1, t_2, \ldots t_n) =$$

$$= P[X(t_1) \leq x_1, X(t_2) \leq x_2, \ldots, X(t_n) \leq x_n]$$

for every positive n.

If the random process states are discrete, as in normally the case of Reliability Engineering, the probability measure is given by n^{th} order probability mass function (PMF),

$$P\ (x_1 x_2 \ldots x_n ; t_1 t_2 \ldots t_n) = P[X(t_1)=x_1, X(t_2)=x_2, \ldots X(t_n)=x_n]$$

for every positive n.

A stochastic process is said to have independent increments if for all $t_1 < t_2 \ldots < t_n$, and all $n \geqslant 1$, the n random variables,

$$X(t_1)\ ;\ X(t_2) - X(t_1)\ ;\ \ldots\ ;\ X(t_n) - X(t_{n-1})$$

are independents.

If a random process has independent increment the nth PMF can be described as products of the probability of its increments.

A random process is said to be stationary if

$$F_X(x_1, x_2, \ldots, x_n ; t_1, t_2, \ldots, t_n) =$$

$$F_X(x_1, x_2, \ldots, x_n ; t_1 + u, t_2 + u, \ldots, t_n+u)$$

for all $n \geqslant 1$, and u.

An example of a discrete states, continuous parameter random process is a Poisson process, defined as an independent increment process whose increment are Poisson distributed, as explained in the next section.

3. POISSON PROCESS

Let $X(u, u + t)$ a random process denoting the number of events at the interval $(u, u + t)$, which satisfies the following assumptions:

(i) The number of events in non overlapping intervals are independent r.v.

(ii) The probability of a single event at interval Δt, is proportional to Δt,

$$P_X(1; t, t + \Delta t) = \lambda \Delta t \qquad \text{as } \Delta t \to 0$$

being λ the process intensity.

(iii) The probability of k - events ($k > 1$) at interval Δt

$$P_X(k; t, t + \Delta t) = 0, \qquad \text{as } \Delta t \to 0$$

then X(t) is a Poisson process.

Assumption (i) refers to the independent increment property; (ii) includes the concept that the events are rare and (iii) excludes the possibility of simultaneous ocurrance of more than one event.

At the Poisson process $T = \mathbb{R}+$ and $E = \mathbb{N}$.

To determine the probability of k events at the interval $(u, u+t]$, the application of the three assumptions to $(u, u+t + \Delta t)$ gives

$$\frac{dP(k; u, u+t)}{dt} = \lambda(u+t)\left[P(k-1; u, u+t) - P(k; u, u+t)\right]$$

$$k = 0, 1, \ldots$$

A recursire first order system of linear diferential equations whose solution,

$$P_X(k; u, u + t) = \frac{\Lambda^k}{K!} e^{-\Lambda}$$

where

$$\Lambda = \int_u^{u+t} \lambda(t)dt \quad ; \quad \lambda(t) \geqslant 0$$

If λ is constant the X(t) is a Poisson homogeneous process, and

$$P_X(k; u, u+t) = \frac{(\lambda t)^k}{K!} e^{-\lambda t}$$

Therefore in a homogeneous Poisson process the number of events occurring between $(s, s+t]$, depends on t, the interval lenght, and not on the interval origen.

Example 1

Defects occur along an undersea cable according to a homogeneous Poisson process of rate $\lambda = 0.05\ \mathrm{Km}^{-1}$. Calculate:

a) Probability of zero defects in the first 3 Km.
b) Probability of zero defects in the second 3 Km, assuming no defects in the first 3 Km.
c) Probability of three defects in 6 Km assuming 1 defect in the first 3 Km.

Solution:

a) $P(0\ ;\ 0,3) = P\ (0;3) = e^{-0.05\cdot 3} = \underline{0.8607}$

b) $P(0\ ;\ 3,6) = P\ (0;3) = \underline{0.8607}$

c) $P(2;3) = (0.05.3)^2\ e^{-0.05\cdot 3}\ /\ 2 = \underline{0.0097}$

The probabilities of a nonhomogeneous Poisson process can be calculated as a homogeneous Poisson process using Λ as the average.

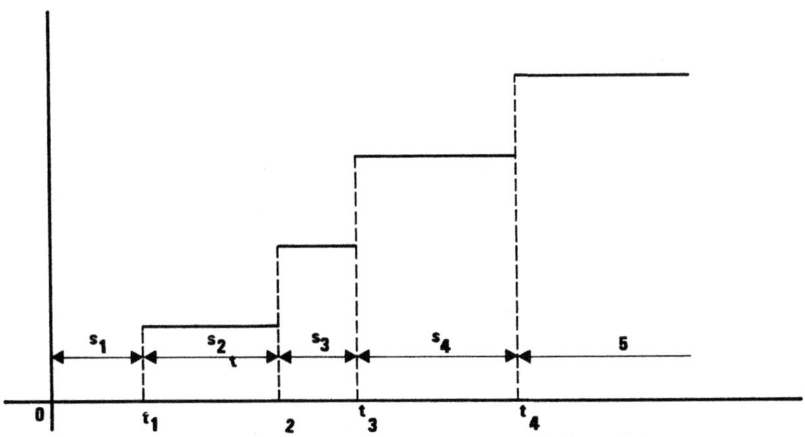

Fig.1 Poisson sample function, showing arrival times -t_n-, and interarrival times -s_n-.

Figure 1 shows a uniform Poisson sample function with indication of the two associated random sequences: arrival times - t_n - and interarrival times - s_n -.

The arrival time T_n is the time of ocurrence of the nth event, so the event $t < T_n \leqslant t + \Delta t$, corresponds to n-1 ocurrences in $(0,t]$, and one in $(t, t+\Delta t]$. Then the probability density function of T_n,

$$f_{T_n}(t)\ \Delta t = P\left[t < T_n < t + \Delta t\right] = $$
$$= P\left[X = n-1;\ 0,t\right] P\left[X = 1\ ;\ t,\ t+\Delta t\right]$$

$$= \frac{(\lambda t)^{n-1} e^{-\lambda t}}{(n-1)!} \lambda \Delta t$$

or

$$f_{T_n}(t) = \lambda \frac{(\lambda t)^{n-1}}{(n-1)!} e^{-\lambda t}$$

being

$$\mu(T_n) = \frac{n}{\lambda} \quad ; \quad \mathrm{Var}\,(T_n) = \frac{n}{\lambda^2}$$

Therefore the arrival times - T_n - have a gamma distribution with parameters λ and n-1.

The interrarrival times - $S_n = T_{n-1} - T_n$ -, measures the sojour time at state n. The s_o, s_1, ... s_n are independent r.v. with an exponential probability density function with parameter λ. For example s_1.

$$f_{S_1}(t)\,\Delta t = P\,[X=0;\ 0,t]\,P\,[X = 1;\ t,\ t+\Delta t] = \lambda e^{-t}\Delta t$$

4. MARKOV PROCESSES

4.1 Definitions

A Markov process is a random process - X(t) - such that given the value of X(t) -state at time t-, the values of X_u for $u > t$, are not influenced by the values of X_s, for $s < t$. Therefore the future state of Markov process does not depend on the past behaviour of the process.

If the number of states are finite or countable infinite, the Markov process is a Markov chain.

The Markov property applied to a Markov chain, can be written as

$$P_X(x_n \mid x_{n-1}, x_{n-2}, \ldots, x_1;\ t_n, t_{n-1}, \ldots, t_1) =$$

$$P_X(x_n \mid x_{n-1};\ t_n, t_{n-1})$$

For all $x_1, x_2, \ldots, x_n$; for all $t_1, t_2, \ldots, t_n$; and for all $n > 0$.

The conditional probabilities represent state transition probabilities.

If $P_x(x_n\ x_{n-1};\ t_n, t_{n-1})$ depends only on time interval, $t_n - t_{n-1}$, the Markov chain is homogeneous.

For an homogeneous Markov chain

$$\lim_{n\to\infty} P_X\,(X_n \mid x_{n-1};\ t_n, t_{n-1})$$

exits, and therefore the process will always have an steady state distribution.

For simplicity we will adopt the following convention:

(i) To label the state space with a set of integers i, preferably strict positive,

$$x(t) = i$$

meaning that the process is at state i if x = i at time t.

(ii) To name probability functions:

$$\text{(a)}\quad P_i(t) = P[X(t) = i]$$

the state probability or unconditional probability that the random process is at state i at time t.
A vector $\underline{P}(t)$ with elements $P_i(t)$ is called the State Probability Vector.

$$\text{(b)}\quad P_{ij}(t,u) = P[X(u) = j \mid X(t) = i]$$

satisfing the conditions,

$$P_{ij} \geqslant 0, \; i,j = 1, 2, \ldots, m$$

$$\sum_{j=1}^{m} P_{ij} = 1 \qquad j = 1, 2, \ldots, m$$

being m the number of states and expresing the last condition that some transition should occur at time t, understanding that a transition has occurred even if the process continue at the same state. Non transition probabilities P_{ii}, $i = 1, 2, \ldots, m$.

<u>Transition probability</u> or conditional probability that the process is at state j at time u given that is at state i at time t, and u > t.

A matrix P (t,u) with elements $P_{ij}(t,u)$ is called the <u>Stochastic or Transition Matrix</u> of the process.

The Markov chain is homogeneous or stationary if transition probabilities depends only on time interval.

$$P_{ij}(t,u) = P_{ij}(u-t)$$

A Markov chain can be a discrete or continuous time Markov process.

For example the Poisson process is a discrete state, continuous time Markov process.

If can be shown that any random process with the independent increment property is a Markov chain. The inverse argument is not true.

A Markov chain can be represented by its state transition diagram (see figure 2), where the nodes represent the process states, and the arrows the possible transitions from state to state.

4.2 Discrete time Markov chains

At a discrete time stationary finite Markov chain $\{X(n),\ n \in \mathbb{N}\}$ transitions between states ocur only at discrete times or steps.

We define X(n) as the state of the system at n-step.

According to definitions,

$$p_{ij}(n) = P[X(n) = j \mid X(0) = i]$$

which means the probabilities for the process goes from state i to state j in n-steps or transitions,

$$\underline{p}(n) = [p_1(n),\ p_2(n), \ldots, p_m(n)]$$

which gives the probabilities for the m states of the process of the process after n-steps.

The n-step transitions probabilities can be expressed, by the two Chapman-Kolmogorov theorems,

$$\text{(I)}\ p_{ij}(n) = \sum_{k=1}^{m} p_{kj}(n-r)p_{ik}(r)$$

$$\text{(II)}\ p_j(n) = \sum_{k=1}^{m} p_{kj}(n-r)p_k(r)$$

which can be proofed breaking down the possible transitions on the r-step, followed by the application of the total probability law and the Markov property.

Therefore a finite state stationary Markov chain is completely defined by its stochastic matrix and it its initial state.

The first Chapman-Kolgomorov theorem express the law of matrix multiplication, so the n-step transition matrix,

$$\mathbf{P}\ (n) = \mathbf{P}\ .\ \mathbf{P} \ldots \mathbf{P} = \mathbf{P}^n$$

and

$$\underline{p}(n) = \underline{p}(o)\ \mathbf{P}^n$$

A Markov chain or its stochastic matrix is ergodic if,

$$\lim_{n\to\infty} \mathbf{P}\ (n) = \mathbf{P}^n_{n\to\infty} = \Gamma$$

exists.

A stochastic matrix is regular if one of its power contains only positive elements. If this is the case, then 1 is an eigenvalue of multiplicity one, and all the other eigenvalues λ_i, satisfy $\lambda_i < 1$.

A regular matrix is ergodic and the rows of $\mathbf{P}(\sim) = \Gamma$ are equal to the left eigenvector associated to the eigenvalue 1. Then the steady state probability state vector of the process $\underline{\pi}$,

$$\underline{P}(\sim) = \lim_{n \to \sim} \underline{P}(o)\, \mathbf{P}^{\,n} = \underline{P}(o)\, \Gamma = \underline{\pi}(\pi_1, \pi_2, \ldots, \pi_n)$$

Therefore the steady state probability state vector is independent of the initial state, and can be calculated using the formula

$$\underline{\pi} = \underline{\pi}\, \mathbf{P}$$

Example 2

Consider a protection system periodically tested. For instance a presure relieve value monthly tested.

Assuming that repair is proceeded only by test, such system can be modelled as a two stage homogenious Markov chain.

Let $X(n) = 1$ means operating; $X(n) = 2$ means failure and P the transition matrix,

$$\mathbf{P} = \begin{bmatrix} 0.99 & 0.01 \\ 0.14 & 0.86 \end{bmatrix}$$

If the system is operating at initial condition, calculate:

a) Probability that the system is operating after 16 steps or periods.

b) Long run probability that the system is at failure condition.

Solution

a) $$\underline{P}(16) = \underline{P}(0) \cdot \mathbf{P}^{\,16} =$$

$$= \begin{bmatrix} 1, 0 \end{bmatrix} \begin{bmatrix} .9383 & .0617 \\ .840 & .1359 \end{bmatrix} = \begin{bmatrix} .9383\ , & 0.0617 \end{bmatrix};$$

b) $$\begin{bmatrix} 1, & 2 \end{bmatrix} \begin{bmatrix} 0.99 & 0.01 \\ 0.14 & 0.86 \end{bmatrix} = \begin{bmatrix} 1, & 2 \end{bmatrix}$$

$$\pi_1 + \pi_2 = 1$$

therefore the long run probabilities,

$$\pi_1 = 0.9333 \qquad \pi_2 = 0.0667$$

It is important to observe that the steady state probabilities are independent of initial conditions.

4.3 Continuous time Markov chain

On a continuous time stationary finite Markov chain $(X(t), t \geq 0)$, transitions between states ocur at any time.

We define X(t) as the state of the system at time t.

The transitions probabilities satisfy,

a) $0 < P_{ij} < 1 \quad ; \quad \sum_{j=1}^{m} P_{ij} = 1$

b) Markov property:

$$P_{ij}(t,u) = P[X(t) = j | X(u) = i]$$

c) Homogeniety property

$$P_{ij}(t,u) = P_{ij}(t-u)$$

d) Chalman-Kolgomorov theorem:

$$\mathbf{P}(t+u) = \mathbf{P}(t) \cdot \mathbf{P}(u)$$

e) Continuity of $\mathbf{P}(t)$ or stochastic continuity, postulating $\lim_{t \to o} P_{ij}(t) = \delta_{ij}$

If the process satisfies (e), $\mathbf{P}(t)$ is differentiable, and the limits

$$\lim_{t \to o} \frac{P_{ij}(\Delta t)}{\Delta t} = q_{ij} \qquad i \neq j$$

$$\lim \frac{1-P_{ij}(\Delta t)}{\Delta t} = -q_{ii}$$

exist and define the process on an infinitesimal time interval.

The stochastic process is conservative if

$$q_{ii} = -\sum_{j=1}^{m} q_{ij} \qquad i \neq j$$

$q_{ij}\Delta t$ represents the probability of a system transition from state i to state j at time interval Δt, $\Delta t \to 0$.

q_{ii} represents the non transition rate at state i.

If the number of states -m-, is finite the Markov chain is always conservative; but an infinite Markov chain can be non conservative.

Assuming a conservative process, a transition rate matrix or Markov chain generator

$$\mathbf{A} = \lim_{\Delta t \to 0} \frac{\mathbf{P} - \mathbf{I}}{\Delta t} = \{ q_{ij} \}$$

exist being **I** the unit matrix.

We can write the state probability at time t +Δt in turn of the state probability at time t,

$$\underline{P}\,(t+\Delta t) = \underline{P}(t)\ \mathbf{P}\ (t+\ \Delta t)$$

Rearranging

$$\frac{\underline{P}(t+\Delta t) - \underline{P}(t)}{\Delta t} = P(t) = \frac{\mathbf{P}\ (t+\Delta t) - \mathbf{I}}{\Delta t}$$

to obtain,

$$\frac{d\underline{P}(t)}{dt} = \underline{P}(t)\ .\ \mathbf{A}$$

The solution of the differential equation in the time domain,

$$\underline{P}(t) = \underline{P}(o)\ \exp \mathbf{A}\ t$$

and in the s-domain

$$\underline{P}(s) = [\underline{P}(o)]\ [s\ \mathbf{I} - \mathbf{A}]^{-1}$$

The steady state can be obtained by setting to zero the system equation,

$$\frac{d\underline{P}(t)}{dt} = \underline{P}(t).\ \mathbf{A} = 0$$

Example 3

The simplest Markov chain is a Poisson Process of λ intensity.

Write the transition matrix and transition rate matrix of a Poisson proces $(X(t),\ t \geqslant 0)$ of λ intensity.

Solution

The state space : $E = \mathbb{N}$

(i) Transition probabilities

$$P_{ij}(t) = P\ [X(t) = j \mid X(t) =]i \ = e^{-\lambda t}\ .\frac{(\lambda t)^{j-i}}{(j-i)!}\ ; \quad j \geqslant i$$

$$= 0 \qquad ; \quad j < i$$

The stochastic property (e) is satisfied.

$$\lim_{t\to o} P_{ij}(t) = \delta_{ij}$$

(ii) Transition rates:

$$q_{ij} = \lim_{t\to o} \frac{P_{ij}(\Delta t)}{\Delta t} = e^{-\lambda\Delta t} \frac{(\lambda\Delta t)^{j-1}}{t(j-i)!}$$

$$\text{If } j=i+1 \text{ , } q_{ii+1} = \lambda$$
$$j>i+2 \text{ , } q_{ij} = 0$$

which represents the assumptions of a Poisson process -see point 3-.

$$q_{ii} = \lim_{t\to o} \frac{1-e^{-\lambda\Delta t}}{\Delta t} = -\lambda$$

The transition rate matrix,

$$\mathbf{A} = \begin{bmatrix} -\lambda & \lambda & . & . & . \\ . & -\lambda & \lambda & . & . \\ . & . & . & . & . \\ . & . & . & . & . \end{bmatrix}$$

5. SYSTEMS AVAILABILITY

5.1 Application to maintained systems

It has been defined the system availability as the probability that the system is able to perform its intended function.

Depending on time interval there are three types of availability:

(i) Instantaneous availability; refered to any time during system life.

(ii) Up-time availability; defined as proportion of time that the system is up during a specific time period.

(iii) Steady state availability or up time availability for a very large period.

The process of failures and repairs of a system is a random process defined by its random function $(X(t=0),t\geqslant 0)$ which indicates the system state. Very often this random function is continuous time, stationary finite Markov chain, where system state depends on its components state and system logic.

For each component is defined a failure rate $-\lambda(t)-$, and a repair rate $-\mu(t)-$, which can be consider constant if the interarrival times are exponentially distributed.

Both, failure rate and repair rate can be understood as a conditional probability; a probability of a failure (repair) at the interval (t, t + dt| if the component is up (down) at time t.

The process satisfies the differential equation,

$$\underline{P}'(t) = \underline{P}(t) \cdot \mathbf{A}$$

being $\underline{P}(t)$ the probability state vector and **A** the transition rate matrix which reflects the components properties $-\lambda_i;\ \mu_i-$ and the system logic.

The system equation can be solved at the s-domain by Laplace trasform or by numerical methods.

It is assumed that in a maintained system the components failures are instantaneously detected and its repairings start at that moment.

5.2 Single component system

We define as state 1, the operating state, and as state 2, the failed state.

Assuming constant rates, the transition rate matrix

$$\mathbf{A}_1 = \begin{bmatrix} -\lambda & \lambda \\ \mu & -\mu \end{bmatrix}$$

and as boundary condition

$$\underline{P}(o) = [1,0]$$

the solution of the process equation,

$$P_1(t) = A(t) = \frac{\mu}{\lambda + u} + \frac{\lambda}{\lambda + u} \quad e^{-(\lambda+\mu)t}$$

$$P_2(t) = \overline{A}(t) = \frac{\lambda}{\lambda + u} \quad 1 - e^{-(\lambda+u)t}$$

and the steady state availability

$$A(\sim) = \lim_{t \to \sim} P_1(t) = \frac{\mu}{\lambda + \mu}$$

$P_1(t)$ gives the probability that the system is at state 1, thought it may have failed and being repaired several times in the time interval $(0,t]$, or the instantaneous availability.

5.3 Two components system

We assume that each component, A and B, has two possible states: operating or failed.

The system state is a combination of components states according the following table:

System state	Components Operating	Components Failed
1	A B	-
2	B	A
3	A	B
4	-	A B

In Markov homogeneous chain the transition rates depends on the initial and final states, which allowed to model dependencies.

5.3.1.Series system.-

The transition rate matrix of the process,

$$\mathbf{A}_2 = \begin{bmatrix} -(\lambda_1+\lambda 2) & \lambda_1 & \lambda_2 \\ \mu_1 & -\mu_1 & 0 \\ \mu_2 & 0 & -\mu_2 \end{bmatrix}$$

being λ_1; μ_1 and λ_2; μ_2 components A and B transition rates.

State 4 is not accesible.

If A and B are identical and there is one repairman, then exist only two system states: 1 and 2, being the transition rate matrix,

$$\mathbf{A}_3 = \begin{bmatrix} -2\lambda & 2\lambda \\ \mu & -\mu \end{bmatrix}$$

equivalent to the transition rate matrix $\mathbf{A}_1$ -single component system-.

5.3.2.Paralled systems.-

In order to model any dependency and repair policy, transition rates are assumed to be dependent of initial and final state, as shown in figure 2.

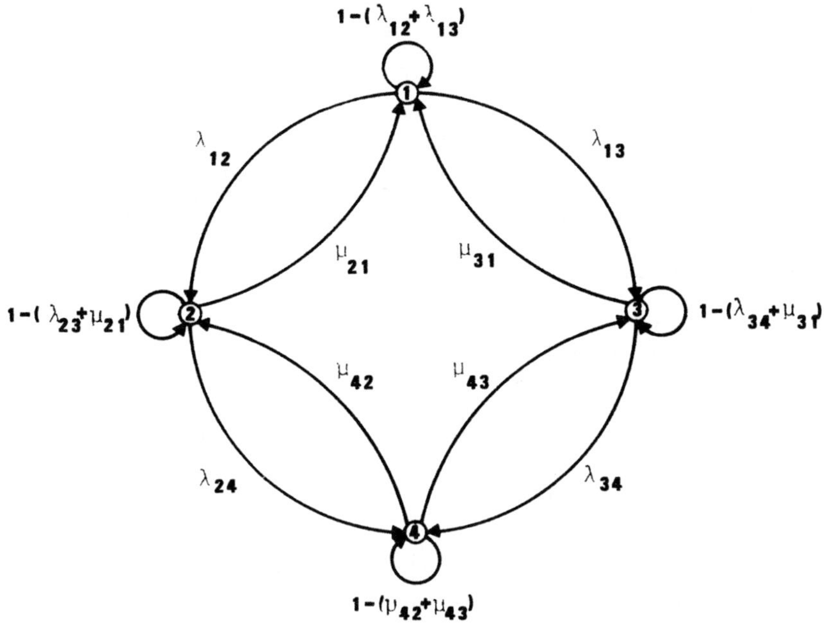

Fig. 2 Transition diagram in two components parallel system.

Where

λ_{12} and λ_{24} represent the failure rates of component A.
λ_{13} and λ_{34} represent the failure rates of component B.
μ_{21} and μ_{42} represent the repair rates of component A.
μ_{31} and μ_{43} represent the repair rates of component B.

The transition rate matrix:

$$\mathbf{A}_4 = \begin{bmatrix} -(\lambda_{12}+\lambda_{13}) & \lambda_{12} & \lambda_{13} & 0 \\ \mu_{21} & -(\lambda_{23}+\mu_{21}) & 0 & \lambda_{23} \\ \mu_{31} & 0 & -(\lambda_{34}+\mu_{31}) & \lambda_{34} \\ 0 & \mu_{42} & \mu_{43} & -(\mu_{42}+\mu_{43}) \end{bmatrix}$$

Since the two components are connected in parallel, the system is operating, at least one component is operating -states 1, 2 and 3-. Therefore the availability at time t,

$$A(t) = P_1(t) + P_2(t) + P_3(t)$$

Assuming that the two components are identical the following cases are of interest:

(i) Two components on line. Failure rate λ; and repair rate μ.
Transition rate matrix:

$$\mathbf{A}_5 = \begin{bmatrix} -2\lambda & 2\lambda & 0 \\ \mu & -(\lambda+\mu) & \lambda \\ 0 & \mu & -\mu \end{bmatrix}$$

(ii) One component off line. It can be on cold or warm standby mode. In cold standby, the standby component can not fail, because it is not subject to stress.
Transition rate matrix:

$$\mathbf{A}_6 = \begin{bmatrix} -\lambda & \lambda & 0 \\ \mu & -(\lambda+\mu) & \lambda \\ 0 & 2\mu & -2\mu \end{bmatrix}$$

In warm standby the standby component can fail, but with an smaller failure rate λ^*.
Transition rate matrix:

$$\mathbf{A}_7 = \begin{bmatrix} -(\lambda+\lambda^*) & \lambda+\lambda^* & 0 \\ \mu & -(\lambda+\mu) & \lambda \\ 0 & 2\mu & -2\mu \end{bmatrix}$$

Example 4

A system is made up of three machines. At most two machines can be operating. When a machine fails it undegoes repair, but only a machine can be in repair. Both operating and repair times are exponentialy distributed with parameters $\lambda=1$ and $\mu=2$. Assuming a cold standby and a perfect switching operation determine:

(i) State space
(ii) Transition diagram
(iii) Transition matrix and transition rate matrix
(iv) Process differential equation
(v) Process intantaneous availability
(vi) Process steady availability

Solution:

(i) The process states can be defined as follow:

State:	Number of machines: Operating	Standby	Repair	Waiting
1	2	1	0	0
2	2	0	1	0
3	1	0	1	1
4	0	0	1	2

(ii) The transition diagram is given in figure 3.

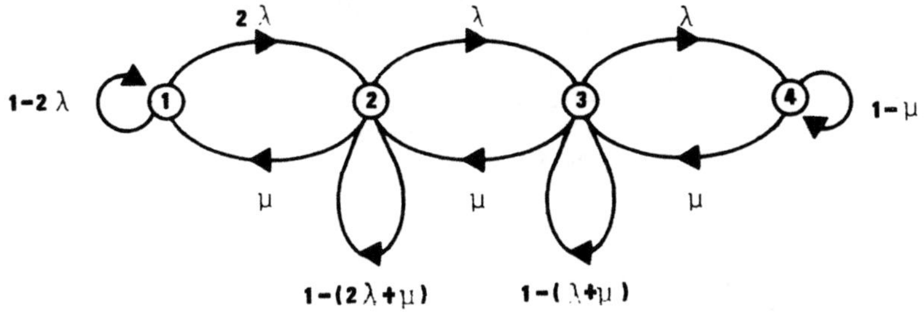

Fig. 3 Transition diagram

(iii) The transition matrix:

$$\mathbf{P} = \begin{bmatrix} 1-2\lambda dt & 2\lambda dt & 0 & 0 \\ \mu dt & 1-(\mu+2\lambda)dt & 2\lambda dt & 0 \\ 0 & \mu dt & 1-(\mu+\lambda)dt & \lambda dt \\ 0 & 0 & \mu dt & 1-\mu dt \end{bmatrix}$$

The transition rate matrix:

$$\mathbf{A} = \begin{bmatrix} -2\lambda & 2\lambda & 0 & 0 \\ \mu & -(\mu+2\lambda) & 2\lambda & 0 \\ 0 & \mu & -(\mu+\lambda) & \lambda \\ 0 & 0 & \mu & -\mu \end{bmatrix}$$

(iv) Process equations:

$$\underline{P}' = \underline{P} \cdot \mathbf{A}$$

or

$$P_1' = -2\lambda P_1 + \mu P_2$$
$$P_2' = 2\lambda P_1 - (\mu+2\lambda)P_2 + \mu P_3$$
$$P_3' = 2\lambda P_2 - (\mu+\lambda)P_3 + \mu P_4$$
$$P_4' = \lambda P_3 - \mu P_4$$

and the boundary condition

$$\underline{P}(o) = (1,0,0.0)$$

Transforming to s-domain the process equation, with $\mu=2$; $\lambda=1$:

$$\begin{aligned} sP_1(s)-1 &= -2\,P_1(s) + 2P_2(s) \\ sP_2(s) &= 2P_1(s) - 4P_2(s) + 2P_3(s) \\ sP_3(s) &= 2P_2(s) - 3P_3(s) + 2P_4(s) \\ sP_4(s) &= P_3(s) - 2P_4(s) \end{aligned}$$

Eliminating P_2, P_3, P_4:

$$P_1(s) = \frac{s^3 + 9s^2 + 20s + 8}{s^4 + 11s^3 + 42s^2 + 28s} =$$

$$= \frac{s^3 + 9s^2 + 20s + 9}{s(s+0.836)\,(s^2+10.164s+33.506)} =$$

$$= \frac{0.286}{s} + \frac{0.14}{s+0.836} + 0.574\,\frac{s + 5.082}{(s+5.082)^2+7.677} +$$

$$+ \frac{1.035}{(s+5.082)^2 + 7.677}$$

and

$$P_2(s) = \frac{-2s^2 + 10s + 8}{s^4 + 11s^3 + 42s^2 + 28s} =$$

$$= \frac{0.286}{s} + \frac{0.082}{s+ 0.836} - 0.367\,\frac{s+ 5.082}{(s+5.082)^2+7.677} +$$

$$+ \frac{21.798}{(s+5.082)^2+7.677}$$

$$P_3(s) = \frac{-4s^3 + 12s^2 + 56s + 16}{(s^2+5s+4)(s^4+11s^3+42s^2+28s)} =$$

$$= \frac{0.286}{s} + \frac{0.366}{s + 0.836} - \frac{0.667}{s+1} - \frac{0.542}{s+4} +$$

$$+ 0.777\frac{s + 5.082}{(s+5.082)^2+7.677} + \frac{0.542}{(s+5.082)^2+7.677}$$

$$P_4(s) = \frac{1}{s} - \left[P_1(s) + P_2(s) + P_3(s)\right]$$

Transforming to the time domain:

$$P_1(t) = 0.286 + 0.14\ e^{-0.836t} + (\ 0.574 \cos 2.771t + $$
$$+ 0.3753 \text{ sen } 2.771t\ )e^{-5.082t}$$

$$P_2(t) = 0.286 + 0.082\ e^{-0.836t} - \ (0.367 \cos 2.771t + $$
$$+ 7.867 \text{ sen } 2.711\ t)e^{-5.082t}$$

$$P_3(t) = 0.286 + 0.366\ e^{-0.836t} - 0.667\ e^{-t} - 0.762\ e^{-4t} + $$
$$+ (0.777 \cos 2.771t + 0.196 \text{sen} 2.771t)e^{-5.082t}$$

(v) Instantaneous availability,

$$A(t) = P_1(t) + P_2(t) + P_3(t)$$

with at least one machine working.

(vi) Steady state availability,

$$A(\sim) = P_1(\sim) + P_2(\sim) + P_3(\sim) = 3 \cdot 0.286 = 0.858$$

The same result can be obtained solving the system:

$$P_1'(\sim) = -2P_1(\sim) + 2P_2(\sim) = 0$$

$$P_2'(\sim) = 2P_1(\sim) - 4P_2(\sim) + 2P_3(\sim) = 0$$

$$P_3'(\sim) = 2P_2(\sim) - 3P_3(\sim) + 2P_4(\sim) = 0$$

$$P_4'(\sim) = P_3(\sim) - 2P_4(\sim) = 0$$

REFERENCES

1. Stark M., Wood J.W. (1986) 'Random Process and Estimation Theory for Engineers', Prentice-Hall.

2. Saiz de Bustamante A. (1985) 'Fundamentals of reliability theory', Proc. ISPRA Course: Reliability Engineering, Kluwer Academic Publishers.

3. Keller A.Z., (1985) 'Markov modelling', Proc. ISPRA Course: Reliability Modelling and Application, Kluwer Academic Publishers.

4. Papazoglou I. (1985) 'Markov approachs to system reliability', Proc. ISPRA Course: Reliability Engineering, Kluwer Academic Publishers.

5. Clarotti C.A. (1986) 'Markov Approach to Calculating System Reliability: Computational problems', Proc. International School of Physics Enrico Perrini, North-Holland.

6. Osäki S. (1985) Stochastic System Reliability Modelling, Wold Scientific.

7. Frankel E.V. (1988) 'System Reliability and Risk Analysis' (2nd edition), Martinus Nijhoff Publishers.

8. Bouleau N. (1988) 'Processas stochastiques et applications', Hermann.

9. Bremanud P. (1987) 'Modélisation markovienne', Ecole Nationale Supérieure de Techniques Avancées (Paris).

THE MONTE CARLO METHOD

J.M. PERLADO
Instituto de Fusión Nuclear
Universidad Politécnica de Madrid
José Gutiérrez Abascal, 2: 28006 Madrid
Spain

ABSTRACT. The Monte Carlo method may briefly be described as the numerical tool of studying an artificial stochastic model of a physical or mathematical process. This paper examines the basic principles of the method, which include the random numbers generation from an uniform density function, its extension for sampling from more complicated probability density functions, and the theory of estimation in straight-analogue ("crude") simulation. The variance reduction techniques are also introduced for improving the efficiency of the Monte Carlo method. A brief comment is given about its use in uncertainty analysis. A clear example of the use of Monte Carlo in the system availability estimation is given.

1.- INTRODUCTION

The **Monte Carlo method** is a numerical method which allows the solution of physical or mathematical processes by describing them through an adequate probabilistic model which represents the random variables involved in the system. There are two critical terms in the method: random sampling and estimation theory. It may be regarded as a branch of numerical analysis/simulation in which random sampling techniques are used, so that the "calculation" involves playing a game of chance. Associated with each "try" in the game of chance is a numerical result or score. The game is designed in such a way that the expected value of the score is that physical or mathematical quantity the player wishes to obtain. Then, if the game is played enough times, according to the central limit theorem, the average score approximates the expected value of the quantity to be calculated.

The name and the systematic development of this method starts about 1944, with the works of **Ulam, Metropolis** and **J. von Neumann**. The first of these applications were in the field of Nuclear Technology and Transport Theory, which have taken large benefit of this technique,

A. G. Colombo and A. Saiz de Bustamante (eds.), Systems Reliability Assessment, 23–43.

contributing definitively to its formal establishment. However, many other areas of physics and engineering are now applying the method of Monte Carlo, among them operational analysis, molecular dynamics, high energy physics and, of course, reliability analysis.

Random sampling methods were used in a number of applications long before the epithet "Monte Carlo" was applied. It is well known that the theory of the statistics of small samples was developed by Gosset ("Student", 1908) with the help in part of sampling experiment. Fermi used sampling methods to study the diffusion of neutrons in matter in 1933, and Kennard (1938) used what would now be called a Monte Carlo method to construct a "theoretical" Brownian motion track using random numbers constructed from digits taken from a telephone book.

The real recognition of testing hypothesis and estimation came when Fisher (1922) gave a new definition of statistics. He established that the object of statistical methods is the reduction of bulky data, Fisher distinguished three basic problems: those of specification of the kind of population from which the data come, of estimation, and of distribution.

In defining the model of a specific system by the Monte Carlo method, a probabilistic scheme, describing adequately the nature of the problem, must be built up, defining the random variables and their probability density functions (pdf) significative to our problem. Also, the desired magnitude to be calculated is represented by other random variable, called **estimator** which, by definition, is that variable whose expected value correspond to the wished quantity.

When calculating with the Monte Carlo method, a large enough number of trials are performed according with the pdf's of the established "*random walk*" in order to obtain the corresponding values of the estimator. The final operation calculates the average of the estimator with the numbers obtained in each trial. Then, if the probabilistic scheme (**random walk**) is defined by the random variables θ_1, θ_2, ... θ_n with pdf's $f_1(\theta_1)$, $f_2(\theta_2)$ ... $f_n(\theta_n)$, and X is the estimator of the magnitud U, to be realised that E(X) = U, the numerical calculation proceeds obtaining N trials of the random variables, $(\theta_{1,1}, \dots \theta_{n,1})$, ..., $(\theta_{1,N}, \dots \theta_{n,N})$, from their respectives pdf's, which gives N values of the estimator X_1, X_2 ... X_N. The magnitude U will be approximated by the average

$$\overline{X} = \frac{1}{N} \sum_{1}^{N} X_i$$

Because of the Monte Carlo method is of statistical nature, the final response is governed by the probability theory and some error will be associated with the value of the estimator. There are two results of the elementary probability theory which have particular relevance for Monte Carlo studies: the law of large numbers and the central limit theorem. The law of large numbers states that the sample mean, with a probability that approach 1 as $N \to \infty$, approaches the population mean, or true mean, then the accuracy of an estimate of a quantity tends to improve when computing on larger and larger samples. The central limit theorem states that the probability, p, that the "exact" or true value of X lies inside the interval ($X-t_p\sigma$, $X + t_p\sigma$) is equal to $2P(t_p)-1$, where P(y) is the normal cumulative distribution function if N is large. The standard deviation, σ, of the arithmetic mean, X, of N independent observations of the random variable is given by

$$\sqrt{\frac{\sum_{i=1}^{N}(X_i-\overline{X})^2}{N(N-1)}} = \frac{1}{\sqrt{N}}\sqrt{\left(\frac{\sum_{i=1}^{N}X_i^2}{N} - \frac{1}{N^2}\left(\sum_{i=1}^{N}X_i\right)^2\right)},$$

$$if N>>1, = \frac{1}{\sqrt{N}}\sqrt{\left(\overline{X}^2-(\overline{X})^2\right)}$$

The precise interpretation of σ in our error estimate is understood showing that the deviation of the arithmetic mean, X, from the "true" mean in a standard deviation (±σ) have a probability of $2P(1) - 1 = 0.68$ ($t_p = 1$), the Monte Carlo result deviates by more than one standard deviation with a probability of 0.32, more than 2σ with 0.046 and more than 3σ with 0.0027.

There is a clasical example that shows the character and limits of the method of Monte Carlo, which is called the Buffon's experiment (1733) and allows to determine an estimation of the value of π. In this paper, other simple mode for calculating π will be presented with the support of a trivial FORTRAN program. As other example, the estimation of a complex surface by the Monte Carlo method is presented, allowing the generalization to more general problems.

Example 1

The number π can be calculated considering that the ratio of the number of points which accomplish that the distance from the central point (0,0) of a circle to them is less than or equal to one, to the total number of randomly sampled points approximates 1/4 of π. A trivial programme represents this simple "area" calculation using a function which chooses the random numbers uniformly distributed in (0,1) using common techniques that will be described below. Random routines are available in all the computers in the corresponding language FORTRAN (RANF), BASIC (RND)...

Example 2

We are interested in the calculation of the surface S of a plane figure defined by some arbitrary boundaries, Figure 1. It is supposed, by simplicity, that the figure is contained in a square of dimension 1.

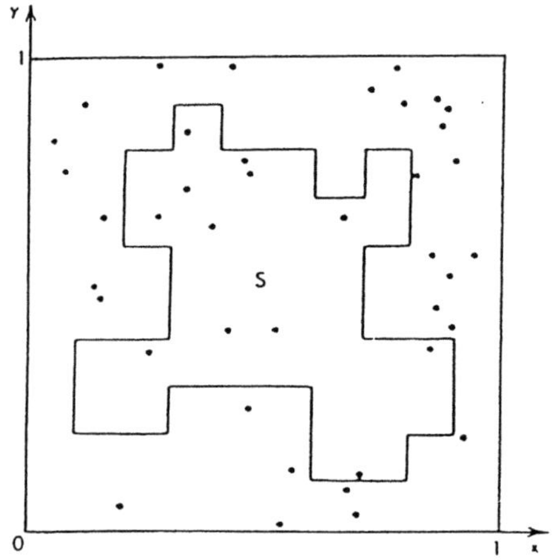

Figure 1. Monte Carlo areas calculation.

A number N of random points are choosen in that square, being N' the number of points inside S. It is obvious that the surface S can be approximated, by geometrical reasons, as the ratio N'/N. The precision in the estimation will increase when the extent of the sample (N) is enlarged. In the example, the extension of the sample is N = 40 with N' = 12. Then, the surface estimation is N'/N = 0.30, while the real value is 0.35.

This is a very simple example of application which is not really practical because many other methods exist for calculating the surface of plane figures. However, it represents the basis for the adequate extension in the calculation of multidimensional volumes, in which resolution Monte Carlo method is the only available solution for many complex problems. Moreover, *this simple case allows the consideration of the Monte Carlo method as the numerical procedure to obtain functional integration in a n-dimensional phase space, at which mathematical formalism can be reduced many of the physical and technical problems.*

2.- TECHNIQUES FOR RANDOM SAMPLING

When running a problem by Monte Carlo methods, it was shown in the **Introduction** that a critical aspect is to obtain a sample of the random variables from their pdfs. These probability density functions could be, depending of the problem, of an arbitrary nature very complex in many applications. It would be shown in the following that the problem of sampling at random from an arbitrary pdf can be conveniently reduced to the problem of obtaining numbers at random from the **canonical distribution.** The canonical distribution represents an uniform distribution, which is the frequency distribution rectangular in the interval (0,1). This particular pdf plays a vital role in Monte Carlo, because it is the simpler and more intuitive way for obtaining **random numbers,** ζ, which, following common usage, *are those numbers independent and uniformly distributed in the interval (0,1).*

In consequence, we have splitted the problem in two parts: random numbers generation and sampling from other different pdf.

2.1.- Random number generation

It is critical to the application of the Monte Carlo method that there should be readily available an adequate supply of numbers "chosen at random". In the early days of the method some effort was made to generate random numbers by means of mechanical, electrical or electronic devices, sometimes involving human agency. These methods proved unreliable in operation and had the distinct disadvantage of being unable to reproduce a previous sequence of numbers which is desirable in Monte Carlo simulations by computer programs.

A classical example of random number machine is ERNIE which uses as random source the discharge of neon tubes, and is dedicated to picking the winning numbers in the British Post Office Premium Savings Bonds Lottery. Other examples are the random generator used to obtain the popular RAND table of 1 million of numbers which applies an standard method of electronic devices with a speed of one per second; the MARK I machine of the University of Manchester, and others.

Except for some special applications, most random numbers for Monte Carlo simulations are now computed by the digital computers using numerical schemes especially designed for this purpose. Random numbers produced in this deterministic manner are often referred as *pseudo* random in order to distinguish them from truly random numbers.

Among other techniques as *mid-square* or *mid-product,* in considering numerical schemes for producing random numbers we can concentrate our attention on a single algorithm known as the **linear multiplicative congruential method,** because the most of the random number generators that are incorporated into computing systems use variants of this method. This algorithm, due to **D. H. Lehmer,** is designed to generate recursively a sequence of random fractions, that is random real numbers uniformly distributed between zero and unity, then each possible number is equally probable.

The important steps in the basis form of the congruential method can be stated in the following way. Choose four numbers:

X_o, the starting value or "seed" ; $X_o \geq 0$
a, the multiplier ; $a > 0$
c, the increment ; $c \geq 0$
M, the modulus ; $M > X_o$, $M > a$, $M > c$

The desired sequence of random number (X_n) is obtained by setting

$$X_{n+1} = aX_n + c \pmod{M}, \quad n \geq 0$$

A sequence produced in this way is called a linear congruential sequence, and the precedent expression is read: X_{n+1} is congruent to ($a X_n + c$) modulus M, and means that X_{n+1} is the remainder of $(aX_n+c)/M$.

It is clear that the numbers produced by that procedure must always be less than M, and therefore when divided by M they must all lie between 0 and 1.

The theoretical justification for the congruential method is based upon results originated in the theory of pure numbers.

The choice of the parameters in the above inequalities is crucial in determining whether or not a *good* random sequence is produced. One important characteristic of any numerical algorithm in generating random numbers is that it produces a cycle of numbers which is repeated endlessly, being the length of the cycle called the *period*. Since many of the Monte Carlo applications requires a very large sequence of random numbers, it is crucial that the period of a random numbers generator should be as large as possible. This, as we shall see, requires that the modulus M should also be a very large number. It is customary to take M to be the largest number that the particular computer can represent in a single *word*. In a binary machine with a word size of p bits, it is convenient to set $M = 2^p$; in a decimal machine also of p bits, M would take the value of 10^p.

2.2.- Sampling from specified distributions.

We need now to establish the relations between the canonical distribution, from which we obtain the random numbers, and any other arbitrary distribution functions which represent our problem. There are, among many proposed techniques, two methods which are extensively used: **inverse cumulative function method** and **rejection technique**.

The **inverse cumulative function method** is probably the most popular technique in Monte Carlo applications, and allows to obtain **analytical expressions** for sampling from the specific *pdf* depending of the random number previously determined. The basis of this method are presented.

If x is a random variable with a probability density function p(x), and $y = f(x)$ is a monotone and increasing function of x with pdf g(y), we can write $P(x \leq a) = P(y \leq f(a))$.

Restating this equality in terms of the respective pdf's gives,

$$P(a) = \int_{-\infty}^{a} p(x)dx = \int_{-\infty}^{f(a)} g(y)dy$$

Now, if we choose g(y) = 1, for 0 ≤ y ≤ 1, then

$$P(\chi) = \int_{-\infty}^{\chi} p(x)dx = \int_{-\infty}^{\zeta} dy = \zeta\binom{random}{number}$$

$$\chi = P^{-1}(\zeta)$$

which is called the **inverse cumulative function.** Of course, this method is really applicable when that inverse function is easily obtained which is the case in many analytical distribution. The inverse cumulative functions for the most common probability density functions are given in Annex 1.

The **rejection technique** was proposed by **J. Von Neumann** and it samples from the direct distribution function through pairs of random number. This technique is particularly useful when no analytical expressions can be built up according to the inversion method. In the more basic form it applies to a *pdf* that is bounded and of finite range, Figure 2.

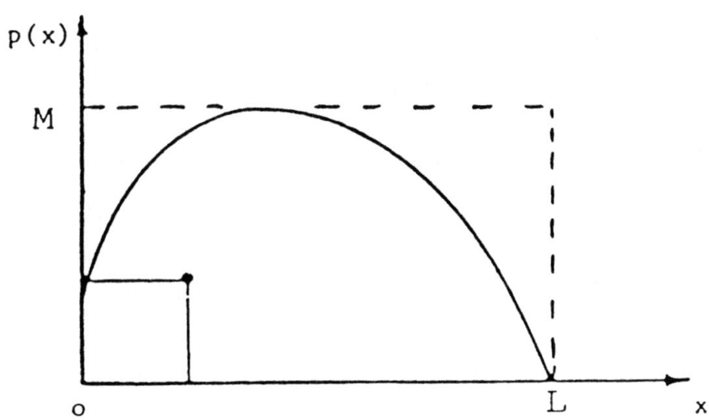

Figure 2. Rejection Technique.

If ζ is a random variable with pdf f(x) defined

$$f(x) = \begin{cases} 0 & -\infty \leqq x \leqq 0 \\ f(x) & 0 < x < L \\ 0 & L \leqq x \leqq \infty \end{cases}$$

and M is the maximum of f(x) in the interval (0,L), the boundaries of the function $f_1(x) = f(x)/M$ should be (0,1). Now, we choose two numbers (t, ρ_2) randomly from the canonical distribution. The conditional probability density function to obtain t when $\rho_2 < f_1(t)$ shall be,

$$h[t|\rho_2 < f_1(t)] = \int_{-\infty}^{f_1(t)} g(t|\rho_2) d\rho_2 = \frac{\int_{-\infty}^{f_1(t)} g(t,\rho_2)\, d\rho_2}{\int_{-\infty}^{f_1(t)} g_1(\rho_2)\, d\rho_2}$$

where

$$g(t,\rho_2) = \begin{cases} \frac{1}{L}, & 0 < t \leq L,\ 0 \leq \rho_2 \leq 1 \\ 0 \end{cases}$$

$$h[t|\rho_2 < f_1(t)] = \frac{\frac{1}{L}\int_{-\infty}^{f_1(t)} d\rho_2}{\frac{1}{L}\int_a^b dt \int_{-\infty}^{f_1(t)} d\rho_2} = \frac{f_1(t)}{\int_0^L f_1(t)\, dt} = f_1(t)$$

Calculating with this technique, two random numbers (t,ρ_2) will be choosen and the value of $f_1(t) = f(t)/M$ will be determined. The value of t should be *rejected* if $\rho_2 > f_1(t)$, and accepted when $\rho_2 < f_1(t)$.

3.- STRAIGHT-ANALOGUE, DIRECT OR "CRUDE" MONTE CARLO

With the basis given in the precedent sections, we are able theoretically to design the Monte Carlo model in order to simulate any system, and to establish the appropriate numerical procedure for running our problem. Three elements are needed to start our calculation:

i) random number generator.
ii) probabilistic scheme of our system, which means to know the dependent random variables and their pdf's.

iii) estimator, or random variable whose expected value gives the desired magnitud of our system.

In reliability analysis, the same conditions must be accomplished. In order to clarify the Monte Carlo application to this particular problem, a simple example of the reliability of a multicomponent system is given below.

We are interested into estimate the average time to failure of a multicomponent system. The time to failure of each element U_i of the system could be considered as a random variable θ_i with a pdf $f_i(\theta_i)$ which shall be known by previous experiments or its technical specifications.

In Figure 3, a schematic working tree is shown with only five elements, two of them (3,4) redundant between them. In this case the estimator of the desired magnitude can be expressed by

$$t = \min [t_1; t_2; \max [t_3; t_4]; t_5]$$

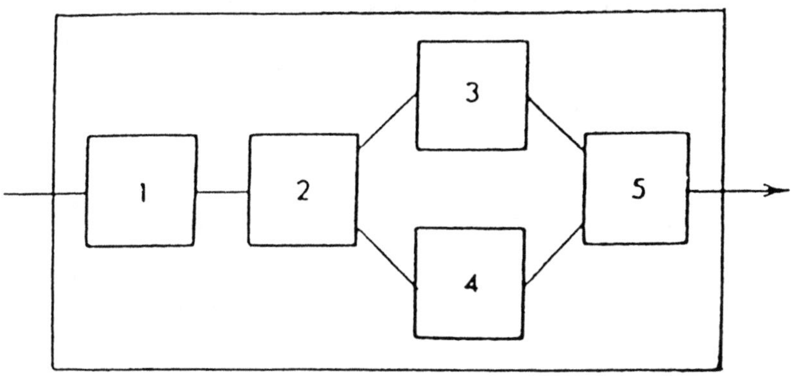

Figure 3. Working tree example in straight-analogue MonteCarlo.

The numerical procedure starts choosing from the pdf's $f_i(\theta_i)$ the values of t_{11}, t_{21}, ... t_{51} which correspond to the first trial of the time to failure for each component. The first value of the estimator t_1 will be

$$t_1 = \min [t_{11}; t_{21}; \max [t_{31}; t_{41}]; t_{51}]$$

The full sample of t should be (t_1, t_2, ... t_N) corresponding to the extent N of the number of trials. According with the theory of Monte Carlo method, the estimate of t (time to failure of the system) can be expressed by

$$E(\theta) \approx \frac{1}{N} \sum_{i=1}^{N} t_i$$

being θ the random variable which represents the estimator of t.

4.- SYSTEM AVAILABILITY BY MONTE CARLO

The use of Monte Carlo method is very common when obtaining the unavailability of a complex system.

System unavailability, as an extension of the component unavailability, can be defined in general as the probability of being in a failed state when required. It is well known that we can distinguish between "point value unavalability" and "probabilistic unavalability". The difference comes because in the first case fixed values are used for the component failure rate, test and maintenance parameters and human error data, and in the probabilistic approach these quantities are treated as random variables. The system unavailability in the probabilistic scheme is a function of these random variables, then it is itself a random variable. Tipically, the process of determining the system unavailability in the probabilistic approach is known as the error or probabilistic range propagation. In fact, each one of the function values (distribution probability function) have associated a probability error. In this approach we are determining "the probability of a probability".

The probabilistic frame is the most common situation in engineering, when the differences in failure, test, maintenance, human rate among the diverse plant facilities are considered. The different component failure, maintenance-test, human error rates are in themself probabilities of ocurring the assign effect in the period of time determined by the normalization unit. In our probabilistic approach, each one of these rates have a experimentally determined density function.

In the simpler case, the system unavailability will be characterized by a boolean logical expression as the mathematical representation of the established fault tree, which depends of the component data, test and maintenance parameters and human error data, assumed as constant probability values. The same expression, but now in terms of the density functions for the components, gives us the unavailability of the system in the probabilistic approach.

A general expression $Y=Y\ (X_1,\ X_2,\ X_i........,\ X_n)$ represents the unavailibility of the system corresponding with the logic of the fault tree. In our case, each one of X_i are the probability density functions of the different events.

Consistently with the established nature of the Monte Carlo method, we can obtain the different values of the function $Y=Y\ (X_1,\ X_2,\ ....,\ X_n)$ sampling different values of X_i, i=1, n, from the given density functions, obtaining finally the mean, standard and distribution function values for a number N of trials. Of course, those function values will be distributed in some specified confidence intervals which respond to the probability of the probability. Remembering the well known concepts of distribution function and confidence limits in statistical theory, the function values corresponding to the increasing confidence limits will be obtained by ordering the y's sampled values in an ascending order, $y_1<y_2<..........<y_n$ to obtain the confidence limits (percentiles) of the distribution Y.

It must be remarked that the Monte Carlo flow chart can be considered as general with the exception of the specific expression of the unavailability, which will correspond to the logical expression representing the fault tree of our system. In terms of programming, this means that a general program can be built with the only exception of the particular unavailability function.

A clear example, which is also used in other reports (WASH-1400 USNRC-1975), will be used here to illustrate the Monte Carlo procedure. The system is represented by the fault tree of Figure 4, where X(I), I=1, 2, 4, 7 represent component failures, X(3) and X(5) non independent operator errors, and X(6) the test and maintenance contribution to system unavailability.

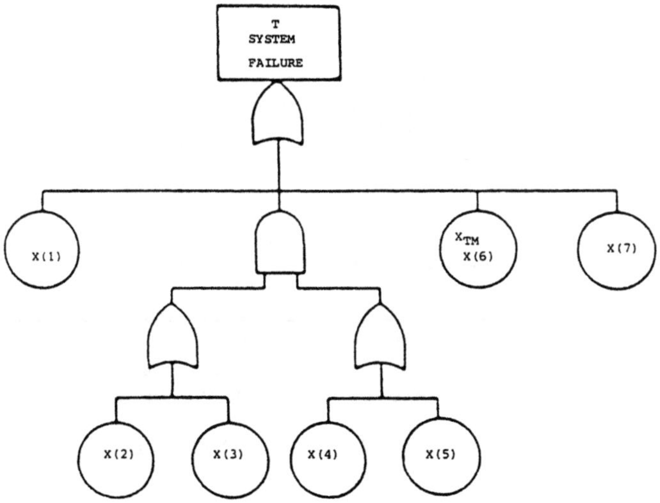

Figure 4. Unavailability example fault tree.

It is supposed that the faults X(3) and X(5) are given to be non independent, which produces a common mode contribution to system unavailability.

The boolean expression for the proposed fault tree is,

$$T= X(1)+ [X(2)+X(3)] * [X(4)+X(5)]+X(6)+X(7)+X_{CM}$$

where, X_{CM} express the common mode contribution of X(3), X(5).

In our probabilistic scheme, this boolean expression can now be interpreted in terms of arithmetical probability by associating each X(i) with the corresponding probability density function and considering for each operation symbol its arithmetic meaning. The system unavailability will be

$$P(T)= X(1)+ [X(2)+X(3)] * [X(4)+X(5)]+X(6)+X(7)+X_{CM}$$

The data used in our example are given in Table 1 assuming a log-normal distribution.

TABLE 1. Example problem data

Fault	Median	Error Factor
Component Failure, X(1)	1.0×10^{-3}	3
Component Failure, X(2)	3.0×10^{-2}	3
Operator Error, X(3)	1.0×10^{-2}	3
Component Failure, X(4)	3.0×10^{-2}	3
Operator Error, X(5)	1.0×10^{-2}	3
Test and Maintenance, X(6)	3.0×10^{-3}	3
Component Failure, X(7)	1.0×10^{-6}	10

The common mode contribution is the square root of the upper and lower bound of the probability combination of X(3) and X(5). The X_{upper} in this case is X(3)=X(5), and the X_{lower} will be the product of both of them (X(3)×X(5)). Then, the common mode distribution can be expressed by

$$X_{CM}=\sqrt{X(3)\times(X(3)\times X(5))}$$

In our particular example, there is other possible simplification when considering that X(7) is much lower than the point value of P(T), which will be simply obtained by substituting in our general expression the point median values. The expression now will be

$$P(T)= X(1)+ [X(2)+X(3)] * [X(4)+X(5)]+X(6)+X_{CM}$$

The coupled error case, when assuming that X(2) and X(4) represent similar coupled faults, and also X(3), X(5), is represented by the expresion,

$$P_{coupled}(T)= \dot{X}(1)+[X(2)+X(3)]^2+X(6)+\sqrt{[X(3)]^3}$$

The flow diagram of a Monte Carlo unavailability program is represented in Figure 5, and the results for the input data and functions here reported are summarized in Annex 2.

5.- VARIANCE REDUCTION TECHNIQUES.

It was shown in the *Introduction* of the method that the accuracy of the estimation is related to the extension of the sample (the number of trials N). In many applications of the Monte Carlo method a very large number of values of the estimator are needed to obtain a reasonable error in the final response. It was also shown that the quantity to measure that error is the associated standard deviation or the **variance** of the estimation. Then, we can read the problem in the following ways:

i) for a fixed extent of the sample, N, to be interested in reducing the associated variance (→ more precision).
ii) for a good enough variance of the estimation, trying to reduce significant the extent of the sample.

There are a significant number of theoretical proposals for variance reduction, but only a few of them are really useful as numerical procedures. Two of them have had more attention in real practice: **correlated** an **importance sampling**. The general philosophy of both of them is to bias the natural probabilistic scheme to an artificial model (changing the associated pdf's and the estimator) in order to obtain a low variance.

The correlated sampling is a particular case of the general *control variates* technique and is based in supposing that the characteristics of the system can be expressed as the difference (or ratio) of two closely related quantities, obtained by solving two very similar problems. Let x be the Monte Carlo estimate of the answer to the first problem and σ_x the corresponding estimated standard deviation in individual samples. Let the addition of a prime (') denote the same quantities for the second of the two closely related problems. Then the estimated standard deviation in the difference, x' - x, for N histories, is given by

$$\sigma_{\bar{x}'-\bar{x}} = \frac{1}{\sqrt{N}}\left[\sigma_x^2 + \sigma_{x'}^2 - 2\,\sigma_x\sigma_{x'}\rho_{xx'}\right]^{1/2} ,$$

where $\rho_{xx'}$ is the correlation coefficient given by

$$\rho_{xx'} = \frac{\overline{xx'} - \bar{x}\,\bar{x}'}{\sigma_x\sigma_{x'}}$$

and

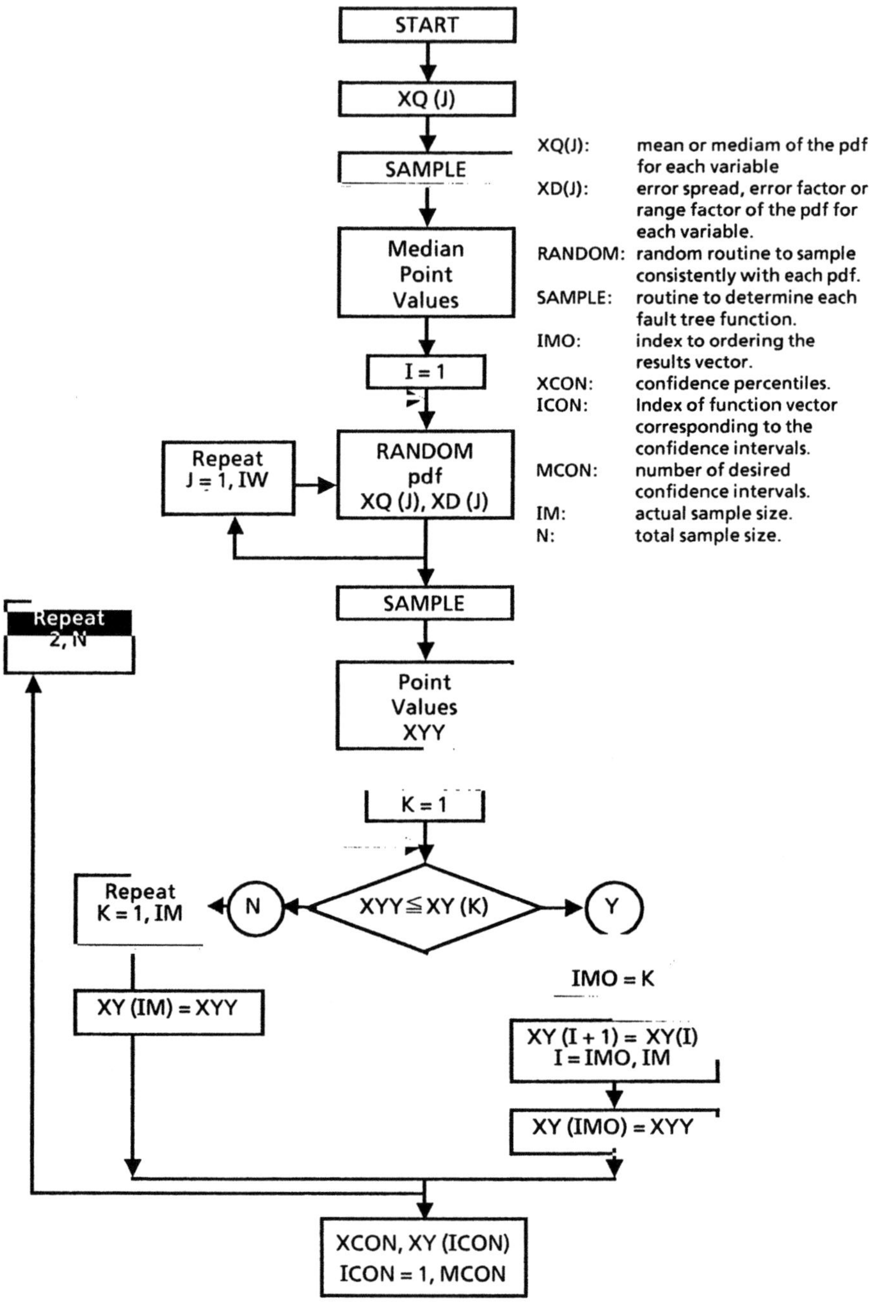

Figure 5.- Flow diagram of Monte Carlo unavailability program.

$$\overline{xx'} = \frac{1}{N} \sum_{i} x_i x_i'$$

These expressions show that the error in the difference can be made very small if the correlation coefficient approaches 1.

The **importance sampling** improves the precision by biasing the natural probabilistic scheme and sampling from the domains of great significance for the desired magnitude. Of course, this bias in the *random walk* supposes a modification in the random variable which represents the estimator in order to accomplish the definition given above. The Monte Carlo applications are in general reduced to integral calculations in a multidimensional space, which read,

$$M \simeq E[\theta(x)] = \int_{-\infty}^{\infty} \theta(x) f(x)\, dx$$

where, in the general formalism, x should represent the vector of random variables, and f(x) the product of the different pdf's of these random variables.

Sampling by importance introduces a change in the pdf, according with the desired significance, without modification in the final response, then

$$M = \int_{-\infty}^{\infty} \frac{\theta(x) f(x)}{g(x)}\, g(x)\, dx = \int_{-\infty}^{\infty} \theta(x)\, g(x)\, dx$$

The new probabilistic scheme should be governed by g(x), and the new estimator correspons to

$$\theta(x) = \frac{\theta(x) f(x)}{g(x)}$$

6.- UNCERTAINTY ANALYSIS

Uncertainty analysis can be an essential requirement in determining the reliability of a calculated quantity. In Monte Carlo calculation, there are two basic types of uncertainties:

i) the statistical uncertainty of the method itself.
ii) uncertainties in the physical parameters being used in the calculation.

A very good review of this theory is given by **H. Rief** from where these brief ideas are taken. An actual review of its practical use in safety analysis is given by **A. Alonso et al.**

Following our knowledge of the Monte Carlo game, a set of responses r_i of the choosen estimator are scored in a Monte Carlo sample represented by the vector **r** which elements are $r_i(i=1, \dots m)$. This response vector depends on a set of input parameter $p_i(i=1,2,\dots m)$ which are elements of the vector **p**, constituting the random variables. This data vector **p** is assumed to be subject to measuring or other uncertainties

$$\mathbf{p} = \mathbf{p}_0 + \delta\mathbf{p}$$

where $\mathbf{p}_0$ is the mean value and the uncertainty $\delta\mathbf{p}$ is characterized by its variance-covariance matrix

$$\mathbf{C} = \langle\delta\mathbf{p}\ \delta\mathbf{p}^T\rangle$$

where the brackets <...> mean a statistical average. Because of the finite sample size, an uncertainty δs_i is introduced,

$$r_i = f_i\ (\mathbf{p}^{(0)} + \delta\mathbf{p}) + \delta s_i$$

Expanding into Taylor series the function f_i and performing a simple algebra, a general expresion for the uncertainty of the response can be obtained

$$\langle\delta\mathbf{r}\delta\mathbf{r}^T\rangle = \mathbf{S}\langle\delta\mathbf{p}\delta\mathbf{p}^T\rangle + \langle\delta\mathbf{s}\delta\mathbf{s}^T\rangle$$

where

$\langle\delta\mathbf{r}\delta\mathbf{r}^T\rangle$: total uncertainty matrix of a Monte Carlo game, combining input data uncertainties with statistical deviations resulting from finite sampling
$\langle\delta\mathbf{p}\delta\mathbf{p}^T\rangle$: uncertainty matrix of the input data
$\langle\delta\mathbf{s}\delta\mathbf{s}^T\rangle$: uncertainty matrix of the Monte Carlo scores
$\mathbf{S} = \langle\partial f_i/\partial p_k\rangle$:sensitivity matrix of calculated responses.

REFERENCES

Hammersley, J.M. and Hadscomb, C.D. (1964), **Monte Carlo Methods,** Methuen.
Kahn, H., (1954), **Applications of Monte Carlo**, AECU-3259, Rand Corporation.
Gelbard, M., Spanier, J., (1969), **Monte Carlo Principles and Neutron Transport Problems**, Addison-Wesley.
Tocher, K.D., (1963), **The art of simulation**, English University Press.
Shreider, A., **Methods of statistical testing. Monte Carlo Method,** Elsevier Pub.
Rubinstein, R.Y., (1981), **Simulation and the Monte Carlo Method**, John Wiley & Sons.
Henley, J. and Kumamoto, H., (1981), **Reliability Engineering and Risk Assesment**, Prentice-Hall.

Beggs, N.J. et. al., (1981), **Statistics for Nuclear Engineering and Scientists**, Bettis Atomic Power Lab., WAPD-TM-1292.
Saiz de Bustamante, A., (1986), **Monte Carlo Methods**, Proceedings of Ispra Course on Reliability Engineering.
Perlado, J.M., (1978), **Método de Monte Carlo y Transporte Neutrónico**, "Montecarlo method and neutron transport", Energía Nuclear, Vol. 22.
Rief, H., (1988) **Monte Carlo Uncertainty Analysis**, in **Uncertainty Analysis**, Ed. Y. Ronen, CRC Press.
USNRC (U. S. NUclear Regulatory Commission) (1975). **Reactor Safety Study: An Assessment of Accident Risk in U. S. Commercial Nuclear Power Plants.** WASH-1400 (NUREG-75/014)
Alonso, A., Ortega, P., et al., (1988), **Análisis de incertidumbres, sensibilidad e importancias.** "Analysis of incertitudes, sensitivities and importances", PIACR-ISI-3601/88. Universidad Politécnica de Madrid.

ANNEX 1.

Exponential distribution

$$f(t) = \lambda e^{-\lambda t}$$
$$F(t) = 1-e^{-\lambda t} = \gamma$$
$$t = -\ell n(1-\gamma)/\lambda$$

or $t = -\ell n\,\gamma/\lambda$

because γ and $1-\gamma$ have the same distribution.

Weibull distribution (two parameters)

$$f(t) = \frac{\beta}{\eta\beta}\, t^{\beta-1}\, exp - \left(\frac{t}{\eta}\right)^{\beta}$$

$$F(t) = 1 - exp - \left(\frac{t}{\eta}\right)^{\beta} = \gamma$$

where β = shape parameter

η = scale parameter

$t = -\eta(\ell n\,\gamma)^{1/\beta}$

Normal distribution

The Central Limit Theorem can be used to generate normally distributed random numbers,

$$x = \sum_{i=1}^{12} \gamma_i \in N(6,1)$$

$$x-6 \in N(0,1)$$

$$\frac{t-\mu}{\sigma} \in N(0,1)$$

then $t = (x-6)\sigma + \mu.$

Log-normal distribution

According ℓn t is distributed normally

$$x = \sum_{i=1}^{12} \gamma_i \in N(6,1)$$

$$\ln t = (x-6)\sigma + \mu$$

$$t = \exp\left((x-6)\sigma + \mu\right)$$

Poisson distribution

To generate a Poisson variate, which is discrete, it is necessary to find the value of x that most closely satisfies

$$\sum_{\lambda=0}^{x} e^{-\gamma} \frac{\lambda x}{x!} = \gamma \ ; \quad \lambda > 0$$

A Poisson random number generator is based on a linear interpolation of two integral values for x, which bracket the γ value, producing a continuum of values. If a true Poisson distribution is desired, x should be an integral approximation of the generated value.

ANNEX 2.

```
      **AVAILAV CODE** TO COMPUTE CONFIDENCE LIMITS OF SYSTEM UNAVAILABILITY
                    BY MONTECARLO METHOD <DENIM>
 0
                              INPUT MEDIANS AND 90% ERROR FACTORS

  ( 1) .1000E-02 .3000E+01 ( 2) .3000E-01 .3000E+01 ( 3) .1000E-01 .3000E+01 ( 4) .300
 0E-01 .3000E+01
  ( 5) .1000E-01 .3000E+01 ( 6) .3000E-02 .3000E+01 ( 7) .1000E-05 .1000E+02 (
 0
           MEDIAN POINT VALUE FOR SYSTEM IS      .6600E-02     FUNCTION NO.=    1
 0
           MEDIAN POINT VALUE FOR SYSTEM IS      .6600E-02     FUNCTION NO.=    2
 0
           MEDIAN POINT VALUE FOR SYSTEM IS      .6601E-02     FUNCTION NO.=    3
       EXAMPLE/SAMPLE=X1+(X2+X3)*(X4+X5)+X6+(X3*X5*X3)**0.5
      FUNCTION PROBLEM NO.=    1
 0
 0                     OUTPUT EVALUATIONS.
    SAMPLE SIZE=.  200   ACCURACY ON 95 PER CENT CONFIDENCE INTERVAL= 2.5 PER CENT
 0  DISTRIBUTION PARAMETERS.   MEAN =     .8728E-02   STANDARD DEVIATION =    .4233E-02
 0
 0                             DISTRIBUTION CONFIDENCE LIMITS
 0                            CONFIDENCE (PER CENT)      FUNCTION VALUE
                                        .5                  .2790E-02
                                       1.0                  .2843E-02
                                       2.5                  .4123E-02
                                       5.0                  .4257E-02
                                      10.0                  .4952E-02
                                      20.0                  .5846E-02
                                      25.0                  .6198E-02
                                      30.0                  .6594E-02
                                      40.0                  .7152E-02
                                      50.0                  .7714E-02
                                      60.0                  .8384E-02
                                      70.0                  .9290E-02
                                      75.0                  .9762E-02
                                      80.0                  .1060E-01
                                      90.0                  .1312E-01
                                      95.0                  .1635E-01
                                      97.5                  .1849E-01
                                      99.0                  .2403E-01
                                      99.5                  .2529E-01
      EXAMPLE/SAMPLE=X1+(X2+X3)**2+X6+X3**1.5
     FUNCTION PROBLEM NO.=    2
 0
 0                     OUTPUT EVALUATIONS.
    SAMPLE SIZE=.  200   ACCURACY ON 95 PER CENT CONFIDENCE INTERVAL= 2.5 PER CENT
 0  DISTRIBUTION PARAMETERS.   MEAN =     .9423E-02   STANDARD DEVIATION =    .5322E-02
 0
 0                             DISTRIBUTION CONFIDENCE LIMITS
 0                            CONFIDENCE (PER CENT)      FUNCTION VALUE
                                        .5                  .2839E-02
                                       1.0                  .3139E-02
                                       2.5                  .3294E-02
                                       5.0                  .3608E-02
                                      10.0                  .4358E-02
                                      20.0                  .5462E-02
                                      25.0                  .5899E-02
                                      30.0                  .6238E-02
                                      40.0                  .7119E-02
                                      50.0                  .8503E-02
                                      60.0                  .9495E-02
                                      70.0                  .1023E-01
                                      75.0                  .1119E-01
                                      80.0                  .1248E-01
                                      90.0                  .1537E-01
                                      95.0                  .1836E-01
                                      97.5                  .2186E-01
                                      99.0                  .3247E-01
                                      99.5                  .3439E-01
      EXAMPLE/SAMPLE=X1+(X2+X3)*(X4+X5)+X6+X7+(X3*X5*X3)**0.5
     FUNCTION PROBLEM NO.=    3
```

```
0
0                       OUTPUT EVALUATIONS.
   SAMPLE SIZE=.  200   ACCURACY ON 95 PER CENT CONFIDENCE INTERVAL= 2.5 PER CENT
0  DISTRIBUTION PARAMETERS.   MEAN =    .8775E-02   STANDARD DEVIATION =    .3838E-02
0
0                          DISTRIBUTION CONFIDENCE LIMITS
0                          CONFIDENCE (PER CENT)     FUNCTION VALUE
                                      .5               .1938E-02
                                     1.0               .2232E-02
                                     2.5               .2868E-02
                                     5.0               .4203E-02
                                    10.0               .4599E-02
                                    20.0               .5551E-02
                                    25.0               .6021E-02
                                    30.0               .6511E-02
                                    40.0               .7308E-02
                                    50.0               .8023E-02
                                    60.0               .8901E-02
                                    70.0               .9776E-02
                                    75.0               .1037E-01
                                    80.0               .1176E-01
                                    90.0               .1439E-01
                                    95.0               .1633E-01
                                    97.5               .1801E-01
                                    99.0               .1917E-01
                                    99.5               .2090E-01
```

AVAILABILITY: A PRACTICAL APPROACH ON AEROSPACE INDUSTRY

S. SANZ FERNANDEZ DE CORDOBA
Aeronautical Engineer, Ph. D.
CASA Engineering Division
Avda. John Lennon, s/n.
28065 - MADRID (SPAIN)

ABSTRACT.

In the present paper, the concepts of reliability, maintenability and availability when used in connection to each other are reviewed.

Mathematical models for availability of single and multiple systems are shown, and the dependence from reliability, maintenability and (for multisystems) the number of units involved is considered.

Differences when dealing with plants or vehicles are reviewed.

1. INTRODUCTION.

Reliability, availability and maintenability as applied to systems are three different concepts which, taken togheter, give a measure of the fitness of a system to perform its intended task during the duty cycle.

However, not all the three concepts are applicable to all systems, nor are they independent from each other: The applicability of the three concepts depend of the definitions of the term systems, task and duty cycle, and the modification of one of them upon a given system, most times implies modification of the other two.

In this presentation, we will try to define each one of the three concepts, analize the implications of the different definitions and the interrelationships which

A. G. Colombo and A. Saiz de Bustamante (eds.), Systems Reliability Assessment, 45–64.

such definitions stablish between the three concepts.
In the presentation, examples taken from the aerospace field will be used to highlit the different concepts.

2. DEFINITIONS.

2.1. BASIC DEFINITIONS.

Before attempting the definition of system reliability, availability and maintenability, three basic concepts have to be defined, namely system, task, and duty cycle. The other concepts will be referred to them, thus it is important to clarify its meaning.

A system is the integration of elements inside a defined physical space which are functionally interrelated.

I must signify that the physical space referred to in the definition, needs not to be physically connected at all times. In this definition, a system of communications through satellite may be defined as the emmiting station, the satellite and the receiving station. However, it needs be physically contained inside closed boundaries, in such a way that we may define in a univoque way wether a physical element belongs or not to the system.

The functional interrelation between the system elements sets a boundary of the possible extension of the system. Naturally. the functional interrelation must be considered as directed to a task, so task must be defined at this point:

Task(s) is the ultimate reason(s) for which the system has been set up.

The definition is intended to cover system which perform multiple task, but also to exclude multiple systems which may be directed to a unique task. Taken together with the previous definition os system, prevent us from extending capriciously the boundaries of a system. Coming back to the previous example of the satellite communications system, the vehicle which sets the satellite into orbit does not belong to the system, since there is not functional relationship between the launcher and any of the other three elements. A failure in the launch vehicle is not a failure of the communications system; it is a failure in building or setting up the system.

Duty cycle, as used here, means either, the predicted life span of the system or a part of the system life such that the system final and initial configuration are very similar.

Since all systems have finite life, no duty cycle may have the same initial and final conditions, but they may be quite close if the duration of the cycle is small when compared with life span.

The above definition of duty cycle is useful for most systems (both, one mission and reparable systems), but there is a notable exception related to mission oriented delivery vehicles, for which the final configuration of the system may be quite different from the initial one. The typical example are modern fighter-bomber aircrafts, for which the duty cycle (usually called mission in military terms) is defined as arriving to the target point, delivering the bombs and returning safely to a non enemy base. Thus, although the aircraft may be (and usually is) rapaired (i.e. the life span is normally composed of several duty cycles) and reffited for a new mission, these stages are not part of the duty cycle, thus the final configuration may be quite different from the initial one. There is a practical reason for defining such a special duty cycle: there is a serious uncertainty about the aircraft lifespan. This may change however during a long attrition war if the industry is not crippled, such as was the case in WWII, where the Germans obtained good satatistical approaches of the life span of their bombers, and redefined the duty cycle in statistical terms.

In all cases the duty cycle of the system is a time, with clear starting and finishing moments, on which the system performs its intended task.

2.2. RELIABILITY, MAINTENABILITY AND AVAILABILITY.

2.2.1. RELIABILITY. When used in the context of performance we are confronted with two different reliability definitions:

System Reliability is defined as the probability of a system working at the designed conditions at (or until) a predetermined time, assuming that at time t=0 the system works as designed.

The above definition, for not reparable systems equals the probability of the system not failing before the given time, while for reparable systems the time is counted only as active or in service time when the predetermined time is longer that the duty cycle. In other words maintenance time is not accounted for in system reliability.

Mission Reliability, as different from System Reliability is defined as the probability which the system has to perform the intended task(s) during the duty cycle, assuming all necessary systems for the task are working at the beginning of the duty cycle.

For one mission vehicles, mission reliability depends directly of system reliability, and they are not really different concepts. But in case of multi-task reparable systems, the relation between both concepts is not so direct. The classical example in aerospace is an unreliable engine starting system for a commercial multi-engine powered aircraft. The duty cycle is normally defined from engine on to engine off conditions, and the possibility of in flight engine re-starting is an added security, no a part of any mission. Furthermore, regulations require to be able to keep flying securely with one half of the engines out. Thus the most unreliable engine starting system has no effect on mission reliability, (it has an effect on availability, as we shall see later). It is also inmediate to see that in a multitask and/or multi-mission systems, some tasks or missions may have a higher reliability with the same system reliability.

Nevertheless, it does exist in general a correlation between system and mission reliability when the system is properly choosen, i. e. we do not include in the system parts which are unnecessary for the mission on which each subsystem is needed.

In the present paper, we will restrict ourselves to system reliability in such case, the reliability can be measured by the parameter defined as:

$$\lambda = \mathrm{MTBF}^{-1} \qquad (1)$$

MTBF = mean time between failures.

The definition of MTBF, depends of the definition of system failures. For complex reparable systems, failures are considered those out of design conditions of the system which need put the system out of service for repair. Failures which may be kept without repair until the next programmed maintenance period, are not considered failures under this definition.

Most complex systems tend to have a constant MTBF over the greatest part of its useful life after the burn-in, so we will assume λ to be constant unless otherwise specified.

2.2.2. MAINTENABILITY. The most used definition of maintenability (or maintenability index) of a system is the total time (or its inverse) under which you may expect a fixed percentage of all failures (normally 60% or 90%) will be repaired.

The definition assumes that the repair time will be normally distributed, so a mean repair rate plus the typical deviation will determine the statistical distribution of the repair time. Instead of using both parameters most designers use only one, namely the "near" mean repair rate, with the 60% criterion (normally used for systems of modular design, which exhibit small typical deviation) or the "near" maximun time for all repairs, with the 90% criterion. Maintenability in then defined by the parameter

$$\mu = MRT^{-1} \qquad (2)$$

MRT = mean repair time

μ exhibits a time behaviour very similar to that of λ, being constant for the greatest part of the useful life after a training period (notice, however, that μ is increasing during the burn-in period).

This definition however is only useful for relatively simple or modular systems like cars or electronic desk calculators. In such systems, the times for servicing or repair depend very little from the origin of the trouble, and statistics on added times of trouble shooting plus repair can be done almost independent of the type of failure. Since, in general, spares and manpower are readely available for such systems, the "mean repair time" gives a real indication of the system maintenability.

When we go to more complex systems, the repair times may be multiplied by high factors from one failure to another. In adition, the general use in such systems of highly reliable equipment, and the implementation of strict preventive maintenance policies, tend to yield very flat frequency distribution curves when a long period of time is considered after the initial burn-in-period. This tendency is further increased in high performance systems, such as modern fighter aircraft, by the tendency to substitute elements which exhibit an unpredicted tendency to fail during the initial service period by more reliable elements and by the difficulty to obtain spares and adequated repair teams for sophisticated equipment parts, all of which introduce random noise in the process.

The flatness of the frequency distribution curve extending over long range of repair times, makes somewhat useless the concept of mean repair time.

What normally develops in such system is groups of repair times, that is, the repair times distribute more or less normally around a certain discrete number of mean times.

Thus, in order to be able to handle maintenability in mathematical terms, the concept of "mean repair time" is substituted in such systems by the concept "maintenance time per service hour". This time is normally measured in man/hours, and represents the number of men work hours needed with the system out of service for each hour of service of the system. Such time is usually well determined, since the "on condition repair time" (i.e. the time to repair unexpected failures) which represent the maximun uncertainty, is only a small part of the total, the biggest part being the "programmed maintenance time", i.e. maintenance needed independent of the system failures and introduced to prevent failures.

The maintenance time per service hour includes both, programmed and on-condition maintenance and for complex systems is normally distributed around a mean maintenance time per service hour.

This form of measuring maintenability is incorrect, since in fact, what is defined in such parameter in the availability of the system, but a mean repair time can be deduced from it, provided it is known the failure rate as we shall see in later paragraphs.

For the time being, lets define the parameter v (maintenance stop proportion) for such systems, as

$$v = \frac{\text{Maintenance time per in service hour}}{\text{Number of people working simoultaneously}} \qquad (3)$$

This parameter v gives the number of hours which the system must be out of service, and must not be confused with μ defined above.

We shall see later that v is in fact related to the steady state unavailability of the system.

2.2.3. AVAILABILITY. Availability is defined as the probability of a system being in service conditions at a determined future time. Clearly for the near future, the availability will depend, among other factors, on the present state (available or not) of the system, but in the far future, the availability of reparable systems will depend only on its failure rate λ and its repair rate μ.

The mathematical model for availability for systems with two states is deduced by Prof. Saiz de Bustamante in his paper "Stochastic Processes, Markov Chains and Systems Availability" in this same volume, so I will just write here the final results. The instanteneous availability (A) and unavailability (U) and steady state availability (As) are given by the equations

$$A = \frac{\mu}{\mu + \lambda} + \frac{\lambda}{\mu + \lambda} \exp\left[-(\lambda + \mu)\ t\right] \qquad (4)$$

$$U = \frac{\lambda}{\mu + \lambda}\left[1 - \exp\left[-(\lambda + \mu)\ t\right]\right] \qquad (5)$$

$$As = \frac{\mu}{\mu + \lambda} \qquad (6)$$

It must be remarked that these equations are derived on the condition that the system has only two states, either in service or failed and being repaired. When the system has a third state, such as being not in service without haveing failed (i.e. during the programmed maintenance as was described en par. 2.2.2), the above equation need certain adjustements to be of practical use. The parameter defined in (3) is related to the steady state unavailability, since obviously

$$Us = \frac{v}{1 + v} \qquad (7)$$

and defining maintenability through v is incorrect as we said in the previous paragraph.

3. MATHEMATICAL MODEL.

3.1. GENERAL CONSIDERATIONS ON MULTI-STATE SYSTEMS.

When we refer to availability, we can definitely say that all systems have at least two states: Available (the system is ready to start its duty cycle and has a chance to complete its task) or non available (either the system can not start its duty cycle, or has not chance to complete its task). But as soon as the system is multitask or a certain degree of degraded performance may be aceptable, intermediate situations arise. Some of them may be defined as states (intrinsic system conditions which appear in discrete forms) while some others look like continous parameters.

Lets take as an example a modern multiengine civil aircraft certified for carrying passangers and cargo. If every thing works correctly, is available. If it has all its engines removed for overhaul is not available. If it has most of its navigation aids failed, it will be unavailable por carrying passangers (by law) but it can still take cargo on VFR (visual conditions flight). This is an intermediate state. But lets assume it has eight different fuel tanks. What is the state when a number from 1 to 7 of the tanks can not be used, while all the remaining element works? The availability in such conditions do not depend at all on the system. It may be available or not depending on the distance you have to go and/or the possibilities of refuelling on the way. So, in this degraded system conditions, availability can not be defined from knowledge of the system.

In order to obtain useful models for these complex systems, it is necessary to reduce those intermediate and undetermined states to better defined conditions, or, plainly, to yes or no states.

3.2. REDUCTION OF STATES IN MAINTENABILITY AND AVAILABILITY.

The usual way of reducing the maintenability to two states (either reparable or not) is to introduce in the definition an external parameter usually cost: When the repair cost goes beyond a certain percentage of the price of a new system, the old system is declared non reparable. Maintenance of the system beyond that point, although it may be done by different reasons (i.e. not having a new system available) is not modelled. Before reaching that point, the system is declared repairable whenever a failure occurs.

Cost is the most common but not the only external criterion used. Another one is time. A systems gets repaired for a number of years and then declared non reparable. This is usually done for systems which have difficult to repair or expensive components which, after a number of years of work become too unreliable, so their failure has to high a probability and is too risky to keep them. This time can measured in years or in cycles. Tipically, guns are discarded after a certain number of cycles (shoots) because the risk of exploding becomes too high.

Whatever the external parameter used, all systems are reduced to two states for modelling: repairable or non repairable, and the model deals with the repairable part (useful life).

A similar exercise is done for availability. The system is reduced to available or non available state during the useful life. In complex system, to do so require several steps.

In the first step, if the system is multitask, each task is separated and studied independent from the others.

Second, for each task a maximun degraded but still admisible duty cycle is defined.

These two initial steps define a unique duty cycle. This duty cycle needs not necessarily be one of the missions the system is designed to fulfill. Most time is in fact a representative mixture of missions, representativeness referred in this case to availability for the real duty cycles.

The third step is choosing the minimum number of subsystem or equipment which has to be in working conditions at the start of the duty cycle for the system having the minimum aceptable chance of performing the defined task. In other words, what it is done in this step is defining the subsystems or equipment which may be inoperative (failed) and are not going to be repaired inmediately. It is important to realize that those failed systems, which we define as not preventing the start of the duty cycle, are going to be repaired eventually but only when either some other equipment essential for the performance of the task fails or when the system is put out of service for programmed maintenance operations. The list of equipment thus defined is variously known as "minimun equipment list" (used in civil aviation) "dispatch list" (used in military aviation), etc., but it is invariably present in all complex systems. The need for such list is clear: Given the levels of redundancy built in such systems, it is both, hopeless and unnecessary to require all the equipment being operative to start the duty cycle.

With the above definitions, the availability of the system on technical terms is uniquely determined. The system is or is not technically available, and there are not intermediate states.

We have used the expression technically available since there is still a fourth step to be performed which is to include the unavailability derived from programmed maintenance. Thus a system technically available may still be unavailable if it has been in service a predetermined time. Once this programmed maintenance is included, availability is perfectly determined.

The above steps, however, have an influence on the definitions of λ and μ, as we shall see next.

3.3. MODIFICATION OF FAILURE RATE λ AND REPAIR RATE μ.

Steps three and four described above amount to a redefinition of the failure rate in terms of availability. In step three, we are discarding a series of failures (those not preventing the start of the duty cycle), and in step four adding as failures the need of programmed maintenance.

In practical life, however, the failure rate is only modified by the first group (i.e. failures not preventing the start of the duty cycle are not counted in the failure rate) while the second group is introduced in the formulation by modifying the repair rate.

The modification of λ is a straight forward operation. The modification of μ for inclusion of programmed maintenance is usually done in one of the two following ways:

a) For relatively simple systems such that the mean repair time (MRT) is known, μ is redefined as

$$\mu = \left(MRT + \frac{MTBF}{TP} \times MPMT \right)^{-1} \tag{8}$$

where

MRT = main repair time

TP = predetermined in service time between programmed maintenance.

MPMT= mean programmed maintenance time (average time the system is out of service because of one programmed maintenance).

MTBF= mean time between failures.

Sometimes, the programmed maintenance is cyclic itself. In that case TP is the longest period in the cycle and MPMT is the added times out of service for all the programmed maintenance operations in the cycle.

b) For more complex systems, where MRT is not defined but rather v as defined in (3) is known, μ is defined as follows

$$\mu = \frac{\lambda}{v} \tag{9}$$

Observe that an alternative definition to (3) for v is v = MPMT/TP with the parameters defined above.

Comparing this last expression with (8) above, we deduce that for complex systems MRT = 0, that is, on condition repair is included in v.

With the definitions of μ given in (8) or (9), and with λ modified as explained above, what we have done is substitute our system by another which is equivalent to ours for availability porpouses with constant λ and μ.

For this system, equations (4) and (5) apply, and are the main mathematical model for the study of availability.

4. INFLUENCE OF λ AND μ ON AVAILABILITY.

4.1. ONE SYSTEM MODEL.

Some important conclusions, not always understood by operators or designers can be reached from eq. (6), the steady state form of model (4). We list some of them:

a) The steady state availability tends to 1 when μ increases and λ remains finite. This is nor surprising. When the mean time to repair is nearly zero, the system is practically always available, even if it has a significant tendency to fail, because of the minimum repair time.

b) Steady state availability tends to zero when either μ decreases towards zero (mean repair time is very large) or λ becomes very large (mean time between failures is very small). This is logical too.

c) Steady state availability is around 0.5 (50% of time available) when μ and λ are on the same order of magnitude, independentily of the absolute value of μ and λ.

This latest conclusion is not generally known or understood. It can be stated in more general terms as follows: Any change in maintenance practices which keep the proportion between μ and λ has no effect on steady state availability of the system.

This means in practice that improving reliability of the system by means of reducing the time between inspections or substitutions does not modify availability of system. And the opposite holds as well: Since increasing the times between inspections and substitutions (increase of μ) has normally the effect of decreasing the MTBF because of poor maintenance (i.e. λ increases as well) this practice does not significantly change availability.

All this means that availability is very much fixed by the design, and the effect of changes in maintenance policies is minimal on it.

But the way of improving availability by improving the design is even more poorly understood in general.

To see how the effect of independent changes in μ and λ affect availability of the system, let us take derivatives of equation (6) with respect to μ and λ writing derivatives in incremental form, we get

$$\frac{\delta As}{\delta \mu} = \frac{\lambda}{(\lambda + \mu)^2} \qquad (10)$$

$$\frac{\delta As}{\delta \lambda} = \frac{-\mu}{(\lambda + \mu)^2} \qquad (11)$$

For a given δAs, the same in both equations, we can combine them to get the expression

$$\lambda \; \delta\mu = (- \delta \lambda) \; \mu \qquad (12)$$

Equation (12) is surprising as can be seen by studying the following extreme cases.

<u>First case</u>: $\mu >> \lambda$ (13)

This condition implies MRT << MTBF; or, in other words, the system has a very satisfactory design with respect to availability, since the repair time is small, the time between failures is large and As tends to 1. From (12) and (13)

$$\delta \mu >> (- \delta \lambda) \qquad (14)$$

Or in other words, a small increase in reliability means a large increase in availability while further improvements in maintenability do not increase significantly availability. Changes or new design in such systems, should concentrate in improving reliability.

Modern transport aircraft belong to this type with typical values of λ of about 0.1 and values of μ ranging from about 0.5 (military transport) to around 1.3 (civil transport).

<u>Second case</u>: $\mu << \lambda$ (15)

This is the case of poor design for availability,

with long repair times, small time between failures and As nearly zero.

From (12) and (15)

$$\delta \mu << \ (-\delta \lambda) \tag{16}$$

Which means that is far more efective to improve availability to concentrate the redesign effort on improving maintenability than on improving reliability.

Sophisticated modern combar aircraft belong to this category, with tipical values of λ around 0.4 and values of μ of around 0.2.

Notice that we are speaking of absolute values (i.e. increasing in 0.01hour-1 μ versus decreasing 0.01hour-1λ).

In relative terms, the increase in availability do not depend of the values of μ and λ, since equation (12) can be written as

$$\frac{-\delta \lambda}{\lambda} = \frac{\delta \mu}{\mu} \tag{17}$$

which means that equal percentage improvements on μ and λ have equal improvement on availability.

Written in the form (17), we have an alternative way of expressing the most effective way of improving availability: The most effective way to improve availability is to act over the parameter for which the possible improvement represents a large percentage of change in the parameter current value.

4.2. MULTI SYSTEM MODEL

An important part of the complexity of many systems is the redundancies included in their design. Those redundancies are in general built to insure the availability of one or more of the redundant parts as a mean to keep the complete system available. Such is the reason for including up to four independent hydraulic system in modern passanger aircraft, on which main control surfaces are hydraudically actuated.

In other cases, redundancies are built to allow repair being done while keeping the system available. This case is not common in aerospace, but it is fairly common in other industrial areas, such as chemical industry or utilities companies.

A third, and very important case, is having a number of complete system to guarantee that a part of them will be available at all times.

This later case is very common when dealing with fairly complex system with relatively low individual availability, such as advanced fighter aircrafts.

In all cases, the problem is the same, and can be stated as finding the number of independent systems needed to ensure, within a certain probability margin, that a determined number of them will be available at any time in the future.

The most simple case can be worked out assuming we have identical units with constant failure rate and repair rate. This model, although simple, applies fairly well to many practical cases such as tactical aviation units, aeroclubs, regional charter/taxi fleets, etc. For this type of groups, the steady state availability of m units out of n, may be modelled as a random function with a binomial distribution. The probability of success of any single outcome at a determined time in the far future is the probability of a determined unit being available at that time, or As as given in equation (6), while the probability of failure is the probability of that unit not being available at that time or (1 - As).

Thus, the probability of having m or more units available out of n can be written as

$$Am = \sum_{x=m}^{n} \binom{n}{x} \frac{\mu^{x} \lambda^{n-x}}{(\mu + \lambda)^{n}} \tag{18}$$

While the probability of having exactly x units available and (n-x) non available (i.e. needing repair) is

$$A = \binom{n}{x} \left(\frac{\mu}{\lambda + \mu}\right)^{x} \left(\frac{\lambda}{\lambda + \mu}\right)^{n-x} \tag{19}$$

The importance of correctly using the binomial distribution model for availability instead of the simplified approach independent random events, so common in reliability analysis, must be emphasized since availability, mainly in high performance aircraft rarely reaches the levels of reliability, and the orders of magnitude never go smaller than 0.01.

As an example, the probability of having available 1 or more fighters from a group of 5, when the independent availability of each fighter is 0.15, becomes 55.63%. If we use the simplified theory that the availability of 1 or more has to be bigger that the availability of 1 will be $A = 5 \times 0.15 = 0.75$, we could deduce that the availability of 1 or more is at least 75% a significant error, since the real availability is 20% lower.

The errors can be much larger when we want individual terms, i.e. the probability of exactly having m working and $n-m$ not working which is an important parameter at the time of sizing the maintenance facilities.

For the case of multisystem model, usually the interest lays on studying the increased availability by including one more unit with the same individual availability versus increased availability with the same number of units by improving individual availability (be it changing μ or λ).

Since n is discrete, no closed form of dAm/dn exists on eq. (18) and the results must be tabulated.

In the next table, availability of m or more units for a fleet of 5 units with individual basic availability of 0.35 is given in the first column. The same availability figures are given when individual availability is increased to 0.4, and when the number of units are increased to 6.

m	n = 5 A = 0.35	n = 5 A = 0.40	n = 6 A = 0.35
1	.8840	.9222	.9246
2	.5716	.6630	.6809
3	.2352	.3147	.3529
4	.0540	.0870	.1174
5	.0053	.0102	.0223

The same case for $n = 15$ units, availability 0.35, with increase of availability to 0.40 and to 16 units, gives the following values.

m	n = 15 A = 0.35	n = 16 A = 0.40	n = 16 A = 0.35
1	.9984	.9995	.9990
2	.9858	.9948	.9902
3	.9383	.9729	.9549
4	.8273	.9095	.8661
5	.6481	.7827	.7108
6	.4357	.5968	.5100
7	.2452	.3902	.3119
8	.1132	.2131	.1594
9	.0422	.0950	.0671
10	.0124	.0338	.0229
11	.0028	.0093	.0062
12	.0005	.0019	.0013
13	.0001	.0003	.0002
14	.0000	.0000	.0000
15	.0000	.0000	.0000

Again, when we compare both tables, we see that the increment in the number of basic units is more effective than the increase on individual availability when the number of units is small, while it is less effective for larger number of units. Unfortunately, no closed form can be obtained in this case thus each particular case has to be numerically calculated.

It is also to be remarked that the sensivity to both parameters increases when we consider the availability of a large number of units compared with the total number of units involved. Thus in the first example, the availability of 5 units or more is doubled when individual availability increases from 0.35 to 0.40, and is multiplied by 4 when the number of units involved is 6. The availability of 1 or more units is only increased in about 4% in both cases.

5. IMPROVING AVAILABILITY OF SYSTEMS.

5.1. GENERAL CONSIDERATIONS ON VEHICLES VS. PLANTS.

Reliability principles and techniques are quite similar no matter whether we are dealing with a termo-electric plant, a car or a interplanetary space ship. In all cases, we are dealing with failures probabilities and effects, and the techniques of component reliability, path duplication and separation, etc, work the same. And the final result we are looking for, namely the ability of the system to perform its task without failures or in spite of failures, applies to them all.

As soon as we go to maintenability, we can set a first clear cut between systems we can broadly define as vehicles and systems we can define as plants. Plants are designed or can be designed to keep operating at least in a degraded state while they are mantained (repaired) while vehicles are designed so they are out of service when they are repaired. And the reason is clear: vehicles are meant to undergo big and frequent changes of speed, i.e. need going through cycles of accelerations. Energy economy thus demands small masses involved, and this requires small spaces and close packing of equipment, which translates into difficult or impossible accessibility to most equipment while in service. Plants, wether fixed or not (i.e. an airplane carrier ship or an space platform orbiting Earth are as much plants as a thermo nuclear electric power plant), occupy larger spaces, equipment is accessible while in operation, and proper design will prevent stopping the plant while repairs are made.

The difference in maintenability reflects itself in availability. If we are dealing with a plant, and we improve reliability by duplicating a path with a second identical path, fully independent, availability is inmediately improved. In a vehicle, such change may not change availability.

Take as an example a simple series systems with MTBF = 1000 hours and MRT = 100 hours. The availability will be:

$$A = \frac{1/100}{1/100 + 1/1000} = 0.9091$$

If we improve reliability by adding a second identical path, assuming the system is a plant (i.e. repairs in a path can be done while the other is working), the availability of the plant is

$$A = \sum_{x=1}^{2} \binom{2}{x} \left(\frac{10}{11}\right)^{x} \left(\frac{1}{11}\right)^{2-x} = 0.9917$$

a significant improvement.

But if the system is a vehicle, so repairs can not be done while it is working, but it is allow to operate with one of the two paths failed, we have to redefine.

$$\text{MTBF} = 2 \times 1000 = 2000 \text{ hours}$$

an improve in reliability. But, when the second path fails and it is taken to repair, there will be 2 failures (1 on each identical path), thence

$$\text{MRT} = 2 \times 100 = 200 \text{ hours}$$

Thus the availability has not changed at all. Even if we repair both paths simoultaneously (which requires double repair team, thus more money), the MRT will be kept at 100 hours but availability will be:

$$A = \frac{1/100}{1/100 + 1/200} = 0.9523$$

or only 4% increase versus 8% increase in the plant.

This result is only put here to show how the improvement on reliability without affecting maintenability is far more difficult in vehicles than in plants, or, in other words, how availability becomes far more closely tied to the conceptual design in the vehicle case than in the plant case.

In plants, improving reliability by means of using more reliable components, duplication of less reliable paths, and other known techniques, leads in general to improved availability. Improving maintenability, i.e. reducing mean repair time by improving accessibility of components (being careful not to make them more prone to accidental or environmental damage), improving removal of components easiness by fixing them with standard elements, etc, also improve availability since μ and λ are fairly independent in a well designed plant.

With vehicles, we have to be more careful since reliability and maintenability are much more connected to each other.

5.2. IMPROVING AVAILABILITY OF VEHICLES.

General techniques of reliability improvements such as duplication, path separation or concentration and, in fact, most of the effective techniques to improve reliability have little effect on availability (improving or worsening it) unless maintenability is carefully considered at the same time.

At the time of design, improving accessibility of components will grately improve μ without, in general, affecting λ. But this is not always the case if placing components in accessible parts make them more prone to accidental or environmental damage, as it is usually the case. It is also usual in vehicles that weight considerations and mass center requirements impose location of components, thus it is not possible to locate them just for maintenance reasons.

A careful consideration has to be given in complex vehicles to the real origin of MRT. As said before, many times the biggest component can be external to the system, for example, trouble shooting, due to need of complex external equipment of very low availability. In those cases, the designer may accept intrinsic deterioration of λ and/or μ in the system, but increase availability after all. Such is the case when built in test are included in the system. Intrinsic reliability and maintenability are deteriorated (the BIT has failures and needs repair), but availability may be improved because of not needing low availability external test equipment.

REFERENCES.

1) P.T. O'CONNOR "Practical Reliability Engineering" John Wiley & Sons, 2nd edition, 1985.

2) "Training Program For Reliability, Probability and Safety Analysis" Douglas Aircraft Company, FAA Contract No. DOT-FA75AC-5123.

3) MIL-STD-781.

4) Standard Mathematical Tables.

5) P.D.T. O'CONNOR ET AL "Reliability Engineering", Hemisphere Publishing Corp. 1988.

RELIABILITY IN NUCLEAR POWER PLANT OPERATIONS

Julio González
NUCLENOR, S.A.
Hernán Cortés, 26
39003 SANTANDER
SPAIN

ABSTRACT. Nuclear Power Plant design and operations are strongly conditioned by the operating licence conditions and regulations. The present paper tries to emphasize some areas where reliability methods are of value to improve plant operations whithin the bounds defined by legally applicable nuclear regulation.

1. INTRODUCTION

Reliability technics do have a substantial contribution to make to nuclear power plant design and operation improvement in the two interrelated areas of Availability and Nuclear Safety.

A measure of Availability is the ratio of the effective number of operating hours to the total number of hours over a certain period of time, or the ratio of the electric energy generated to the energy that could had been generated over the same period of time under ideal conditions.

By Nuclear Safety we understand the freedom from nuclear danger or risks. Risk has been defined in quantitative terms as the expected magnitude of a certain type of damage per unit time to an individual or to a community as consecuence of a certain activity. It is obvious that every nuclear plant must strive for high availability and low risk, and that both concepts as previously defined are of a statistical nature.

Nuclear power generation is a heavily regulated industrial activity. There exists a large body of rules and regulations that apply to the design as well as to the operation of nuclear power plants. It can be said that, by and large, regulations are prescriptive in nature and that they do not relate in any clear manner to probabilistic ideas.

Light-water reactors licenced under USA regulations have to be designed so as to be able to undergo a variety of accidents -called De-

A. G. Colombo and A. Saiz de Bustamante (eds.), Systems Reliability Assessment, 65–74.

sign Basis Accidents- without serious damage to the nuclear fuel and, therefore, without undue radiological consecuences. The total rupture of the largest pipe of the reactor coolant system together with the failure of any single piece of equipment of the emergency core cooling system ("single failure" criterium), under a total loss of electrical power supply to the plant from off-site sources, is considered to be the maximum credible Design Basis Accident. This unlikely scenario plays a central role in the design and operation of light-water reactor plants. Most safety systems are evaluated in relation to their capacity to mitigate this particular accident. Many operating limits -applicable under normal conditions- refer back, one way or another, to regulatory constraints imposed on the maximum Design Basis Accident consecuences.

Early probabilistics studies of nuclear power plant accidents have clearly shown that one can find scenarios of higher probability than the design basis accidents and with similar potential to be the cause of radioactive releases from nuclear power plants. The limited experience derived from accidents and transients that have occurred so far seems to point in the same direction.

After the accident of the Three Mile Island plant in 1979 many power stations were the subject of probabilistic safety studies of the type that we will discuss in the following section. This type of studies are considered today as a valuable method to analyze certain aspects of nuclear power plant design and operation that applicable regulations treat in a rather arbitrary manner or that lie outside the regulatory bounds and scope.

In the following section we discuss in a general manner the reliability methods applied, so far, to the analysis of nuclear power plants. In the last section we discuss three specific areas where probabilistc methods do have a clear and interesting application. These topics are Technical Specifications, testing procedures and preventive maintenance.

2. THE PROBABILISTIC SAFETY ASSESSMENT (PSA) IN NUCLEAR POWER PLANTS (NPP)

2.1 THE OBJECTIVES AND USES OF A PSA

The PSA is an analytical technique for integrating diverse aspects of design and operation in order to assess the risk of a particular nuclear power plant and to develop an information base for analyzing plant-specific and generic issues. In achieving these objectives, a PSA serves a variety of purposes.

An assessment of the plant-specific risk provides a measure of potential accident risks to the public as well as insights into the adequacy of plant design and operation.

The assesment of the adequacy of plant design and operation is achieved by identifying those sequences of potential events that dominate the risk, and establishing which features of the plant contribute the most to the frequency of such sequences. These plant features may consist of potential hardware failures, common-mode failures, human errors during testing and maintenance, or procedural inadequacies leading to human errors, etc. Thus, a probabilistic analysis can reveal, features of a plant that may merit closer attention and so provides a focus for improving safety.

The other objective achieved by PSA is the development of an information base for analyzing plant-specific and generic issues. This information base identifies dominant accident sequences and plant features contributing significantly to risk; as well as containing the models of the plant developed during the study. Knowledge of the most probable severe accidents could assist in developing strategies for coping with accidents beyond the current design-basis accidents. This information could thus provide a focus for training operators to deal with such accidents. Emphasis could be placed on diagnosing the most-probable severe accident sequences and on providing information and guidance to the operators on how to cope with such accidents.

Information developed in the assessment could help in making decisions about the allocation of resources for safety improvements, by directing attention to the features that dominate plant risk.

The plant models developed in the assessment can serve a wide spectrum of uses. They can be used to assess the safety significance of operational occurrences at the plant; they can also be used to assess the applicability and significance of occurrences at other plants. The models provide a basis for evaluating alternative design changes to improve safety.

The utility may well find the information and models developed in the study to be useful in training personnel. The analysis draws together diverse aspects of plant design and operation into an integrated model that could provide plant operators and utility engineers with a different perspective that could prove useful in the training of both.

2.2 MAIN TASKS OF A PSA

The fundamental tasks of a study of this nature are:

A) A description of the so called Initiating Events. These are the events that require some sort of protection or mitigation function to be carried out.

B) A description of possible accidents. That is, a description of the combinations of failures that will result on a serious accident following an initiating event.

C) A reliability analysis of all important plant systems in relation to accident mitigation.

D) A treatment of potential human errors that may contribute to the likehood of a given type of accident or that could be the direct cause of it.

E) A quantitative treatment of the reliability models that represent potential accidents.

A discussion of the tasks of a PSA can be found in many publications about PSA methods and guidelines. A brief description can be found in the author's contribution to the Proceedings of the ISPRA course held in Madrid in 1986.

2.3 PSA LIMITATIONS

The results of a PSA are uncertain to some extent due to the limitations of the models and data that go into the study. There are three important areas where improvements are considered necessary in order to refine PSA analysis. The following is a brief description of ongoing activities in Spain in these areas.

2.3.1 Plant simulation

Since 1984 there has been a continuous activity in Spain to assimilate and improve PSA methods. These activities are funded by the Research and Development program of the electrical utility sector. Plant simulation is a specific area of work within this program.

Plant behaviour under all types of transients and accidents has to be better known. The general objective of this project is to obtain a code package that will allow to simulate in a realistic manner plant response to the spectrum of initiating events considered in a PSA.

This simulation tool must have the following basic characteristics:

A) Simulate plant behaviour until a safe state is reached or core degradation starts.
B) Include models for all plant systems that could be used in response to an accident or transient.
C) Include features to take into account operator action in the course of a transient.

An improved capacity to simulate plant behaviour will permit a better analysis of accident sequences, which is a basic task of a PSA study.

2.3.2 Human reliability

Another area in need of well accepted models and data is human reliability.

It is generally accepted that the fundamental aspect of human reliability with respect to PSA is associated with the adequate diagnostics of plant state rather than with the reliability of the operators in performing specific operations with plant equipment.

A R+D project in this area is underway in Spain. The main objectives of this project are:

A) A definition and classification of the principal factors that influence nuclear power plant operator behaviour in relation to diagnostic and response to plant transients.
B) A data collection campaign at nuclear power plant simulators. The data should be relevant to the validation of models to describe the performance of Spanish plant operators.
C) A methodology to apply human behaviour models to PSA.

2.3.3 Data Banks

Initiating events frequency, equipment failure data and other probability estimates required as PSA input data are not sufficiently well known in some cases.

A data bank that will collect data from all Spanish nuclear power plants is starting to operate at this time. The bank will store data relating to three different aspects of plant operation and energy generation: plant availability, plant events and equipment reliability.

Plant event data is characterized in a way that is relevant to the initiating event analysis task of a PSA.

Equipment reliability data is sent to the Component Event Data Bank of the European Communities (CEDB), based at the JRC in Ispra. This is a reliability data bank designed to be useful for system reliability analysis.

3. PSA APPLICATIONS

3.1 TECHNICAL SPECIFICATIONS

The Technical Specifications is a voluminous and legally binding document that establishes a number of conditions and limitations on the normal operation of a given nuclear power plant.

The technical basis for these limitations and conditions are of such a general and vague nature, that one could make substantial changes on the specific requirements without having to revise the technical basis. For this reason one can say that Technical Specification requirements are set in a rather arbitrary manner.

Technical Specifications do have, however, a considerable bearing on plant operations, and so, there is a growing interest nowadays in finding a reasonable balance between plant availability and safety through revised Technical Specifications.

It seems clear that the traditional nuclear regulatory approach does not have much to contribute to Technical Specifications improvement. Methods based on reliability and risk assessment technics are being studied and proposed at this time.

Technical specifications affect system reliability primarily in the way that they establish Limiting Condition for Operation (LCO) and surveillance requirements for on line test and repair of nuclear power plant safety system components.

Limiting Conditions for Operation refer to the legal constraint on safety system component outage times that are imposed by the regulatory body as part of the reactor operating license. Generally, when a safety system component is removed for repair or test for a period of time there is a period of increased vulnerability concerning the probability that the affected safety system will be available to mitigate an accident. This period of increased vulnerability exists until the component is restored to service. The constraint on the duration of this period, the allowed outage time (AOT), is the aspect of LCO that is of interest here. There are methods that relate measures of risk to the AOT. Only by explicity relating risk to AOT can outage times be constrained by placing limits on risk.

Surveillance requirements refer to the legal requirement to test active, standby safety system components on a periodic basis to verify that they are operational. These requirements are usually expressed as an upper bound on the period between tests. There is no reason why a minimum test period could not also be specified for component testing, as there are also test-caused failures and test-caused degradations that need to be controlled. Risk acceptable test intervals balance these risks in an attempt to achieve an acceptable low, overall risk.

At all times, the assessment of the outage times contributing to a system's unavailability will be easy if a PSA is readily available. On the other hand, adequately judging the acceptable increase of AOTs-caused unavailability is a difficult job, due to the relatively small effect on the NPP-risk as a whole.

Basically, there are two possibilites of judgement:

1. Relative judgement: the increase of system unavailability should be limited by a factor F as compared to system unavailability whith previous AOT.
2. Absolute judgement: the increase of plant risk should be limited in case of changing the AOT.

The first assessment is easy to apply and supplies well defined figures. Its disadvantage is that, for the same safety task, this method could produce shorter AOTs for systems of higher reliability than for those of lower reliability, and does not take into account the importance of the system in the context of the accident sequences.

Basically, the second assessment supplies risk-relevant figures. In this case it must be recognized that the change of risk by a change in the AOTs is small in comparison with the risk uncertainty. In addition, this concept is much more complicated to handle, as, for comparison purposes, a complete plant-specific PSA will have to be available.

3.2 SYSTEM TESTING

Testing of standby components is performed with the frecuency prescribed in the plant Technical Specifications. Testing is done because the implicit underlying failure model for the components is assumed to be one where the component unavailability increases with the component standby time. The component is implicitly assumed to be best modeled by a time-based failure rate. If this were not so, i.e., if the component model were assumed to be a demand model, then testing would not be necessary. In fact, testing would increase component unavailability if the component were removed from service for the duration of the test. Since tested components are implicitly assumed to have a time-based failure rate model, the component unavailability depends on the test period, and the test period must therefore be determined for each such component.

In reality, most components probably have both time-based and demand contributions to their unavailability. Figure 1 represents the failure probability versus test period for a component with a combined demand and time-based failure model. For relatively short test periods, the demand failure model is probably a reasonably good model for most components. However, as the test period lengthens, the time-based model is probably a better representation. There are no scales on this figure, and the details undoubtedly depend on failure probabilities for certain component types.

Testing of safety system components often, but not always, results in removing the component from service for the duration of the test period. When the component is removed from service by a test, it is unavailable to mitigate an accident. Therefore, the test has contributed to the unavailability of the component and the safety system. On the other hand, if the components were not tested, the component's unavailability would steadily increase and approach one (certainty) at some time in the future. Thus, under the assumption of the combined time-based and demand component failure model, testing is required to keep individual component unavailabilities from approaching one. It is necessary to estimate both the component failure contribution and the test contribution to system unavailability.

Component test periods are determined not only by formal tests of the components, but also by incidental "tests" that occur during normal operations. For example, consider the valves in an emergency feedwater system, where monthly valve stroke test is used to test the valve actuators, i.e., to test the "valve fails to open" failure mode, but the "valve plug" failure model is only tested for when the system is actuated. Suppose also, that the feedwater system is required to operate whenever a turbine trip occurs, and the turbine trip frequency is six to ten times per year, even though there is no formal test procedure to test for this failure mode. As a second example, consider a decay heat removal system where the pump tests do not test the complete flow path. If the system is used at refueling, however, then there is a complete flow path test of every component failure modes at refueling or due to anticipated transients must be used in addition to formal tests when determining component test periods.

3.2.1 Completness of tests

Redundancy and diversity in the design of safety systems is required by regulation as a means to ensure safety. However, Technical Specifications requirement for testing do not necessarily guarantee that all components of a system are operable after succesful system testing.

Tests made on the systems do not always check all the components that are in a standby mode. Technical Specifications of the plant require some tests in order to verify that the system completes its function (supplies a certain flow, arrives to a fixed voltage, etc.), but do not require to check every path within the system that can satisfy the same purpose.

The PSA reflects the importance of certain components that appear in the results of the accident sequence quantification, due to its high unavailability value. As it was said before component unavailability increases with the component standby time, and so it is obvius that reducing this time, that is, forcing the tests to cover most of the components of the system, we get lower failure mode and accident probabilities.

Standard reliability methods and PSA are definitely valuable tools to improve system testing procedures for highly reliable systems.

3.2.2 System Testing Strategies

There are two main testing strategies: sequential testing and staggered testing. Considering a two-train system, under the sequential testing strategy, train A would be tested and then restored to service. Immediately after restoring train A, train B would be removed from service, tested, and then restored. This constitutes a system test. At the next test (e.g., one month later), the process would be repeated.

Under the staggered testing strategy, train A would be tested and restored. Two weeks later, train B would be tested and restored. The process would continue in this way.

Components are assumed to fail according to a time-based component failure model, then staggered testing is proved to be an improvement over sequential testing. The optimum scheme is to test the second leg midway between tests of the first leg, and to continue with this staggering scheme.

3.3 PREVENTIVE MAINTENANCE

Plant technical specifications do not consider preventive maintenance explicitly, and, although there appears to be differences in interpretation, it is generally understood that removing safety systems from service to perform preventive maintenance is not allowed.

Preventive maintenance is of importance in relation to various operating aspects as well as on overall plant availability. One of this aspect is the duration of the operating cycle. Nuclear power plants are considering or going to operating cycles as long as two years. The interest of performing preventive maintenance over this period of time is obvious. Thus, there is a growing interest in establishing the bases for an acceptable policy for preventive maintenance.

Preventive maintenance consists of actions performed to retain a system in a specified operating condition. Systematic inspection and performance are coupled with services designed to prevent or substantially reduce incipient failures. Preventive maintenance includes those time elements expended to replace short life parts, to perform clean-up and housekeeping tasks, and to perform routine inspection and alignment. It is possible, as in the case of corrective maintenance, to quantify preventive maintenance actions. This can be accomplished by measuring the time required to perform these tasks.

Once the mean time of preventive maintenance is measured for a type of component, it is estimated the annual mean unavailability that this practice brings to the equipment. That value can be added to the unavailability due to corrective maintenance on the component, so that a sensibility analysis can be made about the variation in the final frequency of accidents due to the increment in the total maintenance unavailability.

It is necessary to point out that the preventive maintenance always results in a smaller number of corrective maintenances, so that the decrement of unavailability due to the concept of maintenance is not the direct sum of both types of maintenances. However it is difficult to evaluate "a priori" in numeric terms, how much the unavailability due to corrective maintenance is going to decrease because of doing preventive maintenance, and a conservative position is adopted summing directly both unavailabilities.

From the comparison between these two kinds of maintenance unavailabilities, can be deduced the benefit of carrying out a preventive maintenance policy or the usefulness of doing it with one or another frequency, because it is obvious that the unavailability due to preventive maintenance should never be greater than the one that would be given by corrective maintenance in case of no sistematic and scheduled action.

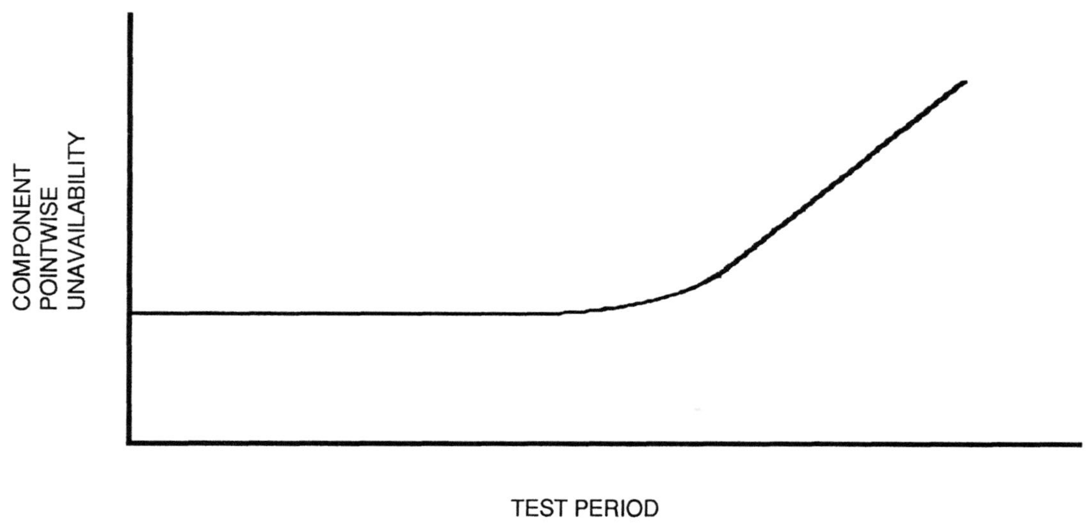

FIGURE 1 COMBINED DEMAND AND TIME-BASED COMPONENT FAILURE MODEL

SECTION 2

ADVANCES ON SYSTEMS RELIABILITY TECHNIQUES

ADVANCES ON FAULT TREE AND EVENT TREE TECHNIQUES

S. CONTINI and A. POUCET
Commission of the European Communities
Joint Research Centre - Ispra Establishment
Institute for Systems Engineering
21020 Ispra (Va) - Italy

ABSTRACT. This paper refers to a previous lecture note /1/ which explains the basics of fault tree and event tree methodologies. The mathematical analysis of fault trees is presented in more detail. Furthermore, a more in depth discussion is given on issues which were only briefly mentioned in the first lecture note, such as

1) the analysis of non-coherent fault trees;
2) the importance calculations;
3) the analysis of event trees.

Although a prior knowledge of the basic issues is needed to understand this note, it was tried to produce a self-contained text for what concerns the treated issues, even if this led to some repetition with respect to the first note.

1. LOGICAL ANALYSIS OF NON-COHERENT FAULT TREES

1.1 Mathematical representation of fault trees

A logic diagram like a fault tree can be represented in a mathematical form by its structure function.

Let $N = \{E_1, E_2, E_3, E_4, \dots, E_n\}$ be the set of primary events of a fault tree. The generic primary event is associated with the failed state of a component, an external system, and so on.

To each primary event E_i, a binary indicator variable y_i can be associated such that:

$$y_i = \begin{cases} 1 \text{ if event } E_i \text{ occurs} \\ 0 \text{ if event } E_i \text{ does not occur} \end{cases}$$

A vector $\vec{Y} = (y_1, y_2, y_3, \dots, y_n)$ can then be used to indicate which primary events occur and which do not.

The binary indicator variable of the Top event can be written as a function of the vector $\vec{Y}$:

A. G. Colombo and A. Saiz de Bustamante (eds.), Systems Reliability Assessment, 77–102.

$$\Phi(\vec{Y}) = \begin{cases} 1 \text{ if the Top event occurs} \\ 0 \text{ if the Top event does not occur} \end{cases}$$

This function is called the **structure function** of the fault tree. The vector $\vec{Y}$ is time-dependent as at any time some event may have occurred (e.g. components failed) and other events have not.
A structure can be either **coherent** or **non-coherent.**
A coherent structure is a structure for which:
1) the structure function is monotone increasing with the vector $\vec{Y}$, i.e.:

$$\Phi(y_1,y_2,\ldots,y_i = 0,\ldots,y_n) \leq \Phi(y_1,y_2,\ldots,y_i = 1,\ldots,y_n)$$

   for all y_i with i = 1,2,...,n. (This means in practical terms that the system must in no way become "better" if some failure occurs);
2) each indicator variable of $\vec{Y}$ is relevant, i.e. there is no y_i such that:

$$\Phi(y_1,y_2,\ldots,y_i = 0,\ldots,y_n) = \Phi(y_1,y_2,\ldots,y_i = 1,\ldots,y_n)$$

   for all y_j with j 0 1,2,...,i-1,i+1,...,n. (This means in practice that every failure must have some relevance for the system).

Fault trees in which the "NOT" operator appears do not fulfill the first conditions and, hence, are not coherent. The analysis of non-coherent fault trees is more complicated than the analysis of coherent trees. In most practical cases, however, the use of the "NOT" operator, either direct or hidden in, e.g., "EXCLUSIVE OR" gates, can be avoided or the "AND-OR-NOT" logic can be written in an approximate "AND-OR" logic. This is the approach described in this section. Whenever applicable, it allows to perform the quantitative analysis on the set of significant minimal cut sets, as if the structure function were coherent.

1.2 Practical Examples of Non-coherent Fault Trees

Example 1:
Consider a two train stand-by system, subject to periodic testing and maintenance. The plant policy prescribes that at a given moment only one of the two trains may be unavailable because of test and maintenance. The fault tree for assessing the unavailability on demand of such a system is given in Fig.1. In this case, the use of the NOT gate allows to obtain a very straightforward representation. It is possible to represent the situation without using the NOT operator (Fig.2), but the model becomes less transparent.

Example 2:
Consider the system represented in Fig.3. Assume that the system has to provide 100% flow to line A, and that, because of a rupture, line B can absorb all flow and thus has to be isolated. The fault tree for

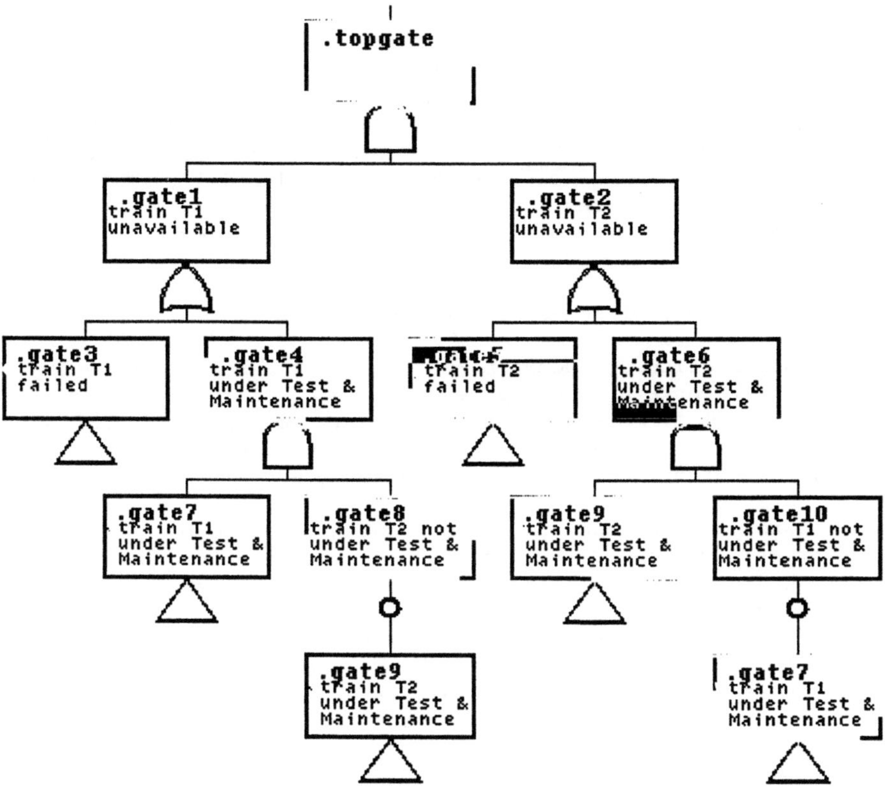

Fig.1: Fault tree for two train stand-by system.

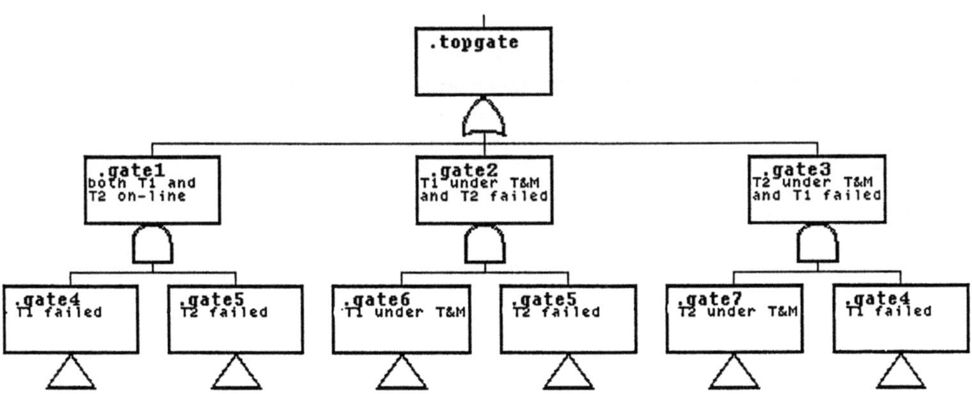

Fig.2: Same as Fig.1 but without using NOT gates.

this system is represented in Fig.4. This fault tree contains mutually exclusive states for, e.g., valve V4 (event V40, "valve V4 was not closed", and event V4C, "V4 is closed") and, hence, is not coherent. It is interesting to note that the fault tree can be made coherent by assuming some boundary conditions on the operation of the system. If valve V4 is assumed to be open and may not be closed for isolating the break, then the fault tree becomes coherent, since the event V40 is removed because is "sure to occur", as well as the event V4C because it becomes "impossible to occur".

1.3 Logical Analysis of Non-coherent Fault Trees

As previously mentioned, non-coherent fault trees are those logical structures that contain negated events, e.g. components in their working state. This means that if a system failure combination contains a negated variable, that combination does not occur if the component, to which the variable is associated, is in its failed state.

Therefore the concept of minimal cut set has to be replaced with that of **prime implicant***. Whereas a coherent structure presents only one s-o-p expression of minimal cut sets, a non-coherent structure may present two or more minimal s-o-p expressions. Consequently, the methods of analysis are very time-consuming, as they are those applied in switching theory for minimizing the combinatorial logical networks /2/ and, in fact, a non-coherent fault tree is equivalent to a complex logical network. In general, most events in a fault tree refer to failure modes of components. Therefore, negated events are often associated with the working state of components, and the corresponding probability is generally very close to 1. In that case, from the quantitative analysis point of view, this probability can be assumed as equal to 1, with an acceptable (conservative) degree of approximation.

If this (realistic) assumption is made, the negated events are only needed for deleting impossible combinations during the analysis of the fault tree. Since these combinations may be generated only by repeated events, the non repeated negated events can be a-priori deleted (i.e. before applying the analysis procedure). The negated-replicated events are taken until they are no more needed for deleting impossible MCS. The final result of this method is the s-o-p expression of MCS as if the fault tree were coherent.

Example: (Fig.5)
F appears only once. $P(\bar{F}) \simeq 1$, therefore the event is replaced by a "sure to occur" event ($P(\bar{F})=1$). A "sure to occur" event can be removed

---

*A prime implicant is a combination (conjunction) of logical boolean variables (some in negated form and the remaining in the affirmative form) that verifies the s-o-p expression, and does not contain any other combination.

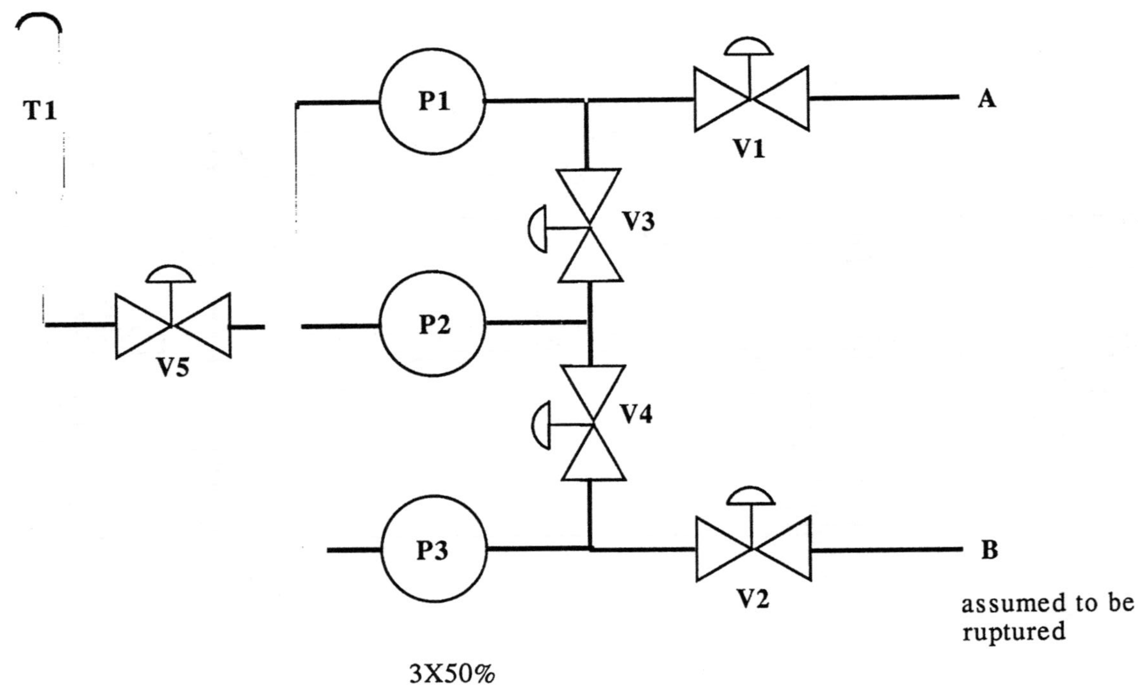

Fig.3: Sample system.

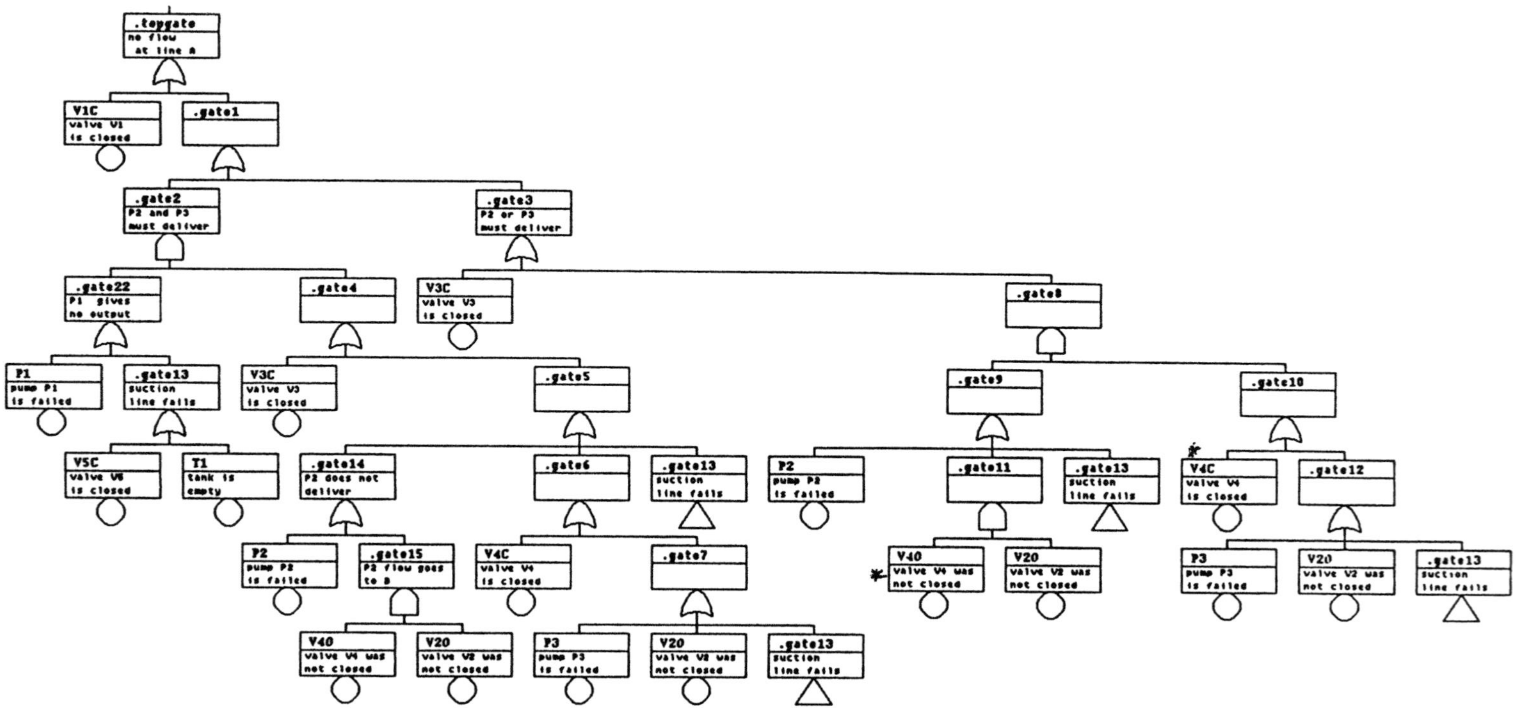

Fig.4: Non-coherent fault tree for the system in Fig.3.

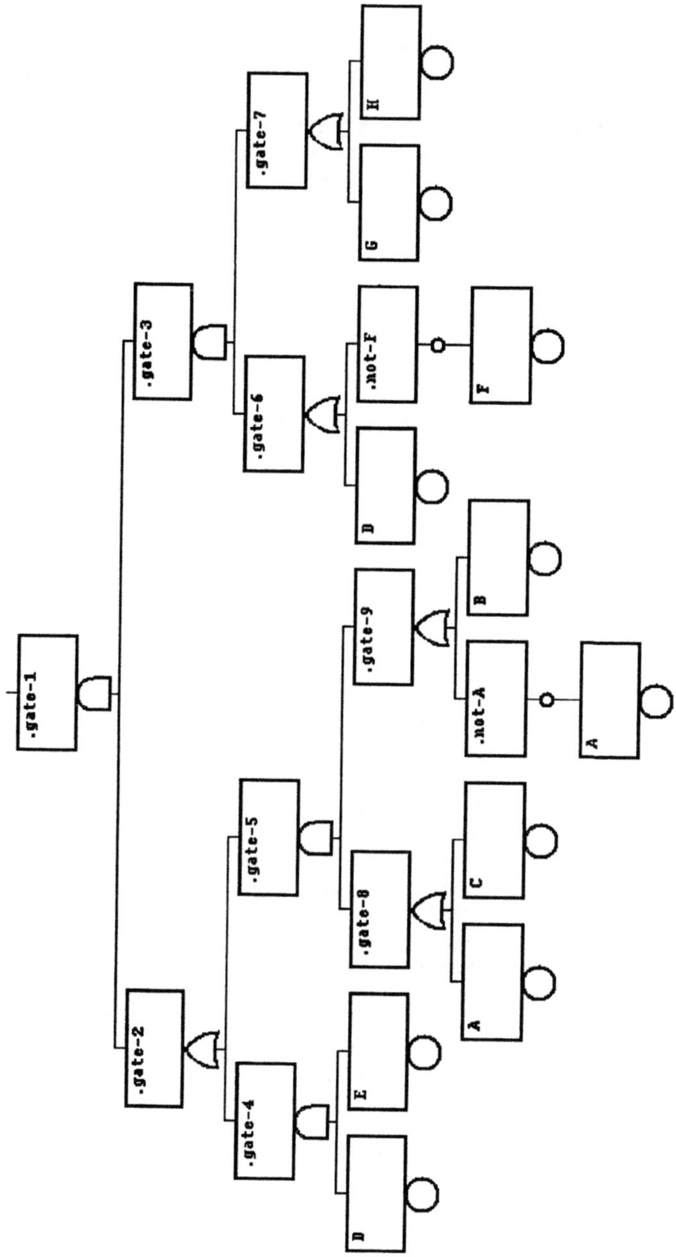

Fig.5: Example of a non-coherent fault tree.

from the fault tree in a proper way:

- if it occurs under an "AND" gate, it is simply removed;
- if it occurs under an "OR" gate, the "OR" gate becomes "sure to occur" (i.e. the event and its father gates until the first "AND" gate will be removed).

The resulting tree is represented in Fig.6.

The event $\bar{A}$ is a repeated event. Therefore it will be needed for removing impossible cut sets. First the tree is analysed as if it were a coherent tree. The list of cut sets is:

G
H
D$\bar{E}$
A$\bar{A}$
A$\bar{B}$
C$\bar{A}$
$\bar{C}$B

The MCS A$\bar{A}$ is impossible and removed from the list. In the remaining list $\bar{A}$ can be eliminated because $P(\bar{A}) \simeq 1$. The final list of MCS is therefore:

G
H
DE
AB
C

2. QUANTITATIVE ANALYSIS OF FAULT TREES

Useful quantitative information on the behaviour of the system can be obtained through the application of the fault tree probabilistic theory /3,4/. This section presents the fundamental equations for determining the unavailability, the expected number of failures, and the unreliability, starting from the component level up to the system level. Afterwards, the calculation of some indexes of components importance will be described.

2.1 Unavailability Calculation

2.1.1 Component unavailability. Consider a new component put into operation at time t = 0. Consider that when it fails, repair processes are started immediately. The history of the component can be graphically represented as shown in Fig.7.

Let f(t) and g(t) be the p.d.f. (probability density functions) of the times to failure and times to repair:

$$f(t) = \lambda(t)\, e^{-\int_0^t \lambda(x)dx} \qquad g(t) = \mu(t)\, e^{-\int_0^t \mu(x)dx}$$

where $\lambda(t)$ and $\mu(t)$ are the failure and repair rates, respectively.

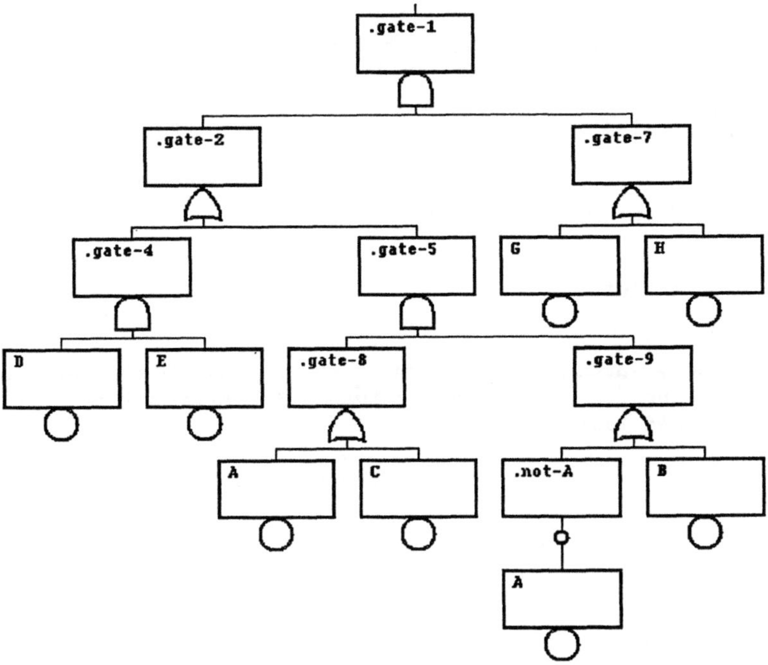

Fig.6: Fault tree in Fig.5 after removing unrepeated negated events.

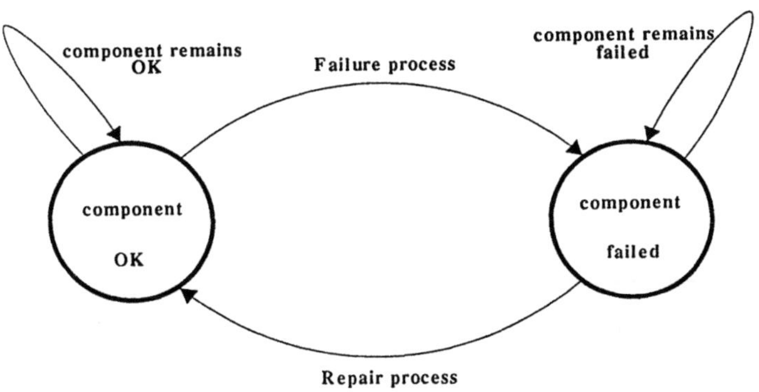

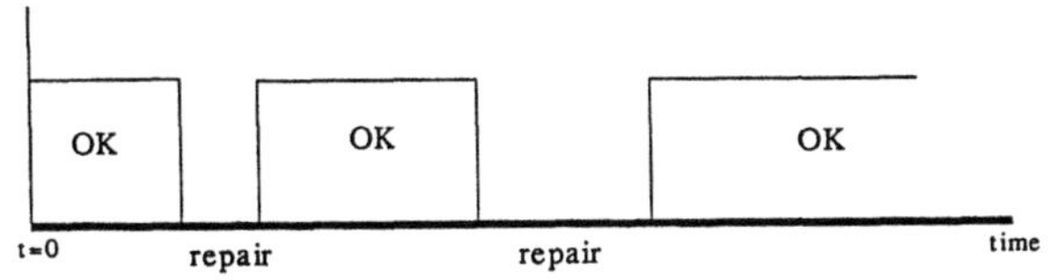

Fig.7: State diagram and state sequence for a repairable component.

The mean time to failure (MTTF) is given by:

$$m = \int_0^\infty tf(t)dt$$

and analogously the mean time to repair (MTTR) is:

$$\tau = \int_0^\infty tg(t)dt$$

The renewal theory allows to determine the unavailability expression of a component subject to whichever type of distributions for the times to failure and times to repair.

It can be shown that the Laplace transform $q^*(s)$ of the component unavailability $q(t)$ is given by the following equation /5/:

$$q^*(s) = \frac{f^*(s) - f^*(s)g^*(s)}{s|1 - f^*(s)g^*(s)|} \tag{1}$$

where $f^*(s)$ and $g^*(s)$ are the Laplace transforms of $f(t)$ and $g(t)$.

The inverse Laplace transformation gives $q(t)$. As a matter of fact the analytical solution is possible only in some simple cases. If $f(t)$ and $g(t)$ are exponential distributions (λ and μ constant), the Laplace transforms have the following expressions:

$$f^*(s) = \frac{\lambda}{s+\lambda}$$

$$g^*(s) = \frac{\mu}{s+\mu}$$

Therefore:

$$q^*(s) = \frac{\mu}{\lambda+\mu}\,\frac{1}{s} + \frac{\lambda}{\lambda+\mu}\,\frac{1}{s+(\lambda+\mu)}$$

and:

$$q(t) = \frac{\lambda}{\lambda+\mu}\,(1-e^{-(\lambda+\mu)t}) \tag{2}$$

Furthermore:

$$m = \int_0^\infty t\ f(t)dt = \frac{1}{\lambda} \qquad \tau = \int_0^\infty t\ g(t)dt = \frac{1}{\mu}$$

The unavailability presents a steady state value which is reached after 3-4 times the MTTR

$$\lim_{t\to\infty} q(t) = \frac{\lambda}{\lambda+\mu} = \frac{MTTR}{MTTF+MTTR} \quad (3)$$

The steady state value can also be obtained through the application of the "final value theorem" which states that for any function $\phi(t)$ which has a Laplace transform:

$$\lim_{t\to\infty} \phi(t) = \lim_{s\to 0} s.\phi^*(s) \quad (4)$$

where $\phi^*(s)$ is the Laplace transform of $\phi(t)$.
Applying this theorem to Eq.(1) the steady state value can easily be obtained without passing in the time domain:

$$\lim_{s\to 0} s\, q^*(s) = \lim_{s\to 0} s\left\{\frac{\lambda}{s(\lambda+\mu)} + \frac{\lambda}{(\lambda+\mu)(s+\lambda+\mu)}\right\} = \frac{\lambda}{\lambda+\mu}$$

The steady state value of the unavailability can also be obtained knowing only the mean values m and τ of f(t) and g(t). It is indeed possible to prove that, for any type of f(t) and g(t), the following equation holds:

$$q_\infty = \lim_{t\to\infty} q(t) = \frac{\tau}{m+\tau} \quad (5)$$

Consider now the case of a not repairable component. An unrepairable component may be conceived as a component with infinite repair time. The availability coincides with the reliability; in fact, saying that the unrepairable component is functioning at a generic time t is equivalent to say that it never failed before t.

For the exponential case, setting $\mu = 0$ (i.e. $\tau = \infty$) in (2) the equation for the unreliability can be obtained:

$$q(t) = 1-e^{-\lambda t} = F(t) \quad (6)$$

Therefore, for not repairable components:

$$F(t) = q(t)$$

2.1.2 Unavailability of a minimal cut set (MCS). Under the hypothesis of components independence, the unavailability of a generic MCS of order n is given by the product of the components unavailability:

$$Q(t) = \prod_{i=1}^{n} q_i(t) \tag{7}$$

In fact, in order that a MCS (parallel system) be failed at time t it is necessary that its components be all failed at t. Obviously, it does not mean that all components fail at the same time t: the product means that when the last component fails the othersare still in the failed state.

Analogously, the steady state unavailability of a MCS is the product of the steady state values of the components:

$$Q = \prod_{i=1}^{n} q_{\infty i} \tag{8}$$

2.1.3 Unavailability of the Top event. Let x_i be a binary variable so that:

$$x_i(t) = \begin{cases} 1 & \text{if the i-th MCS occurs at time t} \\ & \text{(all its components are failed)} \\ 0 & \text{if the i-th MCS does not occur at time t} \\ & \text{(at least one of its components is not failed)} \end{cases}$$

The cut set expression in terms of binary indicator variables is:

$$x_i(t) = \prod_{j=1}^{n} y_j(t)$$

where y_j is the indicator variables of the j-th primary event. The $E(x_i(t)) = Q_i(t)$ is the unavailability of the i-th MCS.

The Top event occurs at a given time t, when at least one of its MCS occurs at that time. If N is the number of MCS, the Top event unavailability $Q_s(t)$ is given by:

$$Q_s(t) = E\left[\coprod_{i=1}^{N} \prod_{j=1}^{n_i} y_j(t)\right] \tag{9}$$

The high number of MCS of real systems makes it impractical to determine the exact value of the Top event unavailability. The application of the inclusion-exclusion principle to the disjunction of the N minimal cut sets gives the bounds that are used in practice:

- upper bound: $$P_{1top}(t) = \sum_{i=1}^{N} Q_i(t) \tag{10}$$

- lower bound: $$P_{2top}(t) = P_{1top}(t) - \sum_{i=2}^{N} \sum_{j=1}^{i-1} P_{ij} \quad (11)$$

in which P_{ij} is the occurrence probability of the intersection of MCS i and j.

The upper bound is very easy to calculate. It does neglect the probability of more MCS's occurring at the same time which is sensible in highly reliable systems and is often called the **"rare event approximation"**.

Other bounds have been proposed for the Top event unavailability. The following, known as the Barlow-Proschan bound, gives a better approximation than Eq.(10), and it gives the exact value when all the MCS are independent (i.e. they do not contain common events):

$$P_{B-P} = 1 - \prod_{i=1}^{N} (1-Q_i(t)) \quad (12)$$

2.2 Calculation of the Expected Number of Failures (ENF)

2.2.1 Expected number of failure of a repairable component. A parameter of paramount importance for a repairable item is the expected number of failures $W_f(t)$ in the time interval 0-t. To determine the ENF, the concept of failure intensity $w_f(t)$ is needed. $w_f(t)$ is the probability that the component fails in the interval t,t+dt. It is useful to point out the difference between the failure intensity $w_f(t)$ and the failure rate $\lambda(t)$. In $w_f(t)$ the component could have been failed and then repaired (one or more times) before t, while the failure rate $\lambda(t)$ requires that the component must be survived until t without any failure (given the component good at t=o).

For determining $w_f(t)$ it is useful to pass through the Laplace transform. In fact, it can be proved that /5/:

$$w_f(s) = \frac{f^*(s)}{1 - f^*(s)g^*(s)} \quad (13)$$

Thus given f(t) and g(t) it is possible to obtain the equation for $w_f(s)$. Unfortunately, only in some particular cases the analytical solution $w_f(t)$ can be found.

The failure intensity $w_f(t)$ presents a steady state value that is reached after about 3-4 times the mean time to repair :

$$w_f = \frac{1}{m+\tau} = \frac{1}{MTTF+MTTR} \quad (14)$$

where m and τ are the mean values of the functions f(t) and g(t), respectively.

The equation in the Laplace domain for the expected number of failures $W_f^*(s)$ can be determined by means of the following formula:

$$W_f^*(s) = \frac{f^*(s)}{s\left[1-f^*(s).g^*(s)\right]} \tag{15}$$

Comparing (13) and (15) it can be observed that:

$$W_f^*(s) = \frac{1}{s}\, w_f^*(s) \tag{16}$$

In the time domain, Eq.(16) represents the following fundamental equation:

$$W_f(t) = \int_0^t w_f(t)dt \tag{17}$$

If the steady state value is used, then

$$W_f(t) = w_{f_\infty}.t = t/(m+\tau) \tag{18}$$

In these cases $W_f(t)$ is linear in t.

For the exponential distribution of the failure and repair times one obtains:

$$w_f^*(s) = \frac{\lambda\mu}{s(\lambda+\mu)} + \frac{\lambda^2}{\lambda+\mu} \cdot \frac{1}{s+\lambda+\mu}$$

$$w_f(t) = \frac{\lambda\mu}{\lambda+\mu} + \frac{\lambda^2}{\lambda+\mu}\, e^{-(\lambda+\mu)t} \tag{19}$$

Noting with $p(t) = 1-q(t)$ the availability of the component, it can be seen that $w_f(t) = \lambda p(t)$. This holds, however, only in the exponential case. In this case the steady state expression is as follows:

$$\lim_{t\to\infty} w_f(t) = \frac{\lambda\mu}{\lambda+\mu} \tag{20}$$

It is important to compare the expressions for the unavailability, unreliability and mean number of failures for a repairable item. Comparing (1) and (15) it can easily be seen that:

$$q(t) < W_f(t)$$

Furthermore, it can be proved that $W_f(t)$ can be expressed as follows:

$$W_f(t) = \sum_{k=1}^{\infty} F^{(k)}(t)$$

where $F^{(k)}(t)$ represents the probability of the occurrence of at least K failures in the time interval 0,t. Then, since $F^{(1)}(t)$ represents the component unreliability, it follows that:

$$W_f(t) > F(t)$$

Therefore, for repairable components:

$$q(t) < F(t) < W_f(t)$$

In case of a non repairable component (infinite repair time) it can easily be shown that:

$$w_f(t) = \lambda e^{-\lambda t} = f(t)$$

and

$$W_f(t) = \int_0^t w_f(t)dt = \int_0^t f(t)dt = F(t)$$

Therefore, for non repairable components:

$$q(t) = F(t) = W_f(t)$$

2.2.2 Expected number of failures for a MCS. In order that a MCS of order n occurs in the time interval t,t+dt it is necessary that (neglecting, as usual, multiple failures in dt):

a) n-1 events be verified at t, and
b) the last event occurs in t,t+dt.

The probability of the event b) is given by $w_{fj}(t)dt$; the probability of event a) is nothing but the product of the remaining components unavailability, that is:

$$\prod_{\substack{k=1 \\ k\neq 1}}^{n} q_k(t)$$

Therefore, the failure intensity of a generic i-th MCS can be obtained as follows:

$$w_{f_i}(t)dt = \sum_{j=1}^{n_i} w_{fj}(t) \prod_{\substack{k=1 \\ k\neq j}}^{n_i} q_k(t)dt \qquad (21)$$

The ENF of the MCS is therefore given by:

$$W_{fi}(t) = \int_{o}^{t} w_{fi}(t)dt$$

In case of constant failure and repair rates, the expression for $W_{fi}(t)$ can easily be determined by considering, conservatively, $\lambda = w_f$ for any event (i.e. p(t) = 1):

$$W_{fi}(t) = \sum_{j=1}^{n} \frac{\lambda_j}{(h+\gamma)q_j(t)} Q_i(t).t \qquad (22)$$

where

$$\gamma \begin{cases} 1 & \text{if } \lambda_j \text{ is the failure rate of a repairable or inspected component} \\ 0 & \text{if } \lambda_j \text{ is the failure rate of an unrepairable component} \end{cases}$$

h is the number of unrepairable components belonging to the MCS

2.2.3 Expected number of failure of the Top event. The ENF of the Top event can be obtained through the integration of the Top event failure intensity:

$$W_{ftop}(t) = \int_{o}^{t} w_{ftop}(t)dt$$

Due to the difficulty in determining the exact values, bounds are used in practice. The upper bound for the failure intensity is given by the sum of the failure intensities of the MCSs:

$$w_{ftop}(t) \leq \sum_{i=1}^{N} w_{fi}(t) \qquad (23)$$

Consequently, the ENF for the Top event is given by:

$$W_{ftop}(T) \leq \sum_{i=1}^{N} W_{fi}(T) \qquad (24)$$

2.3 Calculation of the unreliability

2.3.1 Unreliability of a component. The unreliability F(t) of a single component can be calculated by means of the following well known equation:

$$F(t) = 1 - \exp(-\int_{o}^{t} \lambda(t)\, dt)$$

In case of constant failure rate:

$$F(t) = 1 - \exp(-\lambda t)$$

2.3.2 Unreliability of a minimal cut set. It is useful to distinguish between the case where all components are unrepairable from the case where the on-line repair of components is allowed.

When all components are unrepairable, the unreliability can be calculated by means of the same equations described in 2.1 for unavailability assessments. This case will not be further considered.

More complex is the case where components can be repaired during the mission. The exact unreliability of a MCS can certainly be calculated through the application of the markovian technique. In fault tree analysis, however, use is made of conservative bounds, some of which are briefly described below.

A bound largely applied is the Expected Number of Failures $W_f(t)$. In fact, as already mentioned, it can be proved that the expected number of failures $W_f(t)$ is expressed as follows:

$$W_f(t) = \sum_{k=1}^{\infty} F^{(k)}(t) \qquad (25)$$

Since $F^{(1)}(t)$ is the unreliability of the cut set (probability of failure by t), then:

$$W_f(t) > F(t)$$

Moreover, being $F^{(2)}(t)$ corresponding approximately to $F^{(1)}(t)$ squared, it can be realised that for "sufficiently reliable" components $W_f(t)$ represents a good bound for F(t).

Another bound for F(t) has been proposed by Brown:

$$F(t) = 1 - \exp(-A\ t) \qquad (26)$$

where: $$A = \frac{Q_\infty \cdot \sum_{i=1}^{n} \mu_i}{1-Q_\infty} \quad \text{and} \quad Q_\infty = \sum_{i=1}^{n} q_{\infty i} = \prod_{i=1}^{n} \frac{\lambda_i}{\lambda_i+\mu_i}$$

It can easily be seen that Eq.(26) holds for MCS made up of repairable components only, while $W_f(t)$ has not this limitation.

2.3.3 Unreliability of the Top event. Similarly to the unavailability case, for the Top event unreliability only conservative bounds can be assessed. The Top event unreliability $F_{top}(t)$ is generally obtained by summing up the unreliabilities of the MCSs:

$$F_{top}(t) \leq \sum_{i=1}^{N} F_i(t) \tag{27}$$

2.4 Importance Calculations

One of the most interesting aspects of a fault tree analysis is the possibility to identify weak parts in the system, susceptible for subsequent design improvement /6/.

The logical analysis of the fault tree gives already some indication; if lower order (e.g. first order) minimal cut sets are present, these could give a qualitative indication of particularly critical components. However, more quantitative measures of criticality and importance are developed below.

Assume a system fault tree expressed by its structure function $\Phi(\vec{Y})$ with $\vec{Y} = \{y_1, y_2, \ldots, y_i, \ldots, y_n\}$ the vector of the indicator variables of the n events in the fault tree. Event is said to be critical if the system is in a state such that $\Phi(\vec{Y})$ becomes 1 if event i occurs, or:

$$\Phi(1_i, \cdot) - \Phi(0_i, \cdot) = 1$$

with $\Phi(1_i, \cdot) = \Phi(y_1, \ldots, y_{i-1}, 1, y_{i+1}, \ldots, y_n)$

and $\Phi(0_i, \cdot) = \Phi(y_1, \ldots, y_{i-1}, 0, y_{i+1}, \ldots, y_n)$

Example: Consider a parallel system with three components.

The structure function: $\Phi(\vec{Y}) = y_1 . y_2 . y_3$, and, therefore:

$$\Phi(1_1, \cdot) = 1.y_2.y_3 = y_2.y_3$$

$$\Phi(0_1, \cdot) = 0.y_2.y_3 = 0$$

Hence component 1 is critical if $y_2 . y_3 = 1$ or if both components 2 and 3 are failed and, hence, the probability that component 1 is critical is given by the product of the unavailabilities of components 2 and 3.

Birnbaum Importance $I_i^{BB}(t)$: More in general, it can be proved that the probability of component i being critical is given by:

$$\text{Prob}\{\Phi(1_i, \cdot) - (0_i, \cdot)\} = \frac{\delta Q_s(t)}{\delta q_i(t)} \tag{28}$$

where $q(t)$ and $Q_S(t)$ are respectively the component unavailability and the log-event unavailability. This is called the "Birnbaum Importance" measure.

Fussel-Vesely Importance $I_i^{FV}(t)$:

$$I_i^{FV}(t) = \frac{I_i^{BB}(t) \cdot q_i(t)}{Q_s(t)} \qquad (29)$$

The Fussel-Vesely Importance gives the percentage-wise variation of the system unavailability as a consequence of a percentage-wise variation in the unavailability of component (event) i.

Indeed, assume $Q_s(t) = Q_1(t)+Q_2(t)$ in which $Q_1(t)$ is the contribution of the MCS containing event i, $Q_2(t)$ of those not containing event i:

$$\frac{Q_s(t) - Q_s^*(t)}{Q_s(t)} = \frac{(Q_1(t)+Q_2(t)) - (Q_1^*(t)+Q_2(t))}{Q_s(t)} = \frac{Q_1(t)-Q_1^*(t)}{Q_s(t)}$$

with $Q^*(t)$ the unavailability after that $q_i(t)$ has been replaced by $q^*(t) < q_i(t)$.

Now: $$Q_1(t) = \frac{\delta Q_s(t)}{\delta q_{i(t)}} q_i(t) = I_i^{BB}(t) q_i(t) \qquad (30)$$

Hence: $$\frac{Q_1(t)-Q_1^*(t)}{Q_s(t)} = \frac{\delta Q_s(t)}{\delta q_i(t)} (q_i(t)-q_i^*(t))/Q_s(t) =$$

$$= \frac{\delta Q_s(t)}{\delta q_i(t)} q_i(t) \left(\frac{q_i(t)-q_i^*(t)}{q_i(t)}\right)/Q_s(t) =$$

$$= \frac{I_i^{BB}(t)}{Q_s(t)} q_i(t) \left(\frac{q_i(t)-q_i^*(t)}{q_i(t)}\right) = I_i^{FV}(t)\left(\frac{q_i(t)-q_i^*(t)}{q_i(t)}\right)$$

$I_i^{FV}(t)$ is calculated as follows:

$$I_i^{FV}(t) = \frac{\sum_{j=1}^{N} Q_j(t)\alpha_j}{Q_s(t)}$$

$$\text{with} \quad \alpha_j = \begin{cases} 1 & \text{if the i-th component is present in the j-th MCS} \\ 0 & \text{if the i-th component is not present in the j-th MCS} \end{cases}$$

Indeed, from (30) it follows that:

$$I_i^{FV}(t) = \frac{Q_1(t)}{Q_s(t)}$$

Barlow-Proschan Importance: expresses the contribution of failure of the i-th component (event) to the system failure.

Consider component i; failure of i will provide system failure if the system is in a state in which i is critical at time t, and i fails in (t,t+dt).

The probability of this is given by:

$$\frac{\delta Q_s(t)}{\delta q_i(t)} \cdot w_i(t)dt = I_i^{BB}(t).w_i(t)dt$$

where $w_i(t)$ is the failure intensity of the i-th component.

$\int_o^t I_i^{BB}(t).w_i(t)dt$ gives the expected number of system failures due to the failure of i over a time interval (o,t). It is also possible to show that:

$$\sum_{i=1}^{N} \int_o^t I_i^{BB}(t).w_i(t)dt = W_s(t)$$

Now the Barlow-Proschan Importance $I_i^{BP}(t)$ is defined as:

$$I_i^{BP}(t) = \frac{\int_o^t I_i^{BB}(t)w_i(t)dt}{W_s(t)} \quad \text{for } \lambda = \text{constant} \quad \underline{\sim} \quad \frac{\int_o^t I_i^{BB}(t)\lambda_i dt}{W_s(t)}$$

Hence, $I_i^{BP}(t)$ is the fraction of expected number of system failures caused by component (event) i.

Example:

Consider a system with 5 components A, B, C, D, E and the fault tree TOP = AB+CD+CE.

Consider a mission time of $T = 10^4$ h and the following values for the components, cut sets and system:

	$\lambda_i(h^{-1})$	$\mu_i(h^{-1})$	$q_i(T)$	$Q_j(t)$	$Q_s(t)$
A	10^{-5}	10^{-2}	10^{-3}	-	-
B	10^{-5}	0	10^{-1}	-	-
C	10^{-4}	$5x10^{-3}$	$2x10^{-2}$	-	-
D	10^{-4}	$5x10^{-2}$	$2x10^{-3}$	-	-
E	10^{-5}	0	10^{-1}	-	-
AB	-	-	-	10^{-4}	-
CD	-	-	-	$4x10^{-5}$	-
CE	-	-	-	$2x10^{-3}$	-
TOP	-	-	-	-	$2.14x10^{-3}$

The importance of the components yields:

Comp.	$I^{BB}(t)$	$I^{FV}(t)$	$I^{BP}(T)$
A	10^{-1}	$4.6x10^{-2}$	$6.56x10^{-2}$
B	10^{-3}	$4.6x10^{-2}$	$1.29x10^{-3}$
C	10^{-1}	0.95	0.67
D	$2x10^{-2}$	$1.9x10^{-2}$	0.258
E	$2x10^{-2}$	0.93	$2.5x10^{-2}$

This shows that the importance ranking of components is completely different according to the measure (Birnbaum, Fussel-Vesely or Barlow-Proschan) used. Hence, the importance measures must be chosen in function of the type of the system analysed and its application.

3. ANALYSIS OF EVENT TREES

3.1 Basic approaches

Two approaches to event tree analysis are currently used /7/:

1) In the **"large event tree - small fault tree"** approach, the frontline systems performing safety functions and the support systems necessary for the functioning of the front line systems appear explicitly. This allows to model dependencies explicitely in the event tree. The fault trees are "small" because they do not have to model states of the support functions and systems.
   This method is also called **"event tree with boundary conditions"**, since fault trees have to be developed for the frontline systems under various boundary conditions imposed by (partial) failure or success of the various support systems. For example, for a power supply system with two busbars, the following situations may occur: both busbars function, both busbars fail, busbar 1 functions and 2

fails, busbar 2 functions and 1 fails. For the systems which need the power supply, four fault trees will have to be constructed: one for each of the boundary conditions related to the busbar failure or success situations. The event trees may become very complicate in real cases.
Once the (large) event tree and the fault trees are constructed, the quantification of an accident sequence is rather straightforward, since the fault trees can be assumed to be independent.

2) The **"small event tree - large fault tree"** approach is based on complete fault trees, including support system failures. Hence, the fault trees of the event tree are in general not independent. An accident sequence can be quantified by putting the trees concerning its branches as inputs to an "AND" gate (fault tree linking). Since in an accident sequence success of some systems is assumed, these trees are not only fault trees but may also be the complement of a fault tree: a so-called **dual fault tree** (success tree).
Inclusion of the dual fault tree for systems that are assumed to function in any particular event sequence, will eliminate minimal cut sets of the event sequence in which failure causes appear of a system that is defined to be in a success state.
Fault tree linking may lead to very large fault trees containing complemented events. Such trees cannot be analysed easily. It has to be noticed that due to the presence of complemented events the linked tree of an accident sequence is non-coherent.
An alternative to the dual fault tree approach is **"MCS list matching"**. It is based on the fact that a MCS of a system failure in an event sequence that is also a cut set of a system assumed to be in the success state in that event sequence, should be removed.

3.2 Example of Analysis

Consider the simple event tree represented in Fig.8a, in which the front line systems S_1 and S_2 have to intervene when the initiating event E_i occurs. The fault trees of S_1 and S_2 are shown in Figs.8b,c. We are going to briefly describe the main aspects of the different approaches applied to the analysis of the sequence $S_1 \wedge \bar{S}_2$, being their application to the other sequences quite straightforward.

3.2.1 Fault tree linking. According to this approach the sequence $S_1 \wedge \bar{S}_2$ is analysed by means of a fault tree analysis code able to deal with non coherent structures; this is a very general approach, since S_1 and/or S_2 can also be non coherent. However, the difficulty in applying this method is due to the complexity of the fault tree representing the sequence under study.

When the fault trees of S_1 and S_2 are coherent, the MCS matching method can be applied; it allows to analyse the sequence also by means of computer codes able to solve coherent structures only. Indeed:

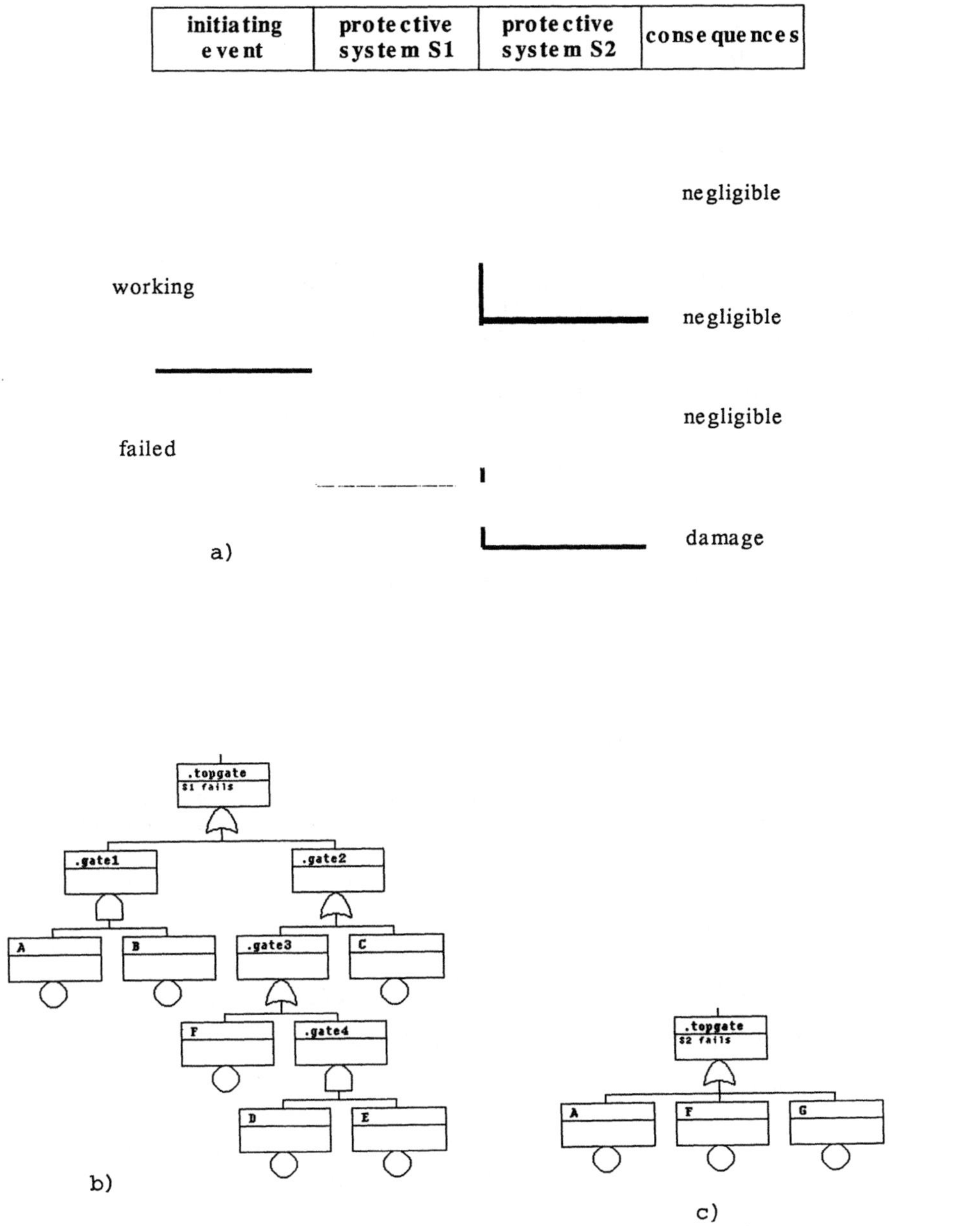

Fig.8: Example of an event tree (8a) and fault trees of the front line systems (8b,c).

$$S_1 = \bigvee_{i=1}^{n_1} K_{1i}$$

$$S_2 = \bigvee_{j=1}^{n_2} K_{2j}$$

where K_{1i} and K_{2j} are the generic MCS of S_1 and S_2, respectively. It can be easily proved that the s-o-p expression of the sequence $S_1 \wedge \bar{S}_2$ is given by the disjunction of the MCS of S_1 that do not verify S_2. In other words a MCS of S_1 is a MCS of $S_1 \wedge \bar{S}_2$ if it does not contain any MCS of S_2.

The advantage of this method is that it is not necessary to complement S_2 in order to analyse the sequence. Therefore, being:

$$S_1 = C + F + A*B + D*E$$

$$S_2 = A + F + G$$

the s-o-p expression of $S_1 \wedge \bar{S}_2$ is C+D*E, since F is common to S_1 and S_2, and A*B in S_1 contains A in S_2.

3.2.2 Event tree with boundary conditions. According to this approach the events A and F, common to S_1 and S_2, are explicitly represented in the event tree that takes the form shown in Fig. 9, where $S_1|.$ and $S_2|.$ have the meaning of S_1, S_2 conditioned to one of the combinations of A,F in their possible states.

The conditioned s-o-p expressions of S_1 and S_2 are represented in Fig.10. The conditioned expressions of S_1 and $\bar{S}_2$ can be obtained directly by complementing S_1 and S_2 for the same states of A,F.

The sequence under examination $S_1 \wedge \bar{S}_2$ is:

$$\bar{A} \wedge \bar{F} \wedge S_1|\bar{A}\bar{F} \wedge \bar{S}_2|\bar{A}\bar{F} = \bar{A} \wedge \bar{F} \wedge (C+DE) \wedge \bar{G}$$

$$\bar{A} \wedge F \wedge S_1|\bar{A}F \wedge \bar{S}_2|\bar{A}F = A \wedge F \wedge 1 \wedge 0$$

$$A \wedge \bar{F} \wedge S_1|A\bar{F} \wedge \bar{S}_2|A\bar{F} = A \wedge \bar{F} \wedge (C+B+DE) \wedge 0$$

$$A \wedge F \wedge S_1|AF \wedge \bar{S}_2|AF = A \wedge F \wedge 1 \wedge 0$$

Obviously, sequences in which at least one conditioned expression is equal to 0 have to be deleted since they are impossible. Therefore, $S_1 \wedge \bar{S}_2 \equiv \bar{A} \wedge \bar{F} \wedge (C+D*E) \wedge \bar{G}$.

Applying the approximation that the probability of negated events are equal to 1, one obtains that:

$$S_1 \wedge \bar{S}_2 = C + D*E$$

which, obviously, is the same result obtained with the MCS matching method.

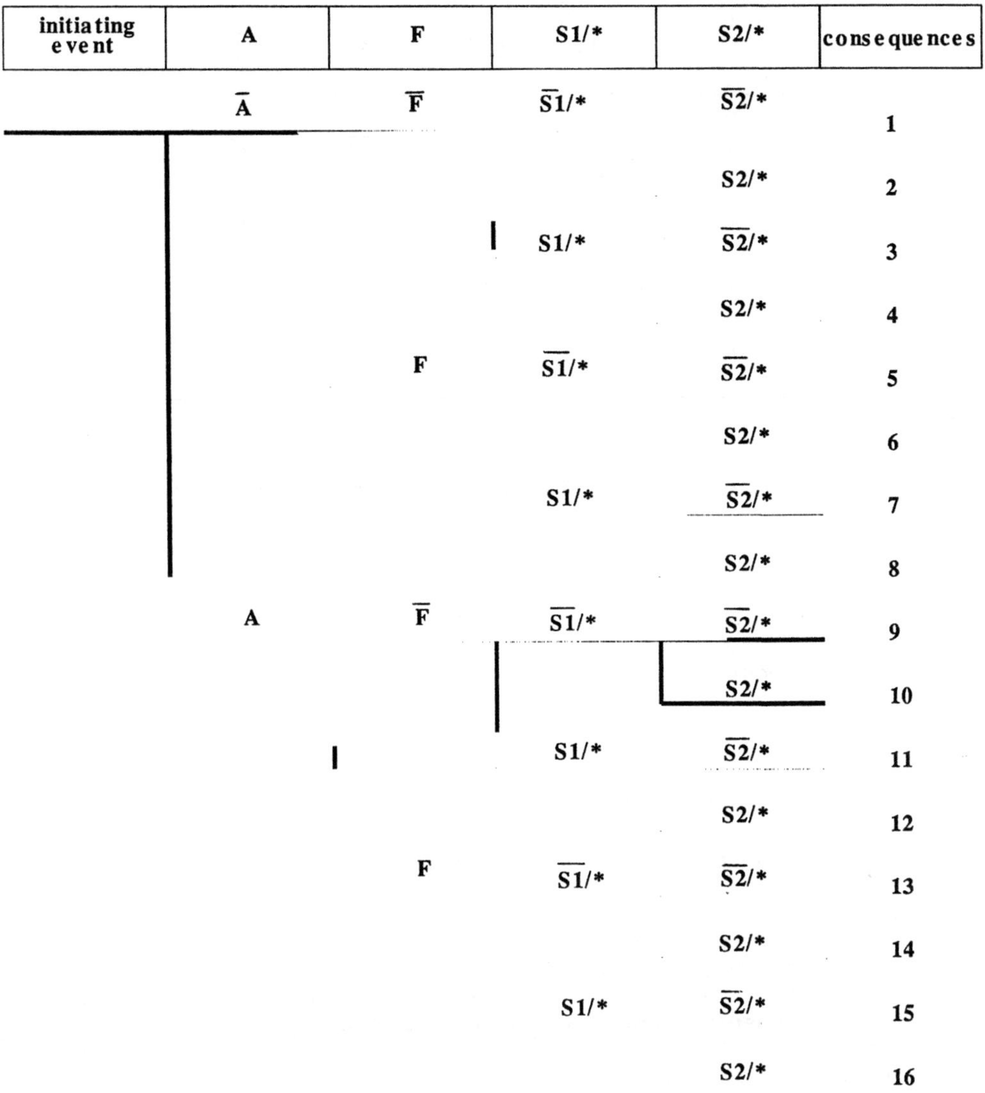

Fig.9: Event tree with common components A,F explicitated.

A	F	S1|.	S2|.	$\overline{S1}$|.	$\overline{S2}$|.
0	0	C+DE	G	$\overline{C+DE}$	$\overline{G}$
0	1	1	1	0	0
1	0	C+B+DE	1	$\overline{C+B+DE}$	0
1	1	1	1	0	0

Fig.10: Truth table for determining the conditioned expressions S_1, S_2.

Once the MCSs of a sequence have been determined, the corresponding frequency can be assessed by means of the equations described in section 2.

REFERENCES

1. Poucet, A., 'Fault tree and event tree techniques', Reliability Engineering, Proc. of the Ispra Course held in Madrid in 1986, edited by A. Amendola and A. Saiz de Bustamante.
2. Mendelson, E., Theory and Problems of Boolean Algebra, McGrow Hill, 1970.
3. Henely, E. and Kumamoto, H., Reliability Engineering and Risk Assessment, Prentice Hall, Y.Y. 1981.
4. Barlow, R. and Proschan, F., Statistical Theory of Reliability and Life Testing, Holt, Rinehart, and Winston, N.Y. 1975.
5. Apostolakis, G., Mathematical methods of probabilistic safety analysis, UCLA-ENG-7464, September 1974.
6. Lambert, H.E., 'Fault tree for decision making in system analysis', UCRL-51829 (Ph.D. Thesis), Lawrence Livermore Laboratory, CA, 1975.
7. U.S. Nuclear Regulatory Commission, PRA Procedure Guide, A Guide to the Performance of Probabilistic Risk Assessment of Nuclear Power Reactors, NUREG CR-2300, 1982.

SYSTEM MODELLING WITH PETRI NETS

Andrea BOBBIO

Istituto Elettrotecnico Nazionale Galileo Ferraris
Strada delle Cacce 91, 10135 Torino, Italy

ABSTRACT. Petri Nets (PN) are a graphical formalism which is gaining popularity in recent years as a tool for the representation of complex logical interactions (like synchronization, sequentiality, concurrency and conflict) among physical components or activities in a system. This notes are devoted to introduce the formalism of Petri nets with particular emphasis on the application of the methodology in the area of the performance and reliability modelling and analysis of systems. The quantitative analysis of the behaviour of systems in time requires the superposition of a stochastic timing mechanism to the classical representation of PN. Timed Petri nets and, in particular, Stochastic Petri nets (SPN) are the object of the second part of the notes. Finally, some fully developed examples enlighten peculiar aspects which differentiate PNs from other modelling techniques usual in reliability analysis. In few words, the goal of these notes is to show that the proposed methodology based on the PN formalism can be conveniently used as a user-friendly language to represent and evaluate complex stochastic systems.

1. Introduction

Petri Nets (PN) are a graphical tool for the formal description of the flow of activities in complex systems. With respect to other more popular techniques of graphical system representation (like block diagrams or logical trees), PN are particularly suited to represent in a natural way logical interactions among parts or activities in a system. Typical situations that can be modelled by PN are synchronization, sequentiality, concurrency and conflict.

The theory of PN originated from the doctoral thesis of C.A. Petri in 1962 [39]. Since then, the formal language of PN has been developed and used in many theoretical as well as applicative areas. Introductory survey papers can be found in [38,3]. Several textbooks on the subject are also available: [37] (where an extended annotated bibliography is contained) [42,13,44]. An yearly Workshop on "Application and Theory of Petri Nets" is held in Europe (the IX edition of the workshop took place in Venice in June 1988).

The classical PNs do not convey any notion of time; in order to use the PN formalism for the quantitative analysis of the performance and reliability of system versus time, a class of Timed PN (TPN) has been introduced. The time variables associated to the PN can be either deterministic variables (leading to the class of models called deterministic PN), or random variables (leading to the class of models called Stochastic PN - SPN). The bibliography on TPN is not as wide as the one on classical PN, however, an extended

A. G. Colombo and A. Saiz de Bustamante (eds.), Systems Reliability Assessment, 103–143.

collection of papers and applications can be found in the proceedings of two international workshops specifically devoted to the use of TPN in performance and reliability evaluation [2,1].

The first part (sections 3,4,5,6 and 7) of these lecture notes is aimed at introducing the classical theory of PN, while the second part (sections 8,9,10 and 11) discusses the stochastic timing of a PN with application in the field of reliability modelling and evaluation.

In particular, Section 2 contains the list of symbols. Section 3 defines the primitive elements of the PN and the execution rules by means of which the dynamic properties of the system are described. Section 4 illustrates typical examples of logical interactions among activities modelled by PN, while Section 5 introduces characteristic properties of PNs. Section 6 shows how a PN can be analyzed through the generation of the *reachability tree*, or by means of matrix techniques. Finally, some possible extensions of the modelling capabilities of classical PNs are considered in Section 7.

The second part of the lecture notes is devoted to illustrate how PNs can be conveniently used as a modelling language for the quantitative analysis of the performance and reliability of systems. The use of PN for this purpose requires that the duration of the activities representing the system operations can be specified and measured; therefore, the first step toward the definition of a suitable modelling framework is the insertion of the notion of time in classical PNs. This topic is addressed in Section 8.

Section 9 examines the class of Stochastic PN (SPN) in which the durations of the activities are exponentially distributed random variables. With this assumption, the dynamic behaviour of the PN can be mapped into a continuous-time homogeneous Markov chain. In this way we cast a natural bridge between SPN and Markov models in reliability analysis. With reference to the above models, we discuss the following topics: the generation of the Markov chain associated to the PN, the assignment of marking dependent transition rates and the partition of PN transitions into immediate and timed transitions. Section 10 shows how SPN models can be naturally used to define interesting measures for the characterization of the system behaviour versus time, and how these measures can be computed from the associated Markov chain. Section 11 illustrates some examples of application of the above methodology in reliability analysis. Section 12 briefly introduces the implementation of simulative techniques in the analysis of SPN.

2. List of Symbols

D^-, D^+, D	-	Input, Output and Incidence matrix
$\underline{e}_j$	-	0-vector with entry j equal to 1
$\mathcal{E}$	-	Execution sequence
$\mathcal{G}_R$	-	Reachability graph
I	-	Input function
$HMSPN$	-	Homogeneous Markov Stochastic Petri Net
$L = \{\lambda_1, \lambda_2, \ldots, \lambda_{nt}\}$	-	Set of firing rates
$M = \{m_1, m_2, \ldots, m_{np}\}$	-	Marking
N	-	Cardinality of the reachability set (state space)
n_p	-	Number of places
n_t	-	Number of transitions

O	-	Output function
PN	-	Petri Net
$P = \{p_1, p_2, \ldots, p_{np}\}$	-	Set of places
$\underline{Q}(t)$	-	State probability vector of the associated Markov chain
$\mathcal{R}$	-	Reachability set
SPN	-	Stochastic Petri Net
$T = \{t_1, t_2, \ldots, t_{nt}\}$	-	Set of transitions
$\mathcal{T}_E$	-	Timed execution sequence
TPN	-	Timed Petri Net
$\underline{U}_p$	-	Unitary vector
v	-	Branching probability in a random switch
$\underline{W}_p$	-	Vector of binary (0, 1) entries
$\underline{X}$	-	Integer vector
$\eta_j(t)$	-	Expected number of firings of t_j in $0 - t$
θ_j	-	Random firing time associated to t_j
λ, μ, γ, ρ	-	Firing rates
Λ	-	Transition rate matrix of the associated Markov chain
ϕ	-	Mean passage time
τ_j	-	Epoch of firing of $t_{(j)}$
$\psi(t)$	-	Expected time spent in a marking in $0 - t$
ω	-	Infinite reproducibility in the reachability tree

3. The Primitive Elements of a Petri Net

For definitions and notation we refer in general to [37]. A Marked PN is a quintuple (P, T, I, O, M), where:

- $P = \{p_1, p_2, \ldots, p_{np}\}$ is the set of n_p places (drawn as circles in the graphical representation);
- $T = \{t_1, t_2, \ldots, t_{nt}\}$ is the set of n_t transitions (drawn as bars);
- I is the transition input relation and is represented by means of arcs directed from places to transitions;
- O is the transition output relation and is represented by means of arcs directed from transitions to places;
- $M = \{m_1, m_2, \ldots, m_{np}\}$ is the marking. The generic entry m_i is the number of tokens (drawn as black dots) in place p_i in marking M.

The graphical structure of a PN is a bipartite directed graph: the nodes belong to two different classes (places and transitions) and the edges (arcs) are allowed to connect only nodes of different classes (multiple arcs are possible in the definition of the I and O relations [37]). Figure 1 is a PN [3].

The dynamics of a PN is obtained by moving the tokens in the places by means of the following execution rules:

- A transition is enabled in a marking M if all its input places carry at least one token;
- an enabled transition fires by removing one token per arc from each input place and adding one token per arc to each output place.

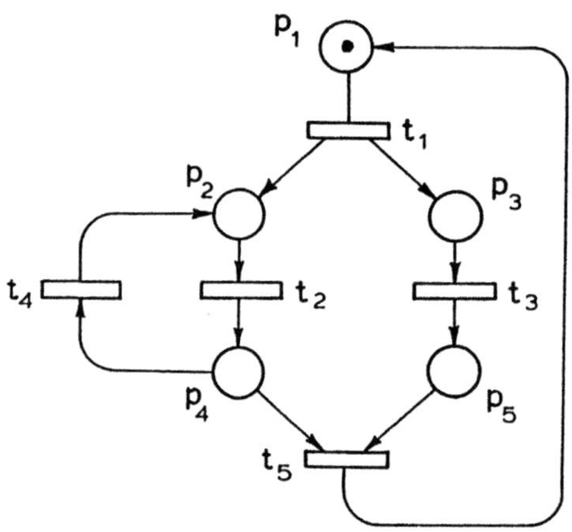

$P = \{p_1, p_2, p_3, p_4, p_5\}$

$T = \{t_1, t_2, t_3, t_4, t_5\}$

$I(t_1) = \{p_1\}$ $\quad O(t_1) = \{p_2, p_3\}$

$I(t_2) = \{p_2\}$ $\quad O(t_2) = \{p_4\}$

$I(t_3) = \{p_3\}$ $\quad O(t_3) = \{p_5\}$

$I(t_4) = \{p_4\}$ $\quad O(t_4) = \{p_2\}$

$I(t_5) = \{p_4, p_5\}$ $\quad O(t_5) = \{p_1\}$

$M_1 = (1, 0, 0, 0, 0)$

Figure 1 - A PN graph with Input and Output functions.

Given an initial marking M_1, the reachability set $\mathcal{R}(M_1)$ is the set of all the markings that can be obtained by repeated application of the above rules.

More formally we can say that t_k is enabled in marking M if:

$$\text{for any} \quad p_i \in I(t_k) \quad , \quad m_i \geq 1$$

Marking M', obtained from M by firing t_k, is said to be *immediately reachable* from M, and the firing operation is denoted by the symbol $(M - t_k \rightarrow M')$. The token count in M' is pictorially represented in Figure 2, and is given by the following relationship:

$$M'(p_i) = \begin{cases} M(p_i) + 1 & \text{if} \quad p_i \in O(t_k)\,, \; p_i \notin I(t_k) \\ M(p_i) - 1 & \text{if} \quad p_i \notin O(t_k)\,, \; p_i \in I(t_k) \\ M(p_i) & \text{otherwise} \end{cases}$$

Let us examine the generation of the reachability set of the PN of Figure 1 given the initial marking $M_1 = (1,0,0,0,0)$. In M_1 the only enabled transition is t_1; firing

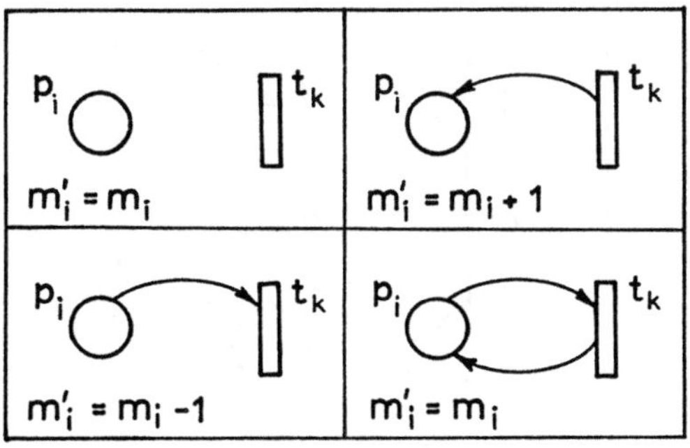

Figure 2 - Token number modification in place p_i subsequent to the firing of transition t_k.

of t_1 removes the token from p_1 and puts a token both in p_2 and p_3 producing the new marking $M_2 = (0,1,1,0,0)$. In M_2 the transitions t_2 and t_3 are both enabled and can fire concurrently. Firing of t_3 leads to $M_3 = (0,1,0,0,1)$ and subsequent firing of t_2 leads to $M_4 = (0,0,0,1,1)$. In M_4 transitions t_4 and t_5 are both enabled, but the firing of either disables the other; the two transitions are in conflict. Firing of t_4 in M_4 produces marking M_3, while firing t_5 in M_4 produces the intial marking M_1. Note that a different firing sequence can be activated from marking M_2, by letting t_2 firing first and obtaining marking $M_5 = (0,0,1,1,0)$. With this, all the possible firing sequences have been examined, and the reachability set $\mathcal{R}(M_1)$ of the net of Figure 1 turns out to contain 5 elements M_1, M_2, M_3, M_4 and M_5.

4. Petri Nets and the Modelling of Systems

PN used for modelling real systems are sometimes referred to as *Condition/Events* nets. Places identify the conditions of the parts of the system (working, idle, queueing, failed), and transitions describe the passage from one condition to another (end of a task, failure, repair ...). An event occurs (a transition fires) when all the conditions are satisfied (input places are marked) and give concession to the event. Occurrence of the event modifies in whole or in part the status of the conditions (marking). The number of tokens in a place can be used to identify the number of resources lying in the condition denoted by that place. The following examples illustrate typical situations of interaction of activities arising in system modelling.

4.1. CONCURRENCY (OR PARALLELISM)

In the PN of Figure 3 transitions t_1 and t_2 are enabled simultaneously; the firing of one of them does not modify the state of the other. The activities modelled by the two transitions run concurrently. In reliability modelling, the PN of Figure 3 can represent two components $\mathbf{C}_1$ and $\mathbf{C}_2$ in parallel redundancy; in this case, places p_1 and p_3 represent the working

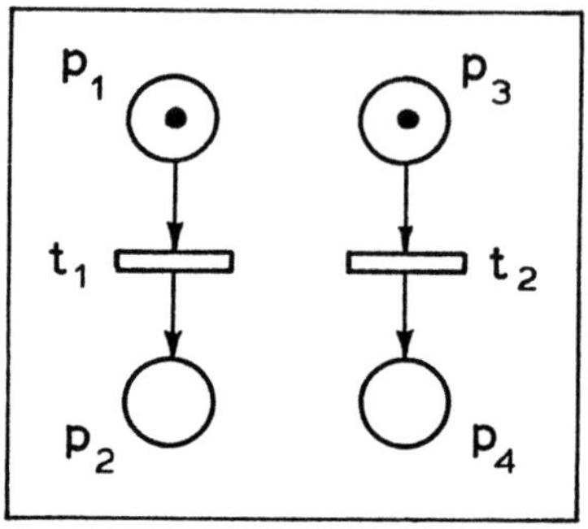

Figure 3 - PN modelling two parallel activities.

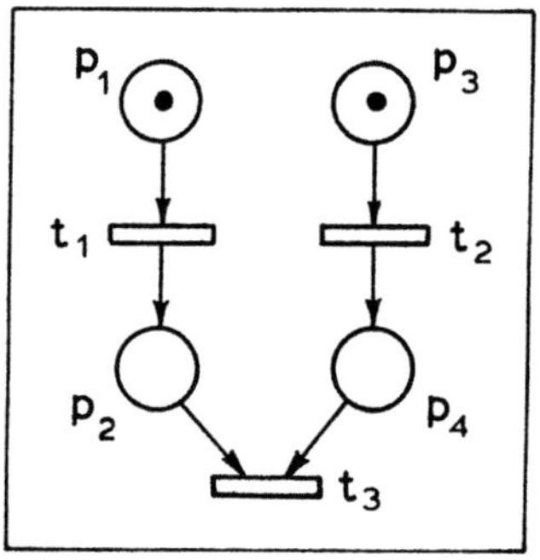

Figure 4 - A PN modelling two parallel activities with synchronization.

condition, p_2 and p_4 the failed condition and t_1 and t_2 the event of failure of $\mathbf{C}_1$ and $\mathbf{C}_2$ respectively.

4.2. SYNCHRONIZATION

In Figure 3 the activities modelled by t_1 and t_2 run concurrently; however, if they represent routines of a parallel program, both should be terminated before the program execution can proceed. The synchronization activity is modelled in Figure 4 by means of transition t_3 whose firing requires a token both in p_2 and p_4.

4.3. LIMITED RESOURCES

A typical factor influencing the performance of distributed systems (multiprocessor sys-

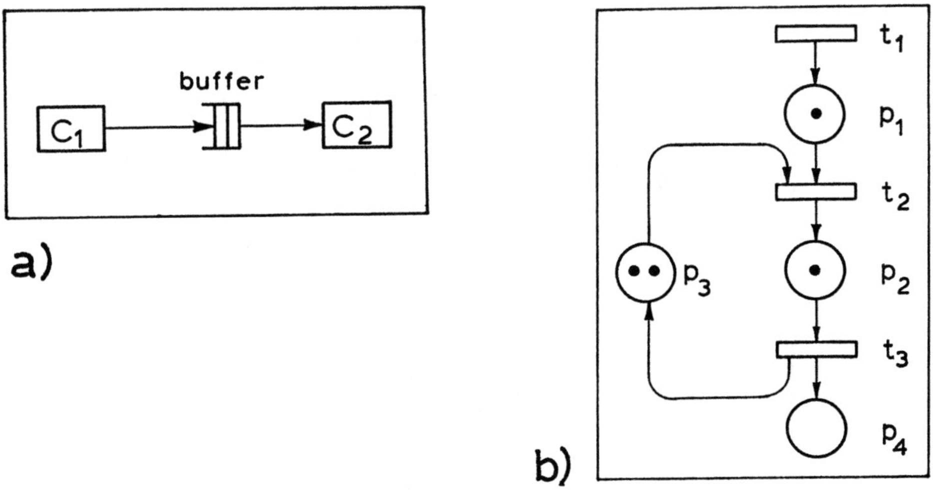

Figure 5 - Block diagram and PN of a buffer with finite size.

tems, flexible manufacturing systems and so on) is the limited number of available resources. Exhaustion of the resources prevents the activities to proceed and blocks the system. Modelling and analysing systems with blocking is a difficult task in almost all modelling frameworks [15,46]. A PN representation of a buffer with limited size is shown in Figure 5b (Figure 5a shows the corresponding block diagram representation). Place p_3 models the number of free buffer positions whereas p_2 the number of filled positions; note that the sum of tokens in p_2 and p_3 is constant and models the total number of available buffer positions (three positions in the figure). Transition t_2 models the filling of one buffer position and can fire if a position free (at least one token in p_3) exists and a task is available to be stored (a token in p_1). Transition t_3 is enabled when at least one buffer position is filled, and firing of t_3 moves one token from p_2 to p_3.

4.4. SEQUENTIALITY (THE PRODUCER/CONSUMER PROBLEM)

A producer produces objects that are put into a buffer from which can be removed and consumed by a consumer. The consuming process must be in sequence with respect to the production process. The PN solution to this problem is reported in Figure 6. A token in p_1 means that the producer is ready to produce. By firing t_1 and t_2 an object is produced (a token is put in the buffer p_5) and the producer is ready again. If the consumer is ready to consume (token in p_3) and an object is in the buffer, transition t_3 can fire removing one token from p_5.

In the PN of Figure 6 the production and the accumulation of objects in the buffer is unbounded. A more realistic situation is obtained by considering a buffer of limited capacity (as in 4.3). The corresponding PN is reported in Figure 7. Place p_6 models the

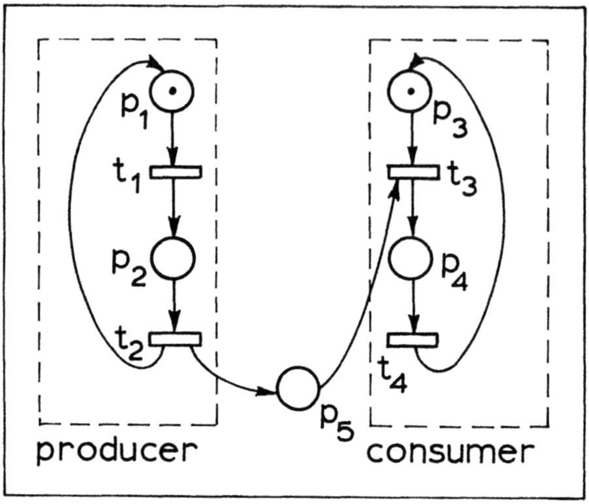

Figure 6 - The producer/consumer problem with unbounded buffer.

free buffer positions and place p_5 the filled buffer positions; the number of tokens in p_5 and p_6 is constant and represents the total available buffer positions. If a single token is assigned to p_6 in the initial marking, we model the situation in which the producer cannot further produce until the consumer has consumed the object in the buffer (a strictly sequential ordering of activities).

4.5. MUTUAL EXCLUSION (CONFLICT)

Two resources $\mathbf{C}_1$ and $\mathbf{C}_2$ are allowed to work in parallel, but are connected to a shared resource $\mathbf{C}_s$ that cannot be accessed by $\mathbf{C}_1$ and $\mathbf{C}_2$ simultaneously (block diagram in Figure 8a). The corresponding PN is in Figure 8b. Places p_1 and p_5 represent $\mathbf{C}_1$ and $\mathbf{C}_2$ working independently; p_2 and p_6 represent $\mathbf{C}_1$ and $\mathbf{C}_2$ requesting access to $\mathbf{C}_s$; p_3 and p_7 represent $\mathbf{C}_s$ busy with $\mathbf{C}_1$ and $\mathbf{C}_2$ respectively. Place p_4 determines which resource can actually access $\mathbf{C}_s$, and prevents places p_3 and p_7 to be marked at the same time; in fact when p_2 and p_6 are both marked, transitions t_2 and t_5 are in conflict. Firing of one of them disables the other. Firing of t_3 or t_6 models the release of the common resource (token back in p_4) and the return to the working condition.

5. Properties of Petri Nets

We enumerate different properties which allow us to classify the primitive elements of a PN or the PN as a whole.

5.1. LIVENESS

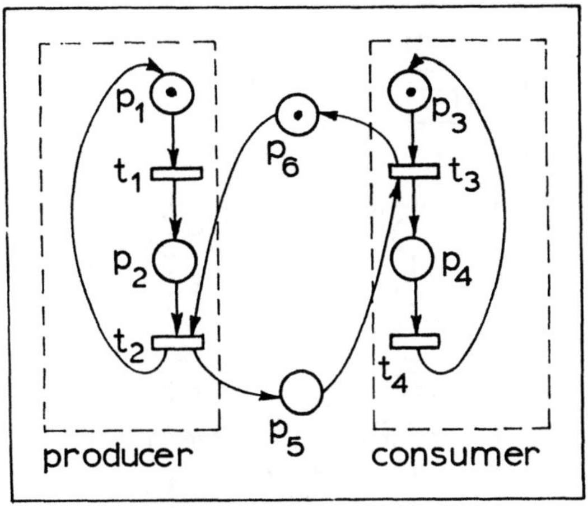

Figure 7 - The producer/consumer problem with finite buffer.

A transition is *potentially firable* in M if there exists a sequence of transition firings which leads to a marking in which the transition is enabled. A transition is *live* if it is potentially firable in any marking of $\mathcal{R}(M_1)$. A transition is *dead* in M if it is not potentially firable; if the PN enters marking M the dead transition cannot fire any more.

5.2. SAFENESS

A place is *safe* if the token count does not exceed 1 in any marking of $\mathcal{R}(M_1)$. A PN is safe if each place is safe. The PNs of Figures 1, 3 and 8b are safe.

5.3. BOUNDEDNESS

A simple generalization of safeness is the concept of boundedness. A place is bounded with bound k, if the token count does not exceed k in any marking of $\mathcal{R}(M_1)$. A PN is k-bounded if each place is k-bounded. The PN of Figure 7 is k-bounded where k is the number of buffer positions. On the contrary, the PN of Figure 6 is unbounded.

5.4. CONSERVATION

A PN is strictly conservative if the total number of tokens is constant in each marking of $\mathcal{R}(M_1)$. The PN of Figure 7 is k-bounded and strictly conservative, while the PN of Figure 8b is safe but not strictly conservative. A subset of places form a *place-invariant* [31] if it is strictly conservative. In the net of Figure 8b the subsets $\{p_1, p_2, p_3\}$, $\{p_5, p_6, p_7\}$ and $\{p_3, p_4, p_7\}$ are place-invariants.

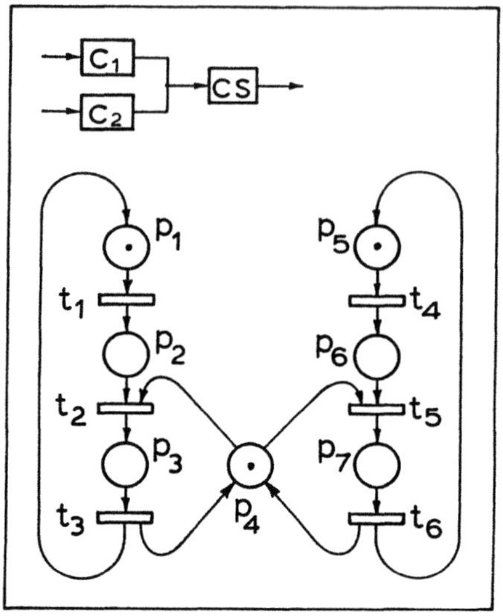

Figure 8 - The mutual exclusion problem: two parallel resources with a common shared block.

6. Analysis Techniques

The success of any model depends on two factors: its modelling power and its decision power. Modelling power refers to the ability to correctly represent the system to be modelled; decision power refers to the ability to analyze the model and determine properties of the modelled system. The modelling power of PN has been examined in the previous sections, and in this section we take into consideration the analysis techniques of PNs.

6.1. THE REACHABILITY TREE AND REACHABILITY GRAPH

The reachability set $\mathcal{R}(M_1)$ of a PN is generated by means of the reachability tree. The initial marking M_1 is the root of the reachability tree. Starting from the root we search for all the enabled transitions; the firing of an enabled transition produces a new marking which is represented as a new leaf in the tree, from which the procedure is iterated.

By properly identifying the frontier nodes of the tree, the generation of the reachability tree involves a finite number of steps [37], even if the PN is unbounded. Let us introduce three kinds of frontier nodes:

- terminal (dead) nodes: nodes in which no transitions are enabled;
- duplicate nodes: nodes which have been already generated in the tree;

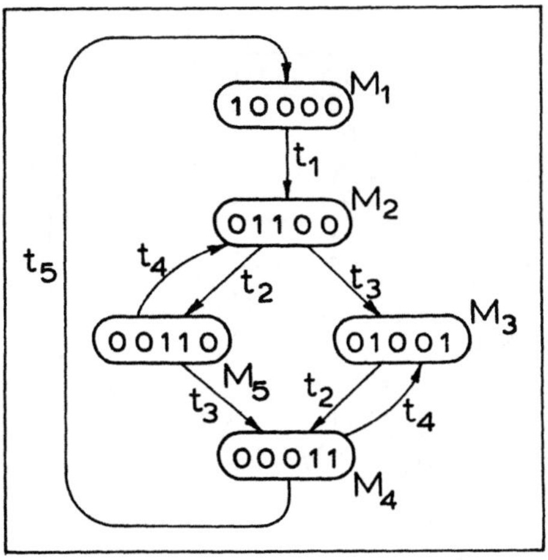

Figure 9 - Reachability graph $\mathcal{G}_R(M_1)$ for the net of Figure 1.

- infinitely reproducible nodes. A marking M'' is an infinitely reproducible node if $M'' \geq M'$ ($m_i'' \geq m_i'$, $i = 1, 2, \ldots, n_p$) for some M' already generated in the tree. Because of the stated relation, the transition sequence from which M'' has been generated starting from M' is surely firable in M''. Thus, the sequence $M' \rightarrow M''$ can be reproduced infinitely often, so that the token count in the places for which $m_i'' \geq m_i'$ can increase indefinitely. We represent the arbitrarily large number of tokens which results from infinitely reproducible nodes by defining a special symbol ω with the following properties:

$$\begin{aligned} \omega + a &= \omega \\ \omega - a &= \omega \\ a &< \omega \end{aligned}$$

  for any positive constant a.

By allowing ω to be a legal symbol in the reachability tree specification, it can be shown that the generation of the reachability tree involves always a finite search algorithm [37]. If the generation of the reachability tree terminates without arriving to infinitely reproducible nodes, the PN is bounded. In this case the reachability set is finite and can be represented as a labelled directed graph whose vertices are the elements of $\mathcal{R}(M_1)$ and such that for each possible transition firing $M_i - t_k \rightarrow M_j$ there exists an arc (i, j) labelled k. The reachability graph associated to a reachability set $\mathcal{R}(M_1)$ will be denoted by $\mathcal{G}_R(M_1)$.

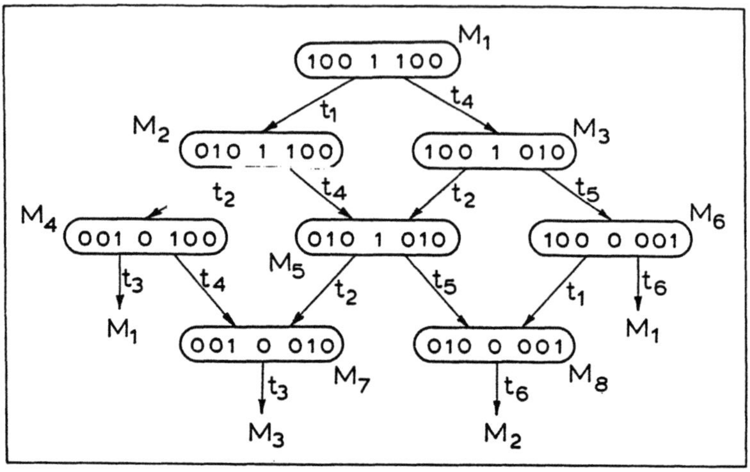

Figure 10 - Reachability graph $\mathcal{G}_R(M_1)$ for the net of Figure 8.

Figure 9 shows the reachability graph of the PN of Figure 1 with initial marking $M_1 = (1,0,0,0,0)$, as discussed in Section 3. Figure 10 shows the reachability graph for the mutual exclusion problem of Figure 8, with initial marking $M_1 = (1,0,0,1,1,0,0)$.

In Figure 11 we have reported the reachability tree of the PN of Figure 6; since the net is unbounded, in order to keep the generation algorithm finite, the symbol ω has been introduced.

If a PN has a finite $\mathcal{R}(M_1)$ all the properties of the net (safeness, liveness, etc..) can be analyzed by inspection of the reachability graph. If the net is unbounded the finite reachability tree representation, by means of the symbol ω, can be an imperfect description of the net (it is possible to find PNs with different properties and behaviours that cannot be distinguished through the reachability tree, due to incomplete information carried by ω [37]).

6.2. MATRIX ANALYSIS

The input and output functions of a PN can be equivalently defined using a matrix notation. Let D^- denotes the input matrix. D^- is a $(n_t \times n_p)$ matrix, whose generic element d^-_{ij} is equal to the number of arcs connecting place p_j with transition t_i. Similarly we define the output matrix D^+ as a $(n_t \times n_p)$ matrix, whose generic element d^+_{ij} is equal to the number of arcs connecting transition t_i with place p_j. The *incidence matrix* D is defined by the following relation:

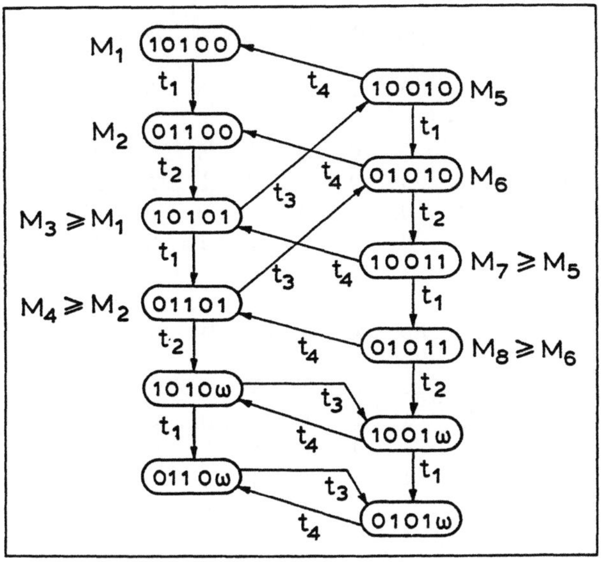

Figure 11 - Generation of the reachability tree for the PN of Figure 6 with unbounded buffer.

$$D = D^{+} - D^{-} \tag{1}$$

The matrices D^-, D^+ and D for the PN of Figure 8b are reported in the following:

$D^- =$

	p_1	p_2	p_3	p_4	p_5	p_6	p_7
t_1	1	0	0	0	0	0	0
t_2	0	1	0	1	0	0	0
t_3	0	0	1	0	0	0	0
t_4	0	0	0	0	1	0	0
t_5	0	0	0	1	0	1	0
t_6	0	0	0	0	0	0	1

$$
D^+ = \begin{array}{c|ccccccc|} & p_1 & p_2 & p_3 & p_4 & p_5 & p_6 & p_7 \\ t_1 & 0 & 1 & 0 & 0 & 0 & 0 & 0 \\ t_2 & 0 & 0 & 1 & 0 & 0 & 0 & 0 \\ t_3 & 1 & 0 & 0 & 1 & 0 & 0 & 0 \\ t_4 & 0 & 0 & 0 & 0 & 0 & 1 & 0 \\ t_5 & 0 & 0 & 0 & 0 & 0 & 0 & 1 \\ t_6 & 0 & 0 & 0 & 1 & 1 & 0 & 0 \end{array}
$$

$$
D = \begin{array}{c|ccccccc|} & p_1 & p_2 & p_3 & p_4 & p_5 & p_6 & p_7 \\ t_1 & -1 & 1 & 0 & 0 & 0 & 0 & 0 \\ t_2 & 0 & -1 & 1 & -1 & 0 & 0 & 0 \\ t_3 & 1 & 0 & -1 & 1 & 0 & 0 & 0 \\ t_4 & 0 & 0 & 0 & 0 & -1 & 1 & 0 \\ t_5 & 0 & 0 & 0 & -1 & 0 & -1 & 1 \\ t_6 & 0 & 0 & 0 & 1 & 1 & 0 & -1 \end{array} \qquad (2)
$$

Let us further introduce the vector $\underline{e}_j$ which is a n_t-dimensional row vector with all the entries equal to 0 except entry j equal to 1. With this notation the execution rules of a PN becomes:

- a transition t_j is enabled in marking M iff $M \geq \underline{e}_j\, D^-$ (note that $e_j\, D^-$ is the $j-th$ row of D^-);
- firing of t_j in M produces a marking M' given by:

$$M' = M - \underline{e}_j\, D^- + \underline{e}_j\, D^+ = M + \underline{e}_j\, D \qquad (3)$$

From the previous definitions follows that, given a PN with initial marking M_1 and a firing sequence $t_i \rightarrow t_j \rightarrow t_k \rightarrow t_j \rightarrow t_i$, the marking obtained at the end of the sequence is given by the following matrix equation:

$$M_{fin} = M_1 + (\underline{e}_i + \underline{e}_j + \underline{e}_k + \underline{e}_j + \underline{e}_i)\, D \qquad (4)$$

By means of the matrix representation, the following properties of PN can be inspected.

6.2.1. Reachability - A marking M' is reachable from M if an integer vector $\underline{X}$ exists such that (see equation 3):

$$M' = M + \underline{X}\, D \qquad (5)$$

Equation (5) provides a necessary but not sufficient condition; all markings reachable from M are solution of equation (5) but not viceversa; for any integer vector $\underline{X}$ a solution to equation (5) exists, but the transition firing sequence represented by $\underline{X}$ can be non-firable. Furthermore, note that the solution of (5) is not affected by the order of transition firings (but only by the number), while the semantics of the net is strongly affected by the order: changing the order a legal sequence can become non-firable.

6.2.2 Conservation - Given a conservative PN, and a n_p-dimensional column vector $\underline{U}_p^T$ with all the entries equal to one, for any marking $M' \in \mathcal{R}(M_1)$ the following relation should hold:

$$M_1\, \underline{U}_p^T \;=\; M'\, \underline{U}_p^T \tag{6}$$

Thus, from equation (5):

$$M_1\, \underline{U}_p^T \;=\; M_1\, \underline{U}_p^T \;+\; \underline{X}\, D\, \underline{U}_p^T \tag{7}$$

since (5) must be satisfied for any $\underline{X}$, it follows:

$$D\, \underline{U}_p^T \;=\; 0 \tag{8}$$

Equation (8) is a necessary and sufficient condition for conservation.

6.2.3. Place Invariant - Let $\underline{W}_p$ be a vector of binary entries (either 0 or 1); we find all vectors $\underline{W}_p$ for which [31]:

$$D\, \underline{W}_p^T \;=\; 0 \tag{9}$$

the places p_i $(i = 1, 2, \ldots, n_p)$ for which $w_i = 1$, form a *place invariant* (a conservative subset of places). With reference to the incidence matrix D of Equation (2), it is easily verified that the following vectors are solution of Equation (9):

$$\begin{aligned} \underline{W}_p^{(1)} &= [\,1\,1\,1\,0\,0\,0\,0\,] \\ \underline{W}_p^{(2)} &= [\,0\,0\,0\,0\,1\,1\,1\,] \\ \underline{W}_p^{(3)} &= [\,0\,0\,1\,1\,0\,0\,1\,] \end{aligned}$$

and therefore, the subsets $\{p_1, p_2, p_3\}$ $\{p_5, p_6, p_7\}$ and $\{p_3, p_4, p_7\}$ are place invariant for the PN of Figure 8b.

7. Extensions

In the use of PN for modelling real systems several authors have found convenient to introduce special constructs either for making the model representation more compact in a given application or for extending the modelling power of the PN formalism. The extensions more often encountered in the literature (and that will be used in the sequel), have been proposed in response to difficulties in modelling priority disciplines by PN. All the extensions mentioned in the sequel are equivalent from the point of view of the modelling power, thus their use depends on the easiness or convenience of the implementation [16].

7.1. INHIBITOR ARCS

An inhibitor arc from place p_j to transition t_k modifies the enabling rules in the sense that the transition can fires only if place p_j does not contain tokens. The inhibition function is usually represented by circle-headed arcs, as in Figure 12 where transition t_k can fire iff p_i contains at least one tokens, but no tokens are present in p_j.

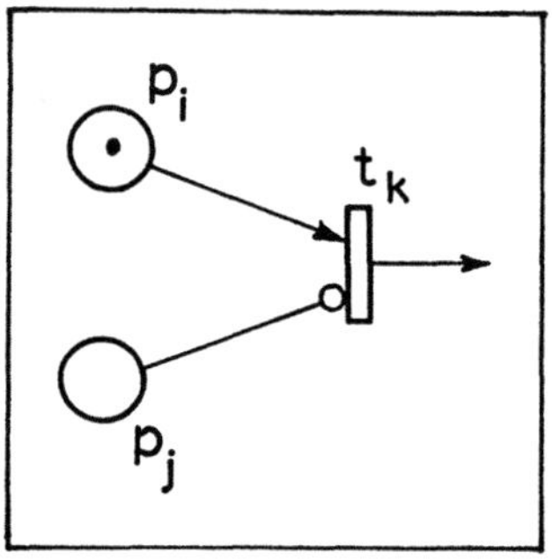

Figure 12 - Inhibitor arc.

In the mutual exclusion problem of Figure 8, the standard PN language does not provide any means to establish precedence rules in the case both resources $\mathbf{C}_1$ and $\mathbf{C}_2$ are simultaneoulsy requesting access to the common resource $\mathbf{C}_s$ (places p_2 and p_5 simultaneously marked). With the insertion of an inhibitor arc from place p_2 to transition t_5 (Figure 13), we model the situation in which, as a conflict arises between $\mathbf{C}_1$ and $\mathbf{C}_2$, $\mathbf{C}_1$ has always the precedence, and blocks (inhibits) $\mathbf{C}_2$ until the common resource is released.

With respect to the reachability graph $\mathcal{G}_R(M_1)$ of the original PN of Figure 8 (reported in Figure 10), the reachability graph of the modified PN of Figure 13 is such that from marking M_5 only transition t_2 can fire while t_5 is inhibited.

7.2. PRIORITY LEVELS

An alternative, but equivalent way to model the same features considered with the introduction of inhibitor arcs, is obtained by attaching to each PN transition a priority level. The standard execution rules are modified in the sense that, among all the transitions enabled in a given marking, only those with associated highest priority level are allowed to fire. In Figure 13 exactly the same precedence policy can be modelled by attaching to transition t_2 a priority level greater than the one attached to t_5. In marking M_5 (see Figure 10) in which both transitions are enabled, only t_2 can fire.

7.3. CONDITIONING FUNCTIONS

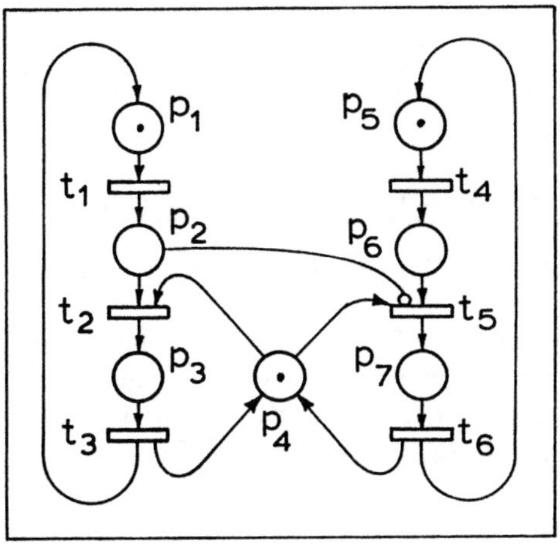

Figure 13 - The mutual exclusion problem of Figure 8, with precedence on the resource $\mathbf{C}_1$.

More complex logical interactions between primitive elements of a PN can be considered by introducing logical *conditioning functions* [20]. Given a marking M, a PN transition is enabled if, beside the normal enabling requirements (including inhibitor arcs and priorities), the conditioning function is *true*. The conditioning functions can be very effective in reducing the graphical complexity of a PN, even if they do not extend the modelling power with respect to inhibitor arcs or priority levels.

7.4. HIGH LEVEL PETRI NETS

In the PN models discussed so far, the individual tokens are indistiguishable. The semantics of the model does not allow to follow the behaviour of an individual token through the net. To overcome this limitation a new class of models has been proposed and discussed. The common characteristic of these models, usually referred to as *high level* PN, is that the position of any single token can be tracked in the PN. Two labelling techniques have been originally proposed: the technique of colouring tokens (coloured PN introduced by Jensen [28]) and the technique of assigning to each token a predicate (*Predicate/Transition* net introduced by Genrich and Lautenbach [24]). However, this class is not further dealt with in the present notes.

8. Timed Petri Nets

An *execution sequence* $\mathcal{E}$ in a marked PN, is a sequence of legal markings obtained by firing a sequence of enabled transitions:

$$\mathcal{E} = \{ (M_{(1)}, t_{(1)}) ; (M_{(2)}, t_{(2)}) ; \dots ; (M_{(j)}, t_{(j)}) ; \dots \}$$

An execution sequence $\mathcal{E}$ can be viewed as a connected path in the reachability graph $\mathcal{G}_R(M_1)$ of the net.

A timed execution sequence $\mathcal{T}_E$ of a marked PN with intial marking $M_{(1)}$, is an execution sequence $\mathcal{E}$ augmented by a non-decreasing sequence of real values representing the epochs of firing of each transition, such that consecutive transitions $(t_{(j)} ; t_{(j+1)})$ in $\mathcal{E}$ correspond to ordered epochs $\tau_j \leq \tau_{j+1}$ in $\mathcal{T}_E$. Thus formally [23,5]:

$$\mathcal{T}_E = \{ (M_{(1)}, t_{(1)}, \tau_1) ; (M_{(2)}, t_{(2)}, \tau_2) ; \dots ; (M_{(j)}, t_{(j)}, \tau_j) ; \dots \}$$

The time interval $\tau_j - \tau_{j+1}$ between consecutive epochs represents the period that the PN sojourns in marking $M_{(j)}$. In the sequel we always assume as initial epoch $\tau_1 = 0$.

> **Definition** - *A Timed PN (TPN) is a marked PN in which a set of specifications are provided and a set of rules are defined such that to each legal execution sequence $\mathcal{E}$ a timed execution sequence $\mathcal{T}_E$ can be univocally associated.*

A variety of timing mechanisms have been proposed in the literature. The distinguishing features of the timing mechanisms are whether the duration of the events is modelled by deterministic variables or random variables, and whether the time is associated to the PN places, transitions or tokens.

Earlier work in timed PN with deterministic timing can be found in [32,43,40,47]. Application of deterministic-PN models are available in different areas, like: communication protocols, performance evaluation, manufacturing. However, in the reliability area stochastic modelling is more appropriate, and therefore we will consider in the sequel only TPN in which the timing mechanism is stochastic; we will refer to this class of models as Stochastic PN (SPN).

SPN were initially proposed in two doctoral thesis [36,35]. In these models, interpreting PN as Condition/Event nets, time was naturally associated with activities that induce state changes, hence with the delay incurred before firing transitions. Although other possibilities have been explored, the choice of associating time with PN transitions is the most common in the literature, and is the only considered in the present notes.

When the random variables associated to PN transitions are exponentially distributed, the dynamic behaviour of the PN can be mapped into a time-continuous homogeneous Markov chain with state space isomorphic to the reachability graph of the PN. This case will be considered in details in the following sections.

Extensions to cover the case of generally distributed transition firing times have been considered in a number of papers [36,21,5,23,26,4]. Releasing the memoryless property of the exponential distribution, in order to univocally associate to each execution sequence $\mathcal{E}$ a timed execution sequence $\mathcal{T}_E$ the concept of SPN *execution policy* needs to be introduced

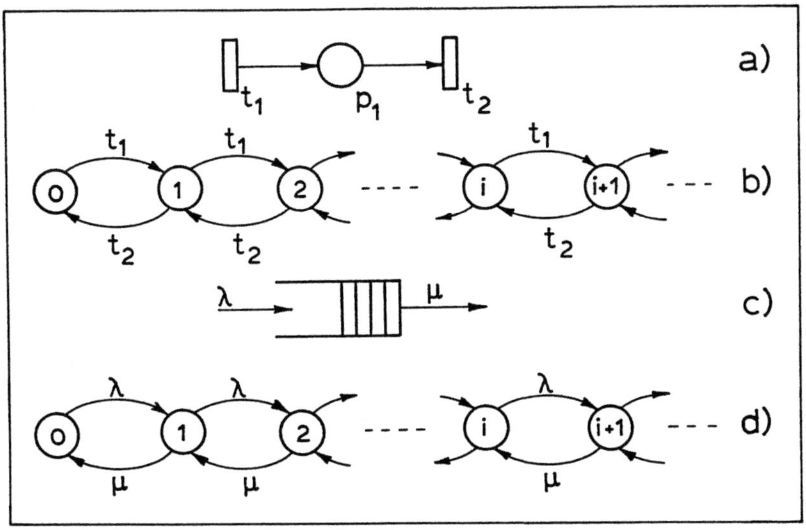

Figure 14 - The $M/M/1$ queue: a) The PN representation; b) The reachability graph; c) The block diagram representation; d) The corresponding Markov transition graph.

[5,4]. The execution policy consists of two parts: the way in which a transition is selected to fire among those enabled in a given marking, and the way in which the time spent is recovered after a transition firing. However, due to the complexity of the semantics of the SPN models, and of the associated stochastic process (both aspects are strictly dependent on the definition of the execution policy), this generalization is no further considered in this paper.

9. Homogeneous Markov SPN (HMSPN)

Let us suppose that the activity modelled by a PN transition takes an exponentially distributed random amount of time to complete once initiated. This means that an exponentially distributed random variable θ_j with parameter $\lambda_j(M)$ is associated to each PN transition t_j. The firing of an enabled transition t_j in marking M becomes a random event which occurs with a time-independent (but possibly marking dependent) firing rate $\lambda_j(M)$. Therefore, knowing the transitions enabled in a given marking and the associated firing rates, we can univocally generate the stochastically timed sequence $\mathcal{T}_E$ from each execution sequence $\mathcal{E}$. In other words, the reachability graph $\mathcal{G}_R(M_1)$ of a marked PN can be univocally mapped into a discrete-state continuous time homogeneous Markov chain, by letting each marking of $\mathcal{G}_R(M_1)$ correspond to a state in the Markov chain, and by substituting the label of the PN transition in each edge of $\mathcal{G}_R(M_1)$ with the firing rate of

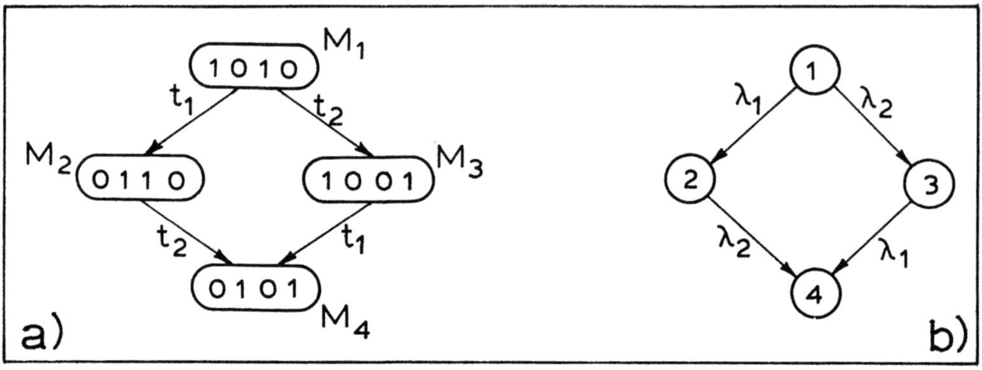

Figure 15 - The reachability graph a) and the corresponding Markov chain b) of the SPN of Figure 3.

the corresponding transition. With this definition we can speak indifferently of *marking* M_i or *state* i.

Example 1 - The M/M/1 queue. Consider the PN of Figure 14a) where it is intended that a transition with no input places is always enabled. The corresponding reachability graph is reported in Figure 14b). The label inside each state is the marking, i.e. the number of tokens in place p_1. Firing of t_1 increases the token count by 1, while firing of t_2 decreases the token count by 1. By associating to transition t_1 the arrival rate λ and to t_2 the service rate μ, the PN of Figure 14a) models the *M/M/1* [29] queueing system. The usual block diagram representation is given in Figure 14c) and the corresponding Markov transition graph is given in Figure 14d). This example is also intended to show how the PN language is suitable to represent queueing systems or queueing networks.

Example 2 - Let the PN of Figure 3 denote the failure process of two components in parallel redundancy; t_1 is the event of failure of component 1 to which a failure rate λ_1 is assigned. Similarly we assigne to t_2 the failure rate λ_2 of component 2. Figure 15a) shows the reachability graph of the net and Figure 15b) the associated Markov chain representing the dynamic behaviour of the net in time.

The probability of the original PN of being in marking M_4 at time t where both components are failed can be computed as the probability of being in state 4 at time t in the corresponding Markov chain.

Example 3 - The reachability graph of the PN of Figure 8 is reported in Figure 10. If all the PN transitions are assigned time-independent firing rates, the reachability graph of Figure 10 is mapped into the Markov chain of Figure 16.

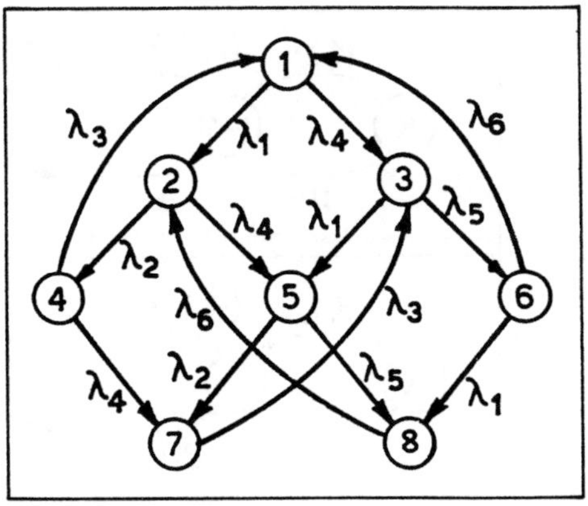

Figure 16 - Markov chain corresponding to the reachability graph of Figure 10.

9.1. FORMAL DEFINITION OF THE MODEL

The Homogenous Markov SPN (HMSPN) is a six-tuple:

$$HMSPN = (P, T, I, O, M, L)$$

where P, T, I, O, M have the same meaning introduced in Section 3, and $L = \{\lambda_1(M), \lambda_2(M), \ldots, \lambda_{nt}(M)\}$ is a set of n_t non-negative real numbers representing the (marking dependent) firing rates of the exponential random variables associated to each PN-transition.

The knowledge of the reachability graph allows us to automatically generate the transition rate matrix Λ of the associated homogeneous Markov chain. Λ is a $N \times N$ matrix, where N is the cardinality of the reachability set $\mathcal{R}(M_1)$.

Let us define $\underline{Q}(t)$ a N-dimensional state probability vector, whose generic entry $q_i(t)$ is the probability of being in state i $(i = 1, 2, ..., N)$ at time t in the associated Markov chain. $\underline{Q}(t)$ is the solution of the standard Markov linear differential equation:

$$\frac{d\,Q(t)}{d\,t} = \Lambda\,\underline{Q}(t) \tag{10}$$

with initial condition $\underline{Q}(0) = [1, 0, 0, ..., 0]^T$. If the steady state probability vector $Q(\infty)$ of the Markov chain exists, it can be calculated from the equation:

$$\Lambda\,Q(\infty) = 0 \qquad \text{with} \qquad \sum_{i=1}^{N} q_i(\infty) = 1 \tag{11}$$

The numerical techniques for the solution of Equations (10) and (11) are outside the scope of the present notes. For a recent survey on methods and techniques for solving equation (10) see [41].

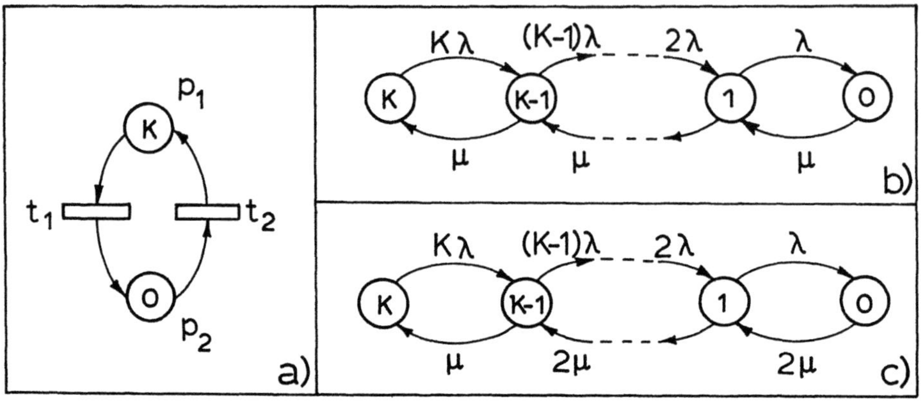

Figure 17 - a) PN modelling K identical parallel components;
b) The associated Markov chain with a single repairman;
c) The associated Markov chain with 2 repairmen.

9.2. MARKING DEPENDENT FIRING RATES

In the formal definition of the HMSPN model the firing rates associated to each transition have been considered as marking dependent. This possibility increases the flexibility of the model and is often used to make the models more compact in the case of the presence of multiple identical resources.

Example 4 - The PN of Figure 17a) has the following physical meaning: place p_1 represents operation; place p_2 non operation; transition t_1 failure and transition t_2 repair. Suppose we have K identical components in parallel redundancy each one with failure rate λ. We can model the system operation by the PN of Figure 17a) with initial marking $M_1 = (K, 0)$ and associating to transition t_1 the marking dependent transition rate $\lambda_{t_1}(M_x) = m_{1x}\lambda$, where m_{1x} is the number of tokens in place p_1 in marking M_x.

Moreover, we can easily model various repair policies: the single repairman policy is modelled by assigning to transition t_2 the repair rate μ. In this case, the Markov chain corresponding to the PN of Figure 17a) is reported in Figure 17b). The independent repair policy can be modelled by assigning to transition t_2 the marking dependent firing rate $\mu_{t_2}(M_x) = m_{2x}\mu$ (as many repairmen as failed components m_{2x}). The case of two repairmen can be modelled by means of more complex logical assignment to the firing rate $\mu_{t_2}(M_x)$ of transition t_2:

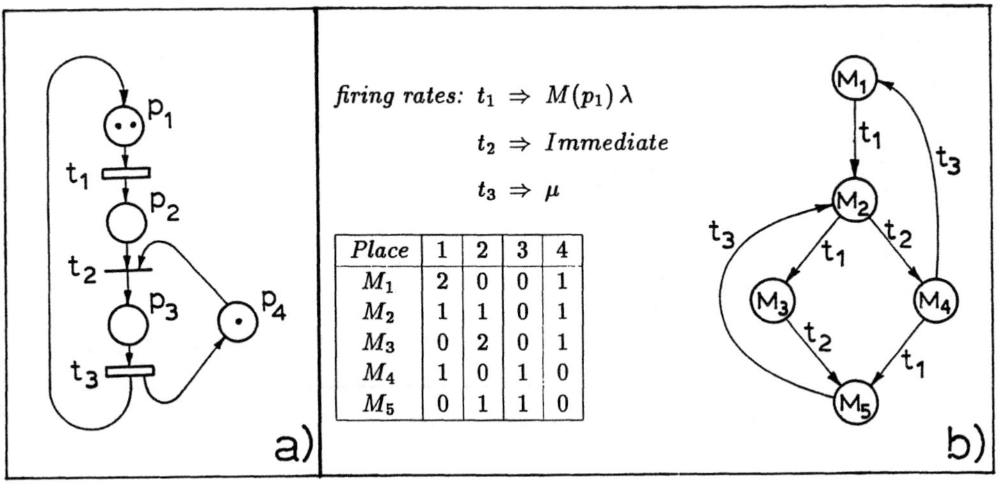

Place	1	2	3	4
M_1	2	0	0	1
M_2	1	1	0	1
M_3	0	2	0	1
M_4	1	0	1	0
M_5	0	1	1	0

Figure 18 - a) Folded PN modelling two identical resources;
b) The associated Markov chain.

$$\mu_{t_2}(M_x) = \begin{cases} 2\,\mu & \text{if} \quad m_2 \geq 2 \\ \mu & \text{if} \quad m_2 = 1 \\ 0 & \text{if} \quad m_2 = 0 \end{cases}$$

In this last case the Markov chain generated by the PN of Figure 17a) is reported in Figure 17c).

Example 5 - In the mutual exclusion problem of Figure 8, if the two resources $\mathbf{C}_1$ and $\mathbf{C}_2$ are identical, we can fold the two simmetric parts of the PN of Figure 8 in the PN of Figure 18a). The stochastic properties of the system are retained by assigning to transition t_1 a firing rate proportional to the number of tokens in p_1. The Markov chain associated to the PN of Figure 18a) is reported in Figure 18b). Note that folding the PN of Figure 18a) corresponds exactly to lumping the Markov chain of Figure 16 into the Markov chain of Figure 18b) (whenever lumpability conditions exist).

9.3. IMMEDIATE AND TIMED PN TRANSITIONS

Many authors [6,11,21] have recognized that the use of SPN for modelling real systems involves the presence of very brief or *fast* transitions, whose duration is short, or even negligible, with respect to the time scale of the problem. Different techniques have been proposed to tackle this problem.

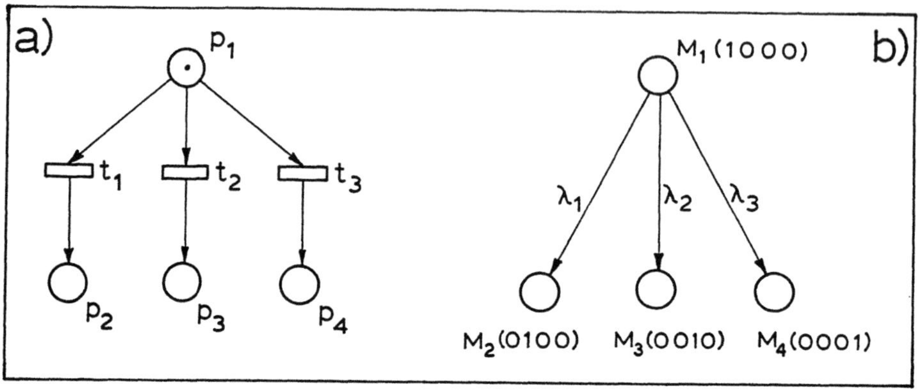

Figure 19 - a) SPN with timed transitions only;
b) The associated Markov chain.

The starting assumption in the GSPN model [6] is that it is desirable to associate a random time only to those transitions which are believed to have the largest impact on the system operation. Transitions are partitioned into two different classes: *immediate* transitions and *timed* transitions. Immediate transitions fire in zero time once they are enabled and have higher priority over timed transitions. Timed transitions fire after an exponentially distributed firing time. In the graphical representation of GSPN, immediate transitions are drawn as thin bars while timed transitions are drawn as thick bars.

Markings (states) enabling immediate transitions are passed through in zero time and are called *vanishing* states. Markings enabling only timed transitions are called *tangible*. Since the process spends zero time in the vanishing states, they do not contribute to the time behaviour of the system so that a procedure can be envisaged to eliminate them from the final Markov chain. With the partition of PN-transitions into a timed and an immediate class, we introduce a greater flexibility at the modelling level without increasing the dimensions of the final state space on which the set of equations (10) or (11) have to be computed.

Given a marking $M \in \mathcal{G}_R(M_1)$ of a GSPN, three different situations may arise:

Situation 1 (Figure 19)
Only timed transitions are enabled (Figure 19a) so that only tangible markings are generated (Figure 19b). The model, in this case, coincides with the HMSPN described in section 9.1.

Situation 2 (Figure 20)
Timed transitions are enabled simultaneously to one immediate transition (Figure 20a). Only the immediate transition is allowed to fire, generating the associate Markov chain of Figure 20b). However, marking M_2 is vanishing and can be eliminated from the chain producing the reduced Markov chain of Figure 20c), in which all the states are tangible.

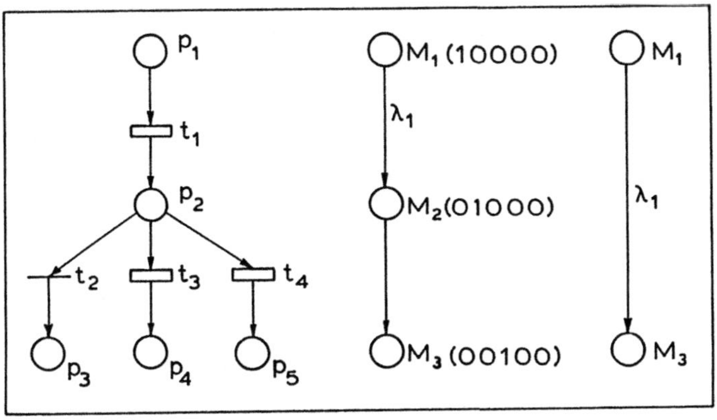

Figure 20 - a) SPN with one immediate transition;
b) The reachability graph;
c) The reduced Markov chain defined over tangible markings.

Situation 3 (Figure 21)
Several immediate transitions are enabled in a marking. In this case in order to define which is the transition that fires first, a probability mass function need to be specified: in the language of GSPN this construct is called a ***random switch***, and the probability mass function the ***switching distribution***. In Figure 21a) the immediate transition t_2 fires with probability v and the immediate transition t_3 with the complementary probability $1 - v$. The equivalent Markov chain is reported in Figure 21b). State M_2 is vanishing and can be eliminated incorporating the switching distribution into the rates leading to state M_2. Elimination of the vanishing state leads to the Markov chain of Figure 21c), which contains only tangible states.

An automatic algorithm can be implemented [6] which recognizes the three situations previously depicted and progressively eliminates vanishing states until a homogeneous Markov chain, defined over tangible states only, is obtained. In this way the reduction procedure becomes completely transparent to the analyst.

The problem of modelling a probabilistic decision, which does not consume time was also considered in [19], by introducing a different construct called *probabilistic arc*. For a comparison of probabilistic arcs with random swithches see [20].

In GSPN [6] only the steady state behaviour of the associated Markov chain is analysed. If the transient analysis is of interest, the use of immediate transitions does not allow to capture the true dynamics of the PN. In this case it is more appropriate to partition the PN-transition into *fast* transitions and *slow* transitions [11]. In this way the transition rate matrix Λ contains rates of very different orders of magnitude, so that the system of differential equations (10) becomes stiff [34]. The increase in the computational load

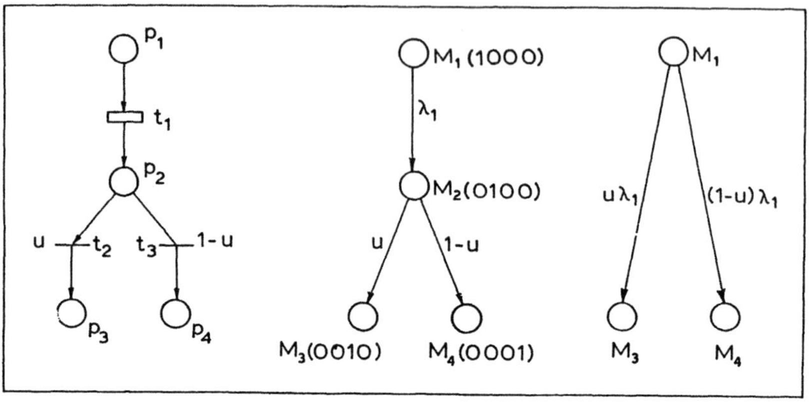

Figure 21 - a) The random switch;
b) The reachability graph;
c) The reduced Markov chain defined over tangible markings.

due to stiffness can be overcome by resorting to a decomposition technique [11,12]. This technique consists in decomposing the transition rate matrix Λ of the associated Markov chain in partitions, such that each partition contains rates of the same order of magnitude. An approximate solution to the original problem is obtained by solving each non-stiff partition in isolation [17]. This technique is incorporated in the package ESP [18].

Example 6 - In the folded PN of Figure 5 (the mutual exclusion problem with identical resources) transition t_2 has only the function of regulating the access to the shared resource $\mathbf{C}_s$ and thus can be modelled by an immediate transition (neglecting, in this case, the access time). The reachability set has 5 states (Figure 18a); markings M_2 and M_3 are vanishing since in these states the immediate transition t_2 is enabled. Eliminating the vanishing states by means of the previous rules leads to the reduced Markov chain of Figure 22 defined over tangible states only.

10. Computation of Measures of Reliability and Performance

A very important point of the time dependent representation of the system behaviour through SPN, is that they allow the user to define in a simple and natural way a large number of different measures related to the performance and reliability features of the system [7,20]. In order to exploit this peculiarity, the input language must be structured for providing a friendly environment for the specification of the output measures. In the sequel we refer in particular to the language of the ESP package [18].

The stochastic behaviour of a SPN is determined by calculating the occurrence probabilities over the states of the reachability set $\mathcal{R}(M_1)$. Therefore, the output measures are

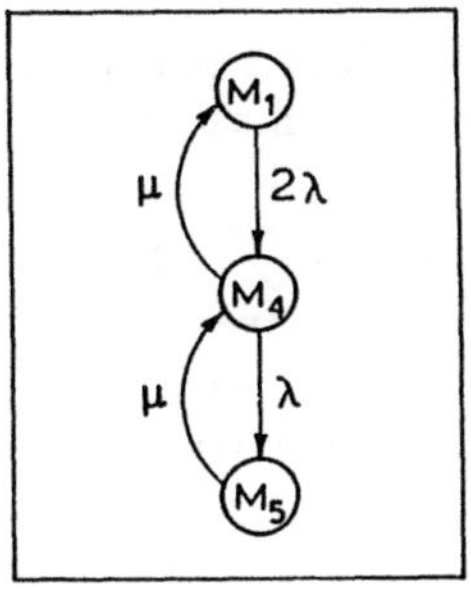

Figure 22 - Markov chain defined over the tangible states of the PN of Figure 5.

defined at the net level, and the numerical computation is carried out automatically by solving the associated equation (10) and by scanning the states in $\mathcal{R}(M_1)$.

Since some of the output measures depend on the integral of the probabilities rather than on the probabilities themselves [25], it is necessary to provide the package with the appropriate computation of the integral of the state probabilities. In the following discussion it is implicitly intended that time t ranges from 0 to ∞, so that all definitions apply to the transient as well as to the steady state solution.

10.1. PROBABILITY OF A GIVEN CONDITION ON THE SPN

By means of logical or algebraic functions of the number of tokens in the PN places, we can specify an output condition (e.g. no tokens in the failed place). We identify in $\mathcal{R}(M_1)$ the subset of places S for which the output condition is true. The output measure

$$Q_S(t) = Prob\,\{condition\ is\ true\ at\ time\ t\,\}$$

is given by:

$$Q_S(t) = \sum_{s \in S} q_s(t) \tag{12}$$

where $q_s(t)$ is the probability of being in state s at time t. For instance, if S is the set of operational states, $Q_S(t)$ in (12) is the usual definition of reliability (or availability).

A very useful case arises when we want to calculate the transient probability that the condition is satisfied for the first time. By using a standard device in the analysis of stochastic processes, we make the states $s \in S$ absorbing, and evaluate this quantity by stopping the process in S. An investigation of the application of SPN for computing the distribution of the completion time as a first marking problem is provided in [10].

10.2. TIME SPENT IN A MARKING

Let S be the subset of markings in which a particular condition is fulfilled. The expected time $\psi_S(t)$ spent in the markings $s \in S$ in the interval $0 - t$ is given by [8]:

$$\psi_S(t) = \sum_{s \in S} \int_0^t q_s(z)\, dz \tag{13}$$

Moreover, it is well known from the theory of Markov chains that as t approaches infinity the proportion of the time spent in states $s \in S$ equals the asymptotic probability:

$$Q_S(\infty) = \sum_{s \in S} q_s(\infty) \tag{14}$$

If S is the set of working states, $\psi_S(t)$ is the expected interval availability [25].

10.3. MEAN PASSAGE TIME

Given that $Q_S(t)$, as calculated in (12), is the probability of having entered subset S before t for the first time, the mean first passage time ϕ_S, has the usual expression:

$$\phi_S = \int_0^\infty [1 - Q_S(z)]\, dz \tag{15}$$

The above formula requires the transient analysis to be extended over long intervals. Of course, in this case, other well known direct techniques can be more effective [14].

10.4. DISTRIBUTION OF TOKENS IN A PLACE

Let p_i be a generic place of the PN. The cumulative distribution function (Cdf) of the number of tokens in p_i at time t is a staircase function in which the amplitude of the k-th step is obtained by summing up the probabilities of all the markings in $\mathcal{R}(M_1)$ containing k tokens ($k = 0, 1, 2, \ldots$) in p_i at time t. The density $f_i(k,t)$ is a mass function equal to the amplitude of the k-th step. The expected value of the number of tokens in place p_i at time t is:

$$E\left[m_i(t)\right] = \sum_{k=0}^{\infty} k\, f_i(k,t) \tag{16}$$

As an example, if place p_i represents identical units queueing up for a common resource the above quantities are the Cdf and the expected value of the number of units in the queue versus time. In reliability analysis a very interesting case arises when place p_i represents failed components. The above quantities provide the Cdf and the expected value of the number of failed components at time t.

10.5. EXPECTED NUMBER OF FIRINGS OF A PN-TRANSITION

Given an interval $(0,t)$ this quantity indicates how many times, on the average, an event modelled by a PN transition has occurred in that interval. Let t_k be a generic PN transition, and let S be the subset of $\mathcal{R}(M_1)$ which includes all the markings $s \in S$ enabling t_k. The expected number of firings of t_k in $(0,t)$ is given by:

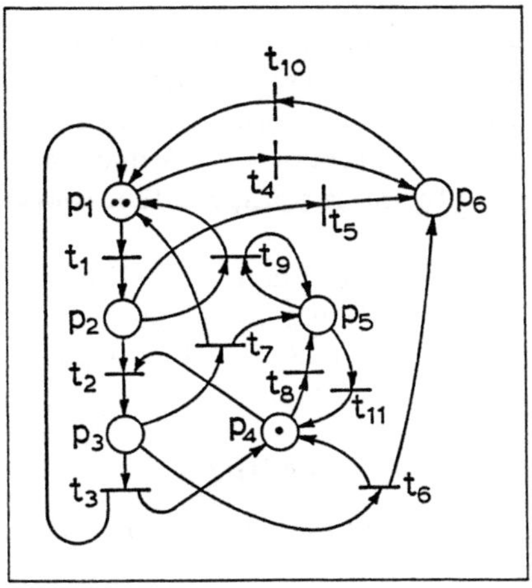

Figure 23 - System of Figure 18 with failures and repairs.

$$\eta_k(t) = \sum_{s \in S} \int_0^t q_s(z)\, \lambda_k(s)\, dz \tag{17}$$

where $\lambda_k(s)$ is the firing rate of t_k in marking s.
In steady state, the expected number of firings per unit time becomes:

$$\nu_k = \sum_{s \in S} q_s(\infty)\, \lambda_k(s) \tag{18}$$

where $q_s(\infty)$ is the steady state probability of state s. As an example, if transition t_k indicates failure (repair) of a component, $\eta_k(t)$ in (17) provides the mean number of failures (repairs) of that component in $(0, t)$.

11. Performance/Reliability Modelling through SPN

Performance-oriented reliability analysis has been the subject of an extensive literature in recent years [9,33,27,45]. We will show, by means of fully elaborated examples, that the SPN language, described in the previous sections, is very suitable to model this class of problems.

11.1. PARALLEL UNITS WITH SHARED RESOURCE

This situation has been depicted in Figure 8, and arises very often in distributed systems. With reference to Figure 8 in a multiprocessor system $\mathbf{C}_1$ and $\mathbf{C}_2$ are independent processors working locally on their private memories and $\mathbf{C}_s$ is a shared global memory which contains common data for the two processors. In a manufacturing system $\mathbf{C}_1$ and $\mathbf{C}_2$ are two working cells connected to the same transportation system or to the same load/unload device $\mathbf{C}_s$.

Assuming $\mathbf{C}_1$ and $\mathbf{C}_2$ to be identical units, the SPN modelling the fault free system operation is reported in Figure 18. Taking into account the failure and repair of each unit the system operation is modelled by the SPN of Figure 23 [11].

TABLE I

Meaning of places and transitions in the SPN of Figure 23

		firing rate
p_1	Unit working independently	
p_2	Unit waiting for access to $\mathbf{C}_s$	
p_3	Unit operating with $\mathbf{C}_s$	
p_4	$\mathbf{C}_s$ free	
p_5	$\mathbf{C}_s$ failed	
p_6	Unit failed	
t_1	Unit requesting access to $\mathbf{C}_s$	$1\ m_1$
t_2	Unit accessing $\mathbf{C}_s$	10^4
t_3	Unit releasing $\mathbf{C}_s$	5
t_4	Unit failure in local mode	$10^{-4}\ m_1$
t_5	Unit failure while waiting	$10^{-4}\ m_2$
t_6	Unit failure when working with $\mathbf{C}_s$	10^{-4}
t_7	$\mathbf{C}_s$ failure while working	10^{-4}
t_8	$\mathbf{C}_s$ failure while free	10^{-4}
t_9	Return to local mode when $\mathbf{C}_s$ failed	10_4
t_{10}	Unit repair	10^{-2}
t_{11}	$\mathbf{C}_s$ repair	10^{-2}

Table I reports the meaning of the places and transitions of Figure 23, and the numerical values assigned to the firing rates associated to each PN transition. With the initial marking M_1 shown in Figure 10, the reachability set $\mathcal{R}(M_1)$ consists in 15 states whose token distribution is reported in Table II. By inspection of Tables I and II the following subsets of states can be recognized:

- States 1,2,5,6,11: fault-free operation of the system.

- States 3,7,13: normal operation of one unit when the other one is in a failed condition.

- States 4,8,12: two units operating and the shared resource failed.
- States 10,14: one unit operating while the other one and the shared resource failed.
- State 9: two units failed.
- State 15: two units and the shared resource failed.

TABLE II

Reachability set and token distribution of the SPN of Figure 23

State	Marking					
	m_1	m_2	m_3	m_4	m_5	m_6
1	2	0	0	1	0	0
2	1	1	0	1	0	0
3	1	0	0	1	0	1
4	2	0	0	0	1	0
5	0	2	0	1	0	0
6	1	0	1	0	0	0
7	0	1	0	1	0	1
8	1	1	0	0	1	0
9	0	0	0	1	0	2
10	1	0	0	0	1	1
11	0	1	1	0	0	0
12	0	2	0	0	1	0
13	0	0	1	0	0	1
14	0	1	0	0	1	1
15	0	0	0	0	1	2

Table III is the literal description of the reachability graph $\mathcal{G}_R(M_1)$ of the SPN; for each state of $\mathcal{R}(M_1)$ on the first column, the enabled transitions and the immediately reachable states (in parentheses) are reported. Substituting the numerical values of the firing rates reported in Table I to the transition labels of Table III the transition rate matrix Λ of the associated Markov chain can be automatically generated.

By considering the access time to $\mathbf{C}_s$ as negligible with respect to the time constant of the system, t_2 and t_9 can be interpreted as immediate transitions. With this assumption it is seen from Table III that the states { 2,5,7,8,12,14 } become vanishing since in these states one of the immediate transitions is enabled. By reducing the state space with the rules of section 9.3, the final Markov chain is defined over a state space containing 9 tangible states.

An interesting performance/reliability measure for this system is the number of units doing useful work at time t, where by useful work we mean the work performed by each unit when operating independently. This measure takes into account the reduction in the system performance due to different effects: the congestion delays due to the sharing of the common resource, the transfer of data or pieces from each unit to $\mathbf{C}_s$ and the failure and repair cycles. By using the definitions of the previous section and looking at Table I, it is seen that this measure coincides with the expected number of tokens in place p_1 and thus can be easily defined at the PN level and computed by means of Equation (16).

TABLE III

Literal description of the Reachability Graph $\mathcal{G}_R(M_1)$

State	Enabled transition and immediately reachable state				
1	1 (2)	4 (3)	8 (4)		
2	1 (5)	2 (6)	4 (7)	5 (3)	8 (8)
3	1 (7)	4 (9)	8 (10)	10 (1)	
4	1 (8)	4 (10)	11 (1)		
5	2 (11)	5 (7)	8 (12)		
6	1 (11)	3 (1)	4 (13)	6 (3)	7 (4)
7	2 (13)	5 (9)	8 (14)	10 (2)	
8	1 (12)	4 (14)	5 (10)	9 (4)	11 (2)
9	8 (15)	10 (3)			
10	1 (14)	4 (15)	10 (4)	11 (3)	
11	3 (2)	5 (13)	6 (7)	7 (8)	
12	5 (14)	9 (8)	11 (5)		
13	3 (3)	6 (9)	7 (10)	10 (6)	
14	5 (15)	9 (10)	10 (8)	11 (7)	
15	10 (10)	11 (9)			

11.2. PARALLEL SYSTEM WITH FINITE INPUT BUFFER

The block diagram of the system is shown in Figure 24. It consists in u identical units $\mathbf{U}_1, \mathbf{U}_2, \ldots, \mathbf{U}_u$ and in an input buffer with b positions $\mathbf{B}_1, \mathbf{B}_2, \ldots, \mathbf{B}_b$ [33].

The GSPN model of the fault free system operation is shown in Figure 25 [7]; the sum of tokens in p_1 and p_2 is equal to b (number of buffer positions; see also Section 4.3), whereas the sum of tokens in p_3 and p_4 is equal to u (number of parallel units). In other words, $\{p_1, p_2\}$ and $\{p_3, p_4\}$ form place-invariants.

The firing rate associated to t_1 is the task arrival rate λ, while the firing rate associated to t_3 is the service rate proportional to the number of active units $m_4\,\mu$, being μ the service rate of a single unit and m_4 the number of tokens in p_4. t_2 is an immediate transition (we neglect the transfer time from the buffer to the service station).

When failures and repairs are considered, the GSPN model becomes as in Figure 26. Heavy lines represent the fault-free operation, light lines failures and dotted lines repairs.

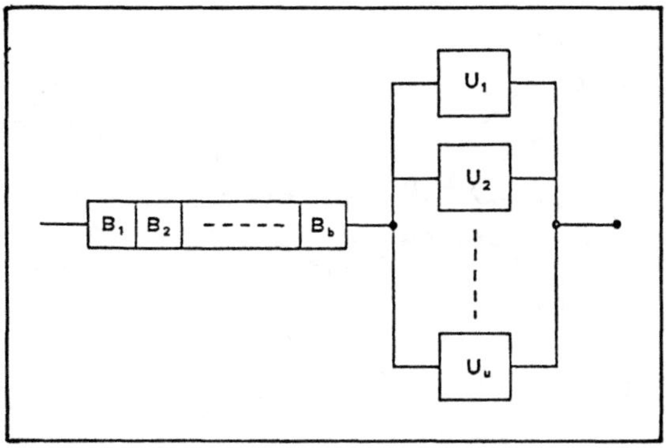

Figure 24 - Block diagram of a parallel system with finite buffer.

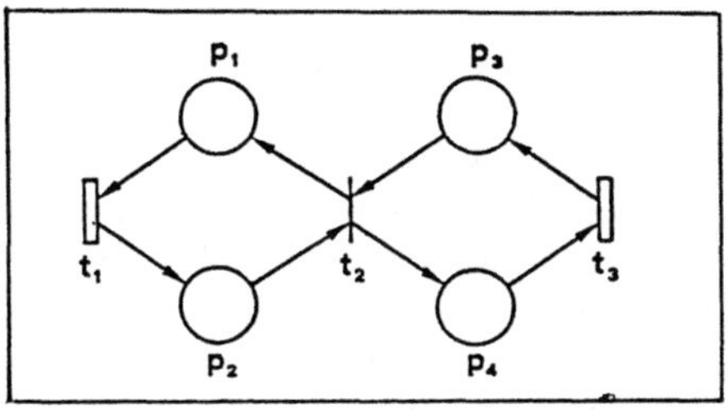

Figure 25 - Fault-free SPN model of the system of Figure 24.

Let us first focus our attention on the failure transitions; with reference to Figure 26 the following hypotheses have been considered:

- Buffer stages fail one at the time, either when free (t_4), or when occupied (t_5), with possibly different failure rates. t_6 and t_7 form a random switch modelling the fact that with probability v_B a buffer stage failure is recovered (the buffer continues to be operational with a storing capacity reduced by one stage), and with probability $1 - v_B$ the failure is not recovered and the buffer locks (inhibitor arc from p_7 to t_2).

- The units $\mathbf{U}_i$ $(i = 1, 2, \ldots, u)$ fail either when idle (t_8) or when active (t_9), with possibly different failure rates. The failure of an idle unit is recovered with probability one, while the failure of an active unit is recovered with probability v_U (random switch t_{10} t_{11}). A task is lost only when an active unit fails.

By slightly modifying the GSPN of Figure 26, different design alternatives or recovery strategies could be accommodated. When repair is considered, t_{12} and t_{13} refer to buffer

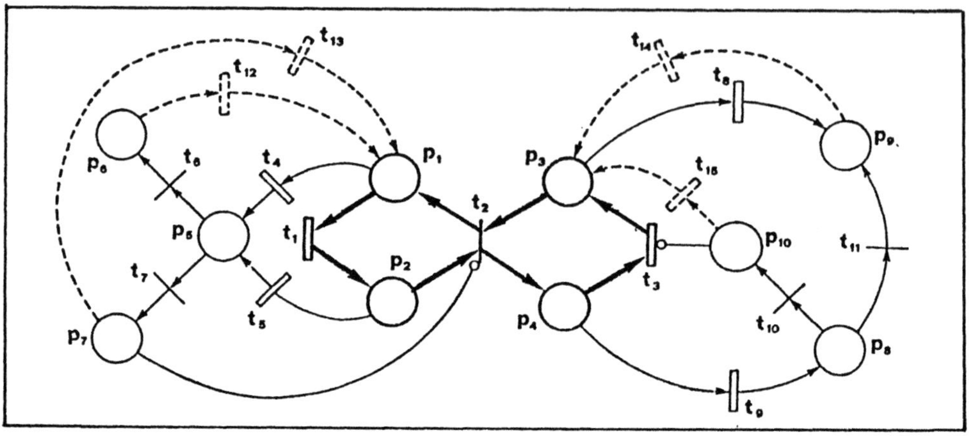

Figure 26 - SPN model of the system of Figure 24 with failures and repairs.

stage repair, and t_{14} and t_{15} to processor repair; the considered model allows us to allocate different repair rates for recoverable and unrecoverable failures.

The meaning of places and transitions in Figure 26 is summarized in Table IV, where the expressions of the firing rates for the timed transitions, and of the switching probabilities for the immediate transitions, are also given.

The initial marking M_1 consists in b tokens in place p_1 and u tokens in p_3. As measures characterizing the system performance and reliability, we define the following.

. *Mean fraction of arrived tasks processed in 0-t.*
The mean number of processed tasks in $0-t$ is given by the mean number of firings of t_3 (Equation 17). The mean number of arrived tasks in $0-t$ is simply $\lambda \cdot t$, having assumed a Poisson arrival process with rate λ. Thus the performance/reliability index $Y(t)$, representing the mean fraction of arrived tasks processed in $0-t$ is calculated as:

$$Y(t) = \frac{\eta_3(t)}{\lambda t} \tag{19}$$

. *Mean number of failures (repairs) in 0-t.*
This quantity is given by $[\eta_4(t)+\eta_5(t)]$ (Equation 17) for buffer stage failure ($[\eta_{12}(t)+\eta_{13}(t)]$ for buffer stage repair), and by $[\eta_8(t)+\eta_9(t)]$ for unit failure ($[\eta_{14}(t)+\eta_{15}(t)]$ for unit repair).

Cdf and mean number of active, idle, failed, units or buffer stages.
These quantities are obtained by applying the procedure of paragraph 10.4 to place p_4 for active units, to place p_3 for idle units, and to places $[p_9 + p_{10}]$ for failed units. Similarly, place p_1 indicates free buffer stages, p_2 filled buffer stages, and $[p_6 + p_7]$ failed buffer stages.

TABLE IV

Meaning of places and transitions in the SPN of Figure 26

p_1	Free buffer stage	
p_2	Occupied buffer stage	
p_3	Idle unit	
p_4	Active unit	
p_5	Failed buffer stage	
p_6	Recovered buffer stage failure	
p_7	Unrecovered buffer stage failure	
p_8	Failed active unit	
p_9	Recovered unit failure	
p_{10}	Unrecovered unit failure	
		firing rate
t_1	Buffer stage becomes occupied	λ
t_2	Transfer from buffer to unit	*immed.*
t_3	Unit ends a task	$m_4\,\mu$
t_4	Free buffer stage fails	$m_1\,\gamma_4$
t_5	Occupied buffer stage fails	$m_2\,\gamma_5$
t_6	Buffer stage failure is recovered	v_B
t_7	Buffer stage failure is not recovered	$(1 - v_B)$
t_8	Idle unit fails	$m_3\,\gamma_8$
t_9	Active unit fails	$m_4\,\gamma_9$
t_{10}	Unit failure is not recovered	$(1 - v_U)$
t_{11}	Unit failure is recovered	v_U
t_{12}	Repair of recovered buffer stage	ρ_{12}
t_{13}	Repair of unrecovered buffer stage	ρ_{13}
t_{14}	Repair of recovered unit	ρ_{14}
t_{15}	Repair of unrecovered unit	ρ_{15}

A numerical example has been run with $u = 2$ and $b = 2$. The reachability set, in this case, comprises 88 tangible states and 84 vanishing states. With reference to Table IV, we have assigned to the parameters the following numerical values (being $w = \lambda/\mu$ the load factor of the system):

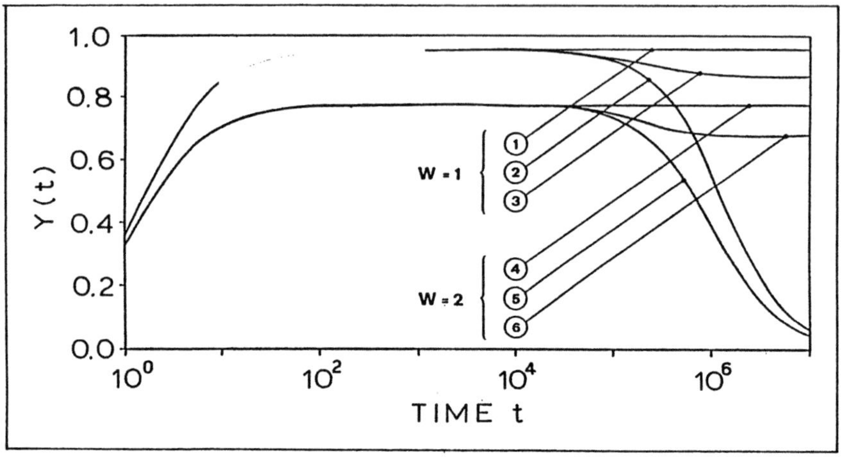

Figure 27 - Mean fraction of arrived tasks processed in $0-t$ versus time.

$$\begin{aligned}
\mu &= 1 \\
\lambda &= w\,\mu \\
\gamma_4 &= \gamma_5 = \gamma_8 = \gamma_9 = \gamma = 1.0\,10^{-6} \\
\rho_{12} &= \rho_{13} = \rho_{14} = \rho_{15} = 10\,\gamma \\
v_B &= v_U = 0.9
\end{aligned}$$

Two cases have been examined with different load factors: $w = 1$ and $w = 2$. The last case represents the ideal load factor since the arrival rate is twice as large as the service rate, but there are two parallel service units. Numerical results have been obtained using the program ESP [18] and resorting to a decomposition technique due to the high spread in the firing rate values. Figure 27 shows $Y(t)$ (Equation 19) as a function of t for the two chosen values of w, and in three different conditions, namely: fault-free operation (curves 1 and 4); with failures (curves 2 and 5); with failures and repairs (curves 3 and 6). Figure 27 shows how the system performance (throughput) is degraded when considering failures and failures/repairs, and can be of valuable support at the design level.

12. Simulative Analysis of SPN

In the previous Sections the SPN was used as a language for generating an associated Markov chain whose transient and ergodic behaviour is obtained by solving Equations (10) and (11) respectively. However, a simulative approach is also possible.

The simulative approach is very simple from a logical point of view, and is easily implementable in a computer program, so that SPN can also be considered as a possible general simulative language. Due to these characteristics, it is conceivable to construct general SPN solvers [20] where both the analytic approach (when feasible and convenient) and the simulative approach are present.

The core of the simulator is that in each marking we need to choose the PN transition which actually fires, among those enabled. This choice is done by generating a random sample from the distribution function associated to each transition and selecting the transition with the minimum firing sample. The simulator clock is updated with the minimum sample and we move to the next marking where a new choice procedure is initiated. The basic algorithm for the generation of a timed execution sequence $\mathcal{T}_E$ (section 8) can be outlined as follows, when the firing times are exponentially distributed:

```
begin
marking ← initial marking
clock = 0
repeat
   for j  := 1 to n_t do
      begin
      if { t_k is enabled } then
         generate a random sample θ_k
      end
   find minimum θ_k
   generate new marking
   clock=clock+min(θ_k)
until { terminating condition is fulfilled }
end
```

The termination criteria are driven by the type of simulation: transient simulation or ergodic simulation [19]. In the transient simulation the user defines exit places or absorbing places; the simulation trial is stopped once a token reaches an exit place. Statistics are gathered by generating random timed execution sequences $\mathcal{T}_E$ through the PN [22].

The ergodic simulation is a regenerative-type simulation [30], in which a return to the initial marking constitutes a regeneration point in the simulation. A trial is defined as the random timed execution sequence $\mathcal{T}_E$ starting and ending with the initial marking.

All the measures defined in Section 10 can be estimated as a result of the simulation approach. The definition of these measures in both transient and ergodic simulation is usually straightforward, and is outlined in [20]. Confidence intervals can be also calculated as a function of the number of trials [19].

The very important fact about the simulative approach, to be noted here, is that in each trial we only generate a single timed execution sequence $\mathcal{T}_E$, so that we do not need to generate and store all the reachable markings at the same time. Moreover, the extension of the simulative approach to the case in which the random variables associated to the PN transitions are generally distributed is, in principle, quite simple. In fact, once the execution policy is specified (i.e. the way in which the SPN keeps trace of the past history; see Section 8), the basic simulation algorithm must be modified by attaching a clock to each PN-transitions. Each time a move is selected, the clocks are updated by recovering the elapsed time as specified by the execution policy, and the following selection is performed by comparing the values of all these clocks. This extension [21,26] is not further considered in the present notes.

13. Conclusion

These lecture notes were intended as introductive material to the use of Petri nets as a general language for the modelling and analysis of the behaviour of complex systems versus time. In the first part, the aim was to show how the semantics of classical PN is suitable to model various kinds of logical as well as physical interactions among components in a system (interactions that are not easily reprensentable in other modelling frameworks).

The second part was more specifically devoted to define the Stochastic PN extension and to present examples taken from the reliability area. Only the case where the stochastic process associated to the SPN is a homogeneous Markov chain has been considered in details. This case arises when the firing times assigned to the PN transitions are exponentially distributed.

From the discussion contained in these lecture notes we can summarize some advantages and disadvantages of the SPN as a modelling tool. The main advantages include: the graphic nature, the conciseness in comparison with state graphs, the possibility of implementing analysis techniques. The graphic nature facilitates the use by non skilled users and allows to implement very friendly graphic editors for the specification of the input net. We finally stress that the use of SPN requires only the specification of the topology of the starting PN, the specification of the firing rates (or of the distribution functions in the general case) associated to the transitions and the specification of the output measures to be computed following the indications provided in Section 10. All the subsequent steps, which consist in:

- the generation of the reachability graph $\mathcal{G}_R(M_1)$;
- the generation of the associated Markov chain;
- the transient and ergodic solution of the Markov chain;
- the evaluation of the relevant process measures;

can be executed in a completely automated way by a computer program, thus making transparent to the user the associated mathematics.

The main disadvantages of SPN arise from the size of the net obtained in modelling very complex distributed systems. In this case the model is difficult to validate at the net level, and the number of reachable markings tends to explode, making analytically intractable the associated Markov chain. It should be recognized, however, that this drawback is common to almost all general purpose modelling techniques.

References

[1] *International Workshop Petri Nets and Performance Models*, IEEE Computer Society Press No. 796, Madison, 1987.

[2] *International Workshop Timed Petri Nets*, IEEE Computer Society Press No. 674, Torino (Italy), 1985.

[3] T. Agerwala. Putting Petri nets to work. *Computer*, 85–94, December 1979.

[4] M. Ajmone Marsan, G. Balbo, A. Bobbio, G. Chiola, G. Conte, and A. Cumani. The effect of execution policies on the semantics and analysis of stochastic Petri nets. *To be published on: IEEE Transactions on Software Engineering*, 1989.

[5] M. Ajmone Marsan, G. Balbo, A. Bobbio, G. Chiola, G. Conte, and A. Cumani. On Petri nets with stochastic timing. In *Proceedings International Workshop on Timed Petri Nets*, pages 80–87, IEEE Computer Society Press no. 674, Torino (Italy), 1985.

[6] M. Ajmone Marsan, G. Balbo, and G. Conte. A class of generalized stochastic Petri nets for the performance evaluation of multiprocessor systems. *ACM Transactions on Computer Systems*, 2:93–122, 1984.

[7] M. Ajmone Marsan, A. Bobbio, G. Conte, and A. Cumani. Performance analysis of degradable multiprocessor systems using generalized stochastic Petri nets. *IEEE Computer Society Newsletters*, 6, SI-1:47–54, 1984.

[8] R.E. Barlow and F. Proschan. *Statistical Theory of Reliability and Life Testing*. Holt, Rinehart and Winston, New York, 1975.

[9] M.D. Beaudry. Performance-related reliability measures for computing systems. *IEEE Transactions on Computers*, C-27:540–547, 1978.

[10] A. Bobbio. Petri nets generating Markov reward models for performance reliability analysis of degradable systems. In *Proceedings of the 4-th International Conference on Modelling Techniques and Tools for Computer Performance Evaluation*, pages 413–431, Palma de Mallorca, 1988.

[11] A. Bobbio, A.Cumani, and R. Del Bello. Reduced markovian representation of stochastic Petri net models. *Systems Science*, 10:5–23, 1984.

[12] A. Bobbio and K.S. Trivedi. An aggregation technique for the transient analysis of stiff Markov chains. *IEEE Transactions on Computers*, C-35:803–814, 1986.

[13] G.W. Brams. *Réseaux de Petri: Théorie et pratique*. Masson, 1983. (in French).

[14] J.A. Buzacott. Markov approach to finding failure times of repairable systems. *IEEE Transactions on Reliability*, R-19:128–134, 1970.

[15] W.M. Chow, E.A. McNair, and C.H. Sauer. Analysis of manufacturing systems by Research Queueing Package. *IBM Journal of Research and Development*, 29:330–341, 1985.

[16] G. Ciardo. Toward a definition of modeling power for stochastic Petri net models. In *Proceedings International Workshop on Petri Nets and Performance Models*, pages 54–62, IEEE Computer Society Press no. 796, Madison, 1987.

[17] P.J. Courtois. *Decomposability: Queueing and Computer System Applications.* Academic Press, New York, 1977.

[18] A. Cumani. Esp - A package for the evaluation of stochastic Petri nets with phase-type distributed transition times. In *Proceedings International Workshop Timed Petri Nets*, IEEE Computer Society Press no. 674, Torino (Italy), 1985.

[19] J. Bechta Dugan. *Extended stochastic Petri nets: applications and analysis.* Technical Report, Phd Thesis, Department of Computer Science, Duke University, 1984.

[20] J. Bechta Dugan, A. Bobbio, G. Ciardo, and K. Trivedi. The design of a unified package for the solution of stochastic Petri net models. In *Proceedings International Workshop on Timed Petri Nets*, pages 6–13, IEEE Comp Soc Press no. 674, Torino (Italy), 1985.

[21] J. Bechta Dugan, K. Trivedi, R. Geist, and V.F. Nicola. Extended stochastic Petri nets: applications and analysis. In *Proceedings PERFORMANCE '84*, Paris, 1984.

[22] G.S. Fishman. *Concepts and methods in discrete event digital simulation.* Wiley, New York, 1973.

[23] G. Florin and S. Natkin. Les reseaux de Petri stochastiques. *Technique et Science Informatique*, 4:143–160, 1985.

[24] H.J. Genrich and K. Lautenbach. System modelling with high level Petri nets. *Theoretical Computer Science*, 13:109–136, 1981.

[25] A. Goyal, S. Lavenberg, and K.S. Trivedi. Probabilistic modeling of computer system availability. *Annals of Operations Research*, 8:285–306, 1987.

[26] P.J. Haas and G.S. Shedler. Regenerative stochastic Petri nets. *Performance Evaluation*, 6:189–204, 1986.

[27] B.R. Iyer, L. Donatiello, and P. Heidelberger. Analysis of performability for stochastic models of fault-tolerant systems. *IEEE Transactions on Computers*, C-35:902–907, 1986.

[28] K. Jensen. Coloured Petri nets and the invariant method. *Theoretical Computer Science*, 14:317–336, 1981.

[29] L. Kleinrock. *Queuing systems, Volume 1: Theory.* Wiley Interscience, New York, 1975.

[30] A.J. Lemoine M.A. Crane. An introduction to the regenerative method for simulation analysis. In A.V. Balakrishnan and M. Thorna, editors, *Lecture Notes in Control and Information Sciences*, Springer-Verlag, 1977.

[31] J. Martinez and M. Silva. A simple fast algorithm to obtain all invariants of a generalized Petri net. In *Proceedings 2-nd European Workshop on Application and Theory of Petri Nets*, Springer-Verlag, 1981.

[32] P.M. Merlin and D.J. Faber. Recoverability of communication protocols - Implication of a theoretical study. *IEEE Transactions on Communication*, COM-24:1036–1043, 1976.

[33] J.F. Meyer. Closed form solution of performability. *IEEE Transactions on Computers*, C-31:648–657, 1982.

[34] W.L. Miranker. *Numerical Methods for Stiff Equations.* Reidel, Dordrecht, 1981.

[35] M.K. Molloy. *On the integration of delay and throughput measures in distributed processing models.* Technical Report, Phd Thesis, UCLA, 1981.

[36] S. Natkin. *Les reseaux de Petri stochastiques et leur application a l'evaluation des systemes informatiques.* Technical Report, These de Docteur Ingegneur, CNAM, Paris, 1980.

[37] J.L. Peterson. *Petri net theory and the modeling of systems.* Prentice Hall, Englewood Cliffs, 1981.

[38] J.L. Peterson. Petri nets. *Computing Surveys*, 9:223–252, 1977.

[39] C.A. Petri. *Kommunikation mit automaten.* Technical Report, Doctoral Thesis, University of Bonn, 1962. (Available in English as: *Communication with automata*, Technical Report RADC-TR-65-377, Rome Air Development Center, Griffis NY, 1966).

[40] C.V. Ramamoorthy and G.S. Ho. Performance evaluation of asynchronous concurrent systems using Petri nets. *IEEE Transactions on Software Engineering*, SE-6:440–449, 1980.

[41] A. Reibman and K.S. Trivedi. Numerical transient analysis of Markov models. *Computers and Operations Research*, 15:19–36, 1988.

[42] W. Reisig. *Petri nets - An introduction.* Springer-Verlag, 1982.

[43] J. Sifakis. Use of Petri nets for performance evaluation. In H. Beilner and E. Gelenbe, editors, *Measuring, modelling and evaluating computer systems*, pages 75–93, North Holland, 1977.

[44] M. Silva. *Las redes de Petri en la Automatica y la Informatica.* AC, Madrid, 1985.

[45] R. Smith, K. Trivedi, and A.V. Ramesh. Performability analysis: measures, an algorithm and a case study. *IEEE Transactions on Computers*, C-37:406–417, 1988.

[46] W. Whitt. Blocking when service is required from several facilities simultaneously. *AT&T Technical Journal*, 64:1807–1856, 1985.

[47] W.M. Zuberek. Timed Petri nets and preliminary performance evaluation. In *Proceedings 7-th Annual Symposium on Computer Architecture*, pages 88–96, 1980.

FAILURE DELAY SYSTEMS RELIABILITY MODELLING

N. LIMNIOS

Division Mathématiques Appliquées
Département Génie Informatique
Université de Technologie de Compiègne
B.P. 649, 60206 Compiègne Cédex
France.

ABSTRACT. In this paper, we introduce the notion of "failure delay systems". A classical assumption in reliability studies of systems, is that time between occurrence of a cut set and system failure is small enough to be neglected : we call this the simultaneity assumption. By considering non negligible time (delay) a large family of systems that we call "failure delay systems" are introduced . For these systems there are two conditions for failure: first, the occurrence of a particular failure, i.e. a cut set, and second, its continuous presence for a given time, fixed or random. We formally define failure delay systems and give the lifetime structure for reliability measures evaluation. We describe next some classes of failure delay systems. A single component system without and with repair are studied and closed-form solutions for lifetime distribution are derived. Finally, as an example of failure delay systems we study the reservoir.

KEY WORDS. Failure delay systems, Reliability related measures, Single component systems, Reservoir.

1. Introduction

A basic assumption in reliabilty studies of systems so far, is to consider that when at a given moment failure conditions are present, like the occurrence of a cut set for example, at this very moment system failure occurs.
In fact, this is not always the case : for a great number of systems the above assumption is not valid. For these systems, when at a given moment failure conditions are present, there still is a time allowance during which if failure conditions cease system failure will not occur. In other words, system failure occurs after an interval of time if and only if failure conditions have not ceased in the meantime.

A. G. Colombo and A. Saiz de Bustamante (eds.), Systems Reliability Assessment, 145–158.

This family of systems we shall call *failure delay systems* ; delay can be either fixed or random.

A good example is a water reservoir of continuous functioning supplied by a water source through a pump, and providing a certain amount of water with fixed rate. A pump failure does not necessarily mean service interruption, not instantaneously; to have this, the pump must remain failed until the reservoir is emptied. In absence of reservoir, evidently, a pump failure would lead to instantaneous interruption of service.
Another example is a reactor cooling circuit ; if circuit failure occurs, the reactor will continue functioning for a time period, after which it will have to stop if the circuit isn't repaired.
Many systems provided with a passive redundancy with life time statistically of the same order as the principal unit's repair time can be considered as failure delay systems.

Ross & Scheetman [8] studied two systems, one with a single component and the other with n components in parallel, each with fixed down time.
Shantikumar [10] generalized the previous model by considering a Markovian parallel system with random safety period at system level.
Pagès & Gondran [7] studied a system with fixed maximum repair time corresponding to the cooling circuit of a nuclear reactor. This system had three states and was modeled by the complementary variables method. It is a particular case of failure delay systems which were called by the authors "almost repairable systems".
Dunbar [1] studied a system with two components and fixed maximum repair time. Its model concerns time-to-first-failure.
Schneeweiss [9] studied the delay operator in fault tree analysis using pseudo-Boolean algebra.

In the following section we define the failure delay systems and give the lifetime structure for reliability measures evaluation; also, some classes of failure delay systems are described. In section 3, a single component system is studied without and with repair. A closed-form solution for lifetime distribution is derived. In section 4, the reservoir case is studied as an example of failure delay systems.

Notations

T	random variable (r.v.) of system's life time
X_i , X	r.v. of up time.
Y_i , Y	r.v. of down time.
τ_i , τ	delay to failure (deterministic or random variable)
φ	system's structure function. = 1, if system is up, = 0, if system is down.
F_L	cumulative density function of r.v. L.
$\tilde{F}_L$	Laplace-Stieltjes transform of r.v. L
R(t) R(n)	reliability function at time $t \in \mathbf{R}_+$ and $n \in \mathbf{N}$ respectively.
A(t), A(n)	availability function.
M(n)	maintainability.

λ	instantaneous failure rate.
μ	instantaneous repair rate.
$\{X_n ; n \in \mathbf{N}\}$	homogeneous Markov chain of reservoir input.
$\{Y_n ; n \in \mathbf{N}\}$	homogeneous Markov chain of reservoir output.
$\{Z_n ; n \in \mathbf{N}\}$	family of r.v. of reservoir level.
$\{W_n = (Z_n, X_n, Y_n) ; n \in \mathbf{N}\}$	3-variate homogeneous Markov chain.
$\mathcal{L}^{-1}(.)$	inverse Laplace transform of (.)
*	convolution.

2. Failure Delay Analysis. Lifetime Structure.

2.1. THE SIMULTANEITY ASSUMPTION

A fundamental assumption in classical system reliability theory, is that of simultaneity. This means that the occurrence of a cut set or an implicant implies an immediate failure of the system. For an 1-out-of-n:F system for example, we have :

$$T = \min_i \{ X_i \}$$

In other words, we admit system failure to occur at the first component failure.This is the simultaneity assumption.
In general, system failure follows the occurrence of a cut set after a time interval $\tau > 0$. According to the system, τ varies from a few fractions of seconds (electric and electronic systems) to a few hours and even more for some industrial processes.
The simultaneity assumption is valid for small τ and permits a considerable simplification and tractability of the relevent mathematical models; for large τ however, it becomes abusive.

2.2. FAILURE DELAY SYSTEMS

When the simultaneity assumption is not valid, we have to consider a time (delay) $\tau = t_1 - t_o$, where t_1 is the instant of system failure and t_o the instant of first occurrence of a cut set, with $t_1 > t_o$ (fig. 2.1).
Thus, failure of a failure delay system is subject to the following two conditions :

(i) *Structural condition* : same as for classical systems, i.e. occurrence of a cut set.

(ii) *Temporal condition* : demands the structural condition to persist for a time $t \geqslant \tau$.

For a 1-out-of-n:F system, we have :

$$T = \min_i \{ X_i \} + \tau$$

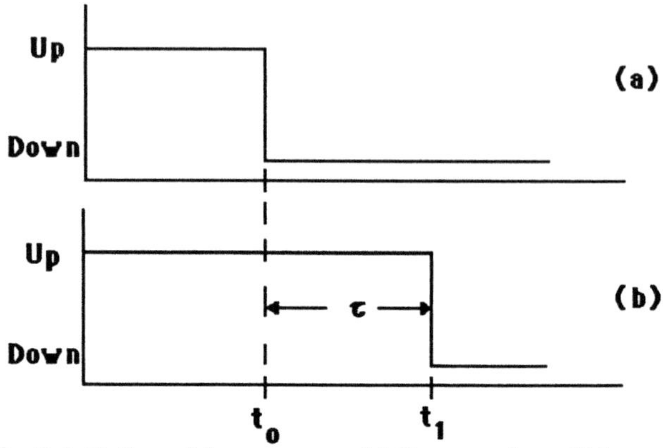

Fig. 2.1 : Failure delay systems. (a) Structural condition occured at time t_o and (b) system failure occured at time t_1, i.e. τ units of time after t_o

Consider a system with a single component, with $\{X_i\ ; i \in \mathbf{N}^*\}$, sequence of random variables where X_i denotes component lifetime in cycle i, and $\{Y_i\ ; i \in \mathbf{N}^*\}$, sequence of r.v. where Y_i denotes component repair time since cycle i.

Let $\{\tau_i\ ; i \in \mathbf{N}^*\}$, sequence where τ_i denotes maximum permitted repair time since cycle i. We say that the system fails for the first time in cycle i, iff : $\tau_i < Y_i$ and $\tau_j > Y_j$, $\forall\, j \in \{1,2,..., i-1\}$.

Then system lifetime will be :

$$T = X_1+X_2+....+X_n + Y_1+Y_2+....+Y_{n-1} + \tau_n$$

with

$$n = \mathit{min}\,\{k \in \mathbf{N}^* : \tau_i \geqslant Y_i, i = 1,....,k-1, \tau_k < Y_k\}$$

Remarks :

1) Generally, we can consider that at time t_0 the component was either functioning or failed.

2) When the system is considered failed during time periods Y_i , i = 1,2,... (failure mode 1, not absorbing), it is called "almost-repairable system", [7].

According to the delay's nature, some classes of systems may be distinguished :

Systems with deterministic delay : τ_i are deterministic variables. This class can be partitionned further into:

- fixed delay : $\tau_i = \tau$, for all i,
- variable delay : τ_i varies according to a deterministic law, i.e. depending on t.

Systems with random delay : τ_i are random variables. This class can be partitionned further into :

- delay with known distribution : when τ_i are i.i.d.with common distribution.
- delay depending on system evolution in (0, t).

2.3. n COMPONENTS SYSTEM

Consider a coherent system with n components and k minimal cut sets. Let X_i, i = 1,2,...,k r.v. denoting minimal cut sets' occurrence times.

Lifetime T for a *non-repairable system* will be :

$$T = \min_i \{X_i\} + \tau$$

For repairable systems we have :

$$T = \min \{t \in \mathbf{R}_+ : t \geqslant t_o + \tau, \varphi(t_o) = 1, \varphi(t_o+u) = 0, \forall u \in [t_o, t]\}$$

3. Failure Delay Systems Modelling

3.1. SINGLE COMPONENT SYSTEM

In this section, we study a single component failure delay system with and without repair, first with fixed and next with random delay. In both cases a closed-form solution for the pdf of system's lifetime is derived. The first case is studied by Ross et al[8], and the second by Shantikumar [10], who derives interesting properties for the lifetime of NBU (New Better than Used).

3.1.1. Non Repairable Component. We shall consider two cases : deterministic and random delay.

For deterministic delay τ, we have :

$$F_T(t) = P\{T \leqslant t\} = P\{X+\tau \leqslant t\} ,$$

hence

$$F_T(t) = \begin{cases} F_X(t-\tau) & \text{for } t \geqslant \tau \\ 0 & \text{for } 0 < t < \tau \end{cases}$$

For random delay τ, we have :

$$F_T(t) = P\{X+\tau \leqslant t\} ,$$

thus

$$F_T(t) = (F_X * F_\tau)(t)$$

where $$(F_X * F_\tau)(t) = \int_0^t F_\tau(t-u)dF_X(u)$$

3.1.2. Repairable Component. We shall consider two cases; delay systems with fixed and random delay.
For a fixed delay we have :

$$T = (X_1+Y^*_1) + \ldots\ldots + (X_{n-1}+Y^*_{n-1}) + X_n + \tau \qquad (3.1)$$

where Y^*_i is repair time of component before system failure.

$$Y^*_i = \begin{cases} Y_i & \text{if } Y_i < \tau \\ +\infty & \text{if } Y_i \geqslant \tau \end{cases}$$

The Laplace-Stieltjes transform of (3.1), gives :

$$\tilde{F}_T(s) = \frac{qe^{-s\tau}\tilde{F}_X(s)}{1-(1-q)\tilde{F}_X(s)\tilde{F}_{Y*}(s)} \qquad (3.2)$$

where

$$\tilde{F}_T(s) = \int_0^\infty e^{-st}\, dF_T(t)\ , \tilde{F}_X(s) = \int_0^\infty e^{-st}\, dF_X(t)\,,\ \tilde{F}_{Y*}(s) = \int_0^\tau \frac{e^{-st}\, dF_Y(t)}{F_Y(\tau)}$$

and $q = P(Y_i \geqslant \tau) = 1 - F_Y(\tau)$.

The following proposition, based on the equation (3.2), gives an interesting result for the evaluation of lifetime distribution in the real domain.

Proposition 3.3 : When X and Y are exponentially distributed with parameters λ and μ respectively, the lifetime pdf is :

$$f_T(t) = \lambda \sum_{n=0}^{r-1} (\lambda\mu)^n e^{-(n+1)\mu\tau} \Phi_n(t-(n+1)\tau)\,, \quad \text{for } t \geqslant \tau$$

$$= 0, \text{ for } 0 \leqslant t < \tau$$

for $r\tau \leqslant t < (r+1)\tau$
with

$$\Phi_o(t) = \frac{\mu}{\lambda+\mu} + \frac{\lambda}{\lambda+\mu} e^{-(\lambda+\mu)t}$$

$$\Phi_n(t) = \frac{1}{(n!)^2} \{ n \sum_{k=0}^{n-1} (-1)^{n+k} \binom{n-1}{k} (1+\frac{\mu}{n-k}t) t^{n-1-k}\ I_{n+k}(t) + \mu I_{2n}(t) \},$$

$$n = 1, 2, \ldots.$$

$I_k(t)$ is given by the following lemma :

Lemma 3.4 : For $t \in \mathbb{R}_+^*$ and $k \in \mathbb{N}$, we have :

$$I_k(t) = \int_0^t u^k e^{-au} du = \frac{k!}{a^{k+1}} \{ 1 - e^{-at} \sum_{j=0}^{k} \frac{(at)^j}{j!} \}$$

For proof of proposition see appendix A.

Numerical example : Consider a single component system with $\lambda = 0.001/h$, $\mu = 0.1/h$ component failure and repair rates, and fixed delay $\tau = 10$ h.We obtain $F_T(50) = 1.46.10^{-2}$

For random delay, we have :

$$T = (X_1 + Y^*_1) + \ldots\ldots + (X_{n-1} + Y^*_{n-1}) + X_n + \tau^*_n \qquad (3.5)$$

with

$$Y^*_i = \begin{cases} Y_i & \text{if } Y_i < \tau_i \\ +\infty & \text{if } Y_i \geqslant \tau_i \end{cases}$$

and

$$\tau^*_i = \begin{cases} \tau_i & \text{if } Y_i \geqslant \tau_i \\ +\infty & \text{if } Y_i < \tau_i \end{cases}$$

The Laplace - Stieltes transform of T, is :

$$\tilde{F}_T(s) = \frac{q\tilde{F}_{\tau*}(s)\tilde{F}_X(s)}{1-(1-q)\tilde{F}_X(s)\tilde{F}_{Y*}(s)} \qquad (3.6)$$

When X, Y, and τ are exponentially distributed with parameters λ, μ, and c, (3.6) gives :

$$\tilde{F}_T(s) = \frac{c\lambda}{s(s+\lambda+\mu+c)+\lambda c}$$

from which we take,

$$f_T(t) = \frac{c\lambda}{s_2 - s_1}(e^{s_2 t} - e^{s_1 t}) \qquad (3.7)$$

where s_1 and s_2 roots of $s^2 + (\lambda+\mu+c)s + \lambda c = 0$. ($s_1, s_2 \in \mathbb{R}_-$, and $s_1 > s_2$)

3.2. STAND-BY AS FAILURE DELAY SYSTEM

Consider a system of two components {1,2}; component 1 is active and component 2 is stand-by. When component 1 fails component 2 assures functioning until complete repair of component 1. In other words, the mission of component 2 is to assure system's functioning while component 1 is being repaired.

Let λ and μ be component 1's failure and repair rate ; suppose that component 2 is not repairable, and let c its failure rate. There will be three system states (Fig. 3.1.) :

state 0 : component 1 up and component 2 stand-by (up state)
state 1 : component 1 down and component 2 functioning (up state)
state 2 : component 1 down and component 2 down (down state).

State 2 is an absorbing state.

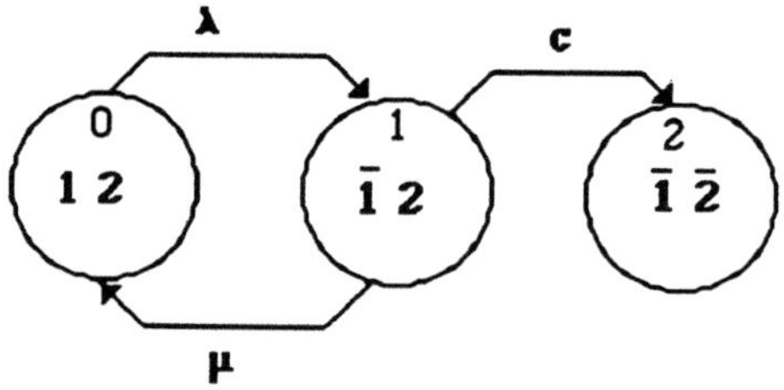

Fig. 3.1 : State diagram for a two components system.

We have the following result :

$$R(t) = P_o(t)+P_1(t) = -s_2\frac{e^{s_1 t}}{s_1 - s_2} + s_1\frac{e^{s_2 t}}{s_1 - s_2} \qquad (3.8)$$

where s_1 and s_2 roots of $s^2 + (\lambda + c + \mu)s + \lambda c = 0$. By derivation of (3.8) one takes relation (3.7).

We need here the following two assumptions, which caracterize random delay systems.

(i) $\lambda << c$, and
(ii) c is of the same order as μ.

Hence,

$$s_1 = -\lambda/2 \text{ and } s_2 = -(\frac{\lambda}{2} + c + \mu)$$

and (3.8) becomes

$$R(t) = \{\frac{\lambda}{2} - (\frac{\lambda}{2} + c+\mu)\, e^{-(c+\mu)t}\}\, \frac{e^{-\lambda t/2}}{c+\mu}$$

MTTF is approximately equal to $2/\lambda$.

4. The reservoir case

4.1. THE RESERVOIR AS A FAILURE DELAY SYSTEM

Systems with reservoirs or storages are a particular case of failure delay systems. Generally, they are conceived to provide to the output a certain autonomy from production: if production stops (following a failure), the system will still be able to fulfill its mission for a certain period of time.
Consider for example a water reservoir of continuous functioning supplied by a water source through a pump, and providing a certain amount of water with a fixed rate. A pump failure does not necessarily mean service interruption, not instantaneously; in order to have service interruption the pump must remain in failed state until the reservoir is emptied. In absence of reservoir, evidently, a pump failure would lead to instantaneous interruption of service.

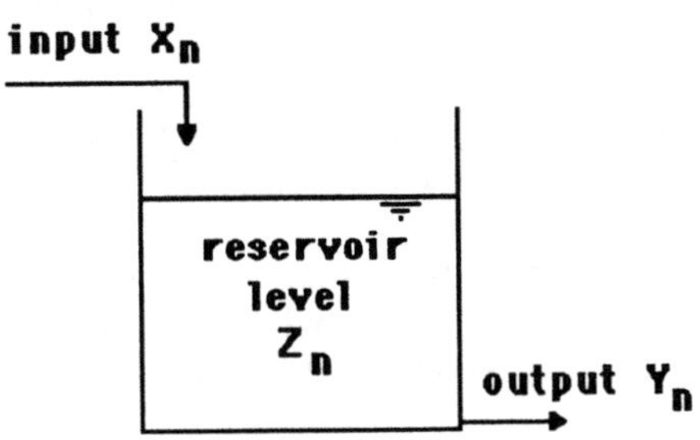

Fig. 4.1. : Reservoir system

Eventhough the reservoir constitutes a real paradigm in the probability field and has inspired numerous theoretical developments [5],[6], only very recently is it considered with respect to reliability; Henley & Kumamoto [2] treat a simplified case by considering time to first failure.
We have developed three general models for the study of such systems [3],[4]. The first one concerns a time dependent Markov input-output reservoir modeled by a 3-dimensional Markov chain; the second the Moran's reservoir modeled by a 1-dimensional Markov chain, for which we have obtained analytical formulation of reliability measures in case of binary input; the third the Moran's reservoir modeled by a 2-dimensional Markov chain.

4.2. RESERVOIR MODELLING

Let $X = \{ X_n ; n \in \mathbf{N}\}$, $Y = \{ Y_n ; n \in \mathbf{N}\}$ and $Z = \{ Z_n ; n \in \mathbf{N}\}$, the random variable families describing the amount of volume units respectively entering, leaving and in the reservoir, respectively at time interval (n, n+1), at the end of this interval and at time t, and E_X, E_y and E_Z the state spaces of X, Y and Z respectively.
Reliability measures can be obtained with respect to the following three events:

$$A = \{ Z_t = 0 , X_t = 0 \text{ or } Y_t = 0, Z_u > 0, X_t > 0, \forall u \in [0,t), Z_o = u \}$$
$$B = \{ Z_t = 0 , X_t = 0, Z_u > 0, \forall u \in [0,t), Z_o = u \}$$

$$C = \{ Z_t = 0, Z_u > 0, \forall u \in [0,t), Z_o = u \}$$

Event A denotes output interruption, event B reservoir emptiness and event C system functioning with zero reservoir level.

Model 1: Markov dependent input-output reservoir, modeled by 3-dimensional Markov chain.
In this model the reservoir is constituted by v pumps on reservoir input and w pumps on output while the reservoir capacity is c volume units. Thus at time $n \in \mathbf{N}$ there are rQ, $r = 0,1,..., [c/Q]$, volume units in the reservoir, and an amount of iQ, $i = 0,1,.....$, v volume units enters the reservoir during the time interval $(n, n+1)$, while at the end of this interval an amount of jQ, $j = 0,1,.....$, w volume units leaves the reservoir.
We have prooved [3] that if X and Y are Markov chains, the process $W = \{ W_n = (Z_n, X_n, Y_n); n \in \mathbf{N}\}$ is equally a Markov chain on the state space $E_z \times E_x \times E_y$.
With this model we can study reliability with respect to all three events A, B and C.

Model 2 : Moran's reservoir modeled by scalar Markov chain.
It is the model of Moran's reservoir in terms of reliability[4]. For this reservoir, output is fixed and equal to 1 volume unit and inputs are independent :

$$P(X_n = j) = p_j, \forall n \in \mathbf{N} ; j = 0,1,2,.......$$

Under these assumptions Moran demonstrates that the process $Z = \{ Z_n ; n \in \mathbf{N}\}$ is a Markov chain on the state space $\{0,1,...,c-1\}$, even when c is infinite.
With this model we can study reliability with respect to event C only.

Model 3 : Moran's reservoir modeled by a 2-dimensional Markov chain.
In order to study event B for Moran's reservoir we have developed the following 2-dimensional Markov chain : $\{(Z_n, X_n) ; n \in \mathbf{N}\}$.
In this model, when state (a,b) $\in \{0,1,...,c-1\}x\{0,1,2,...\}$ verifies $a+b \geqslant c$, it is aggregated with state (a,b+), where $b = c-a$. Thus event B concerns state (0,0).

4.3. NUMERICAL EXAMPLE

In this section we give a numerical example of a reservoir with random input and output.
Consider a reservoir of capacity $c = 5$ volume units, supplied by a pump of flow rate: 1 volume unit /h, of failure rate $\lambda_I = 10^{-3}\ h^{-1}$ and of repair rate $\mu_I = 10^{-2}\ h^{-1}$. Distribution is assured by a pump of flow rate: 1 volume unit/h, of failure rate $\lambda_o = 10^{-4}\ h^{-1}$ and of repair rate $\mu_o = 10^{-2}\ h^{-1}$.
At initial time $t = 0$ we have $Z_o = 5$, $X_o = 1$ and $Y_o = 1$ with probability 1 for reliability and availability. For maintainability we consider at $t = 0$, $Z_o = 0$, $X_o = 0$, $Y_o = 0$ with probability 1.
The event of interest is A (section 4.2.). Reliability, availability and maintainability are presented in fig. 4.2.
We also have:
MTTF = 945.8 h, MTTR = 159.9 h, MUT = 908.5 h, and MDT = 100.5 h.
The steady-state availability is :0.90

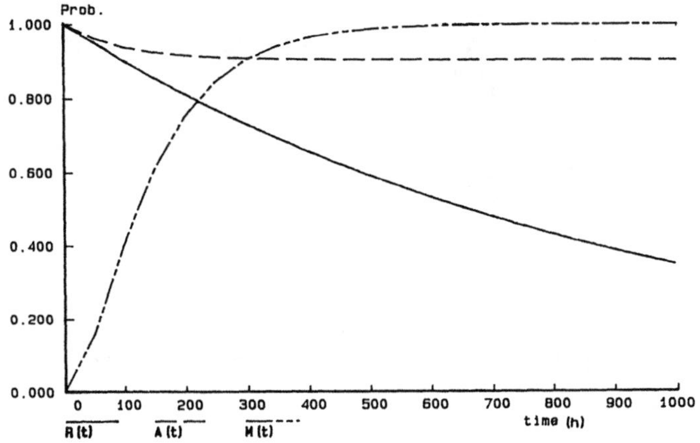

Fig. 4.2 :Reliability, Availability and Maintainability.

5. Conclusions

Failure delay systems constitute a large family in reliability, concerning systems like industrial processes whose delays are due to a relatively long response time, and equally systems like electric ones where failure delays may come from functional or maintenance specifications. A finest modelling is possible in such cases.
A major drawback of delay systems is that standard Boolean algebra cannot be used for their description. However, other algebra can be constructed for their study, i.e. the pseudo-boolean algebra proposed by Schneeweiss [9] which can take into account state variables persistance.
The results presented and in particular the relation (3.3) can be used for the treatment of fault trees with delay operators.

Appendix A : Proof of proposition 3.3.

When X and Y are exponentially distributed with parameters λ and μ respectively, relation (3.2) becomes :

$$\tilde{F}_T(s) = \frac{\lambda(s+\mu)e^{-(s+\mu)\tau}}{s^2+(\lambda+\mu)s+\lambda\mu e^{-(s+\mu)\tau}}$$

We have $\left|\frac{\lambda\mu e^{-(s+\mu)\tau}}{s(s+\lambda+\mu)}\right| < 1$

and thus we can write :

$$\tilde{F}_T(s) = \lambda \sum_{n=0}^{\infty} (-\lambda\mu)^n e^{-(n+1)\mu\tau} \tilde{\Phi}_n(s) e^{-(n+1)s\tau}$$

with

$$\tilde{\Phi}_n(s) = \frac{s+\mu}{[s(s+\lambda+\mu)]^{n+1}}, \qquad n = 0, 1, 2, \ldots$$

thus,

$$F_T(t) = \lambda \sum_{n=0}^{r-1} (-\lambda\mu)^n e^{-(n+1)\mu\tau} \Phi_n(t-(n+1)\tau)$$

Inversion of $\tilde{\Phi}_n(s)$:

$$\Phi_n(t) = \mathscr{L}^{-1}\left(\frac{s+\mu}{[s(s+\lambda+\mu)]^{n+1}}\right) = \mathscr{L}^{-1}\left(\frac{s+\mu}{s^{n+1}}\right) * \mathscr{L}^{-1}\left(\frac{s+\mu}{(s+\lambda+\mu)^{n+1}}\right)$$

where $*$ stands for convolution

$$\mathscr{L}^{-1}\left(\frac{s+\mu}{s^{n+1}}\right) = f_{n-1}(t) + \mu f_n(t)$$

$$\mathscr{L}^{-1}\left(\frac{s+\mu}{(s+\lambda+\mu)^{n+1}}\right) = e^{-(\lambda+\mu)t} f_n(t)$$

thus,

$$\Phi_n(t) = f_{n-1}(t) * f_n(t)\, e^{-(\lambda+\mu)t} + \mu\, f_n(t) * f_n(t) e^{-(\lambda+\mu)t}$$

where $f_n(t) = \frac{t^n}{n!}$

$$\Phi_n(t) = \frac{1}{(n!)^2}\{n\, J_1(t) + \mu\, J_2(t)\}$$

$$J_1(t) = \sum_{k=0}^{n-1} (-1)^k \binom{n-1}{k} t^{n-1-k} \int_0^t u^{k+n} e^{-au} du$$

$a = \lambda+\mu$

$$J_2(t) = \sum_{k=0}^{n} (-1)^k \binom{n}{k} t^{n-k} \int_0^t u^{k+n} e^{-au} du$$

The integrals in the above sums are evaluated by lemma 3.4, and after some elementary algebra operations, we obtain the result.

Appendix B : Reliability measures formulation in terms of discrete homogeneous Markov chains.

Reliability $R(n) = \mathbf{P}_r(0)(\mathbf{M}_{11})^n \mathbf{1}_k$

Availability $A(n) = \mathbf{P}_a(0)(\mathbf{M})^n \mathbf{1}_{s,k}$

Maintainability $M(n) = \mathbf{P}_m(0)(\mathbf{M}_{22})^n \mathbf{1}_{s-k}$

Mean times

$$MTTF = \mathbf{P}_r(0)(\mathbf{I}-\mathbf{M}_{11})^{-1}\mathbf{1}_k$$

$$MTTR = \mathbf{P}_m(0)(\mathbf{I}-\mathbf{M}_{22})^{-1}\mathbf{1}_{s-k}$$

$$MUT = (\boldsymbol{\pi}_1 \mathbf{1}_k)(\boldsymbol{\pi}_1 \mathbf{M}_{21} \mathbf{1}_k)^{-1}$$

$$MDT = (\boldsymbol{\pi}_2 \mathbf{1}_{s-k})(\boldsymbol{\pi}_2 \mathbf{M}_{12} \mathbf{1}_k)^{-1}$$

where

M, the transition matrix with the following partition

$$\mathbf{M} = \begin{matrix} & U & D \\ U & \mathbf{M}_{11} & \mathbf{M}_{12} \\ D & \mathbf{M}_{21} & \mathbf{M}_{22} \end{matrix}$$

U: up states, D: down states

$\mathbf{P}_r(0)$, initial distribution for reliability

$\mathbf{P}_a(0)$, initial distribution for availability

$\mathbf{P}_m(0)$, initial distribution for maintainability

$\boldsymbol{\pi}$, ergodic vector of the chain

s, number of system states

k, number of up system states and s-k number of down states

$\mathbf{1}_k = (1,\ldots\ldots,1)^T$, k-unit vector

$\mathbf{1}_{s,k} = (1,\ldots,1,0,\ldots,0)^T$, s-dimensional vector with the first k elements equal to 1.

Acknowledgments : The author wishes to thank O. Theologou and C. Cocozza-Thivent for their useful comments on this work. Section 4 is part of paper [3] reproduced here with Editor permission.

References

[1] Dunbar, L.C., "A mathematical expression describing the failure probability of a system of redundant components with finite maximum repair time", Reliab. Eng., vol 7, 1984, pp169-179.

[2] Henley, E.J. and Kumamoto, H., Reliability Engineering and Risk Assessment, Prentice-Hall, Inc., Englewood Cliffs, 1981, N.J.

[3] Limnios, N., "Reliability evaluation of reservoir systems", Reliab. Eng. Syst. Saf., vol 23, 1988, pp 65-73.

[4] Limnios, N., "Reliability evaluation of a Moran reservoir ", IEEE Trans. Reliability, december 1989, (to appear).

[5] Lloyd, E.,"Wet and dry water : Remarks on pure and applied mathematics as illustrated by the recent history of stochastic reservoir theory", Bull. Ins. Math. Appl., vol. 10, 1974, pp 348-353.

[6] Moran, P.A.P., The Theory of Storage , Mathuen, 1959, London.

[7] Pagès, A. and Gondran, M., Fiabilité des Systèmes, Eyrolles, 1980, Paris.

[8] Ross, S. and Schechtman, J., "On the first time a separately maintened parallel system has been down for a fixed time", Nav. Res. Log. Quart., vol 26, 1979, pp 285-290.

[9] Schneeweiss, W.G., "Use of fault tree with delayed inputs", IEEE Trans. Reliab.,vol.R30, N°4, Oct. 1981, pp 339-344.

[10] Shanthikumar, J.G., "First failure time of dependant parallel systems with safety periods", Microelectron. Reliab., vol 26, n° 5, 1986, pp 955-972.

THE DYLAM APPROACH TO SYSTEMS SAFETY ANALYSIS

A. AMENDOLA
Commission of the European Communities
Joint Research Centre - Institute for Systems Engineering
21020 Ispra (Va) - Italy

ABSTRACT. Event trees are able to delineate accident scenarios as representative of classes of events to be included into a probabilistic safety assessment (PSA) of a complex system. They are, however, models which hardly give an adequate account of dynamic processes interacting with systems' states. Petri nets or other similar approaches model fairly well timed sequences of events; they fail again in giving an account of the process physics.

The DYLAM technique, which is the subject of the present lecture, presents under these regards significant advantages, especially for developing stochastic models of dynamic systems, and therefore is a powerful aid for the design of protections and decision support systems to assist operators in the control of hazardous processes as well as for systems safety assessment.

1. INTRODUCTION

The present lecture summarizes the principles on which the DYLAM methodology is based; enlightens its possible applications and reviews principal applications that have demonstrated DYLAM adequacy in solving manifold problems connected with safety and reliability assessment of complex dynamic systems. These include:

- identification and analysis of accident sequences in hazardous processes such as nuclear power plants and chemical installations, where DYLAM can substitute or advantageously complement the usual fault tree/event tree approach;
- study of man-machine systems, where DYLAM allows the analyst to adequately model the dynamic interaction of human operators with time evolving processes; and
- control of accident simulators, where DYLAM can drive generation of stochastically changing boundary conditions for the computer codes simulating the process under investigation, and therefore help design of protections and decision support systems.

A. G. Colombo and A. Saiz de Bustamante (eds.), Systems Reliability Assessment, 159–182.

2. EVENT TREES

Event trees have proved to be a valuable tool for assessing major contributions to the overall risk connected with hazardous plants. They are based on an inductive method which starts to consider a possible incident initiator; then it considers the state of the system/components which must first interact with the perturbed process; and successively all other systems which could operate to mitigate the incident or might be damaged by the progression of the incident. From a single initiating event, many paths are in this way derived: some of these lead to process to more or less safe conditions, others to different degraded or accident conditions.

Possible incident courses are described by separate physical calculations and consequence evaluations, i.e. the dynamics of the process, and timings of events and interventions must be studied separately. Even the success conditions for the demanded systems are determined by separate calculations. Once the scenario and the success conditions are determined then the probability to be associated with each path of the event tree is assessed by developing the appropriate fault trees, and by taking into account possible interdependencies.

In a PSA each event tree is representative of a class of events such as a loss of coolant, station blackout, etc., and therefore can be associated with a finite probability value.

During operation any perturbation, any transient is one element of a certain set, i.e. each single loss of feedwater transient in a nuclear power plant is more or less different from another one, and certainly different from the transient assumed as representative of this class in a PSA, because its course is very dependent on the actual state of the process (for instance, power level, time of fuel cycling, etc.) at the time of its occurrence as well as on the actual state of the relevant systems.

Therefore, even during anticipated occurrences, the operator might be confronted with a multiplicity of scenarios similar to (sometimes, however, different from) those assumed in a PSA. His response is also among the factors provoking the reality to be different from the models. This, in fact, might have a wide time variability, since event diagnosis might not be immediate, actions might require communication among plant personnel, inspections, in-site operations, etc. In the meantime, the process evolves according to its physical laws.

The multiplicity of real scenarios makes event trees inadequate to achieve the completeness required when the plant analysis is principally directed to the design of protections /1/ or decision support systems to assist plant operators /2/. In fact, even if, in principle, it would be possible to lay down even more detailed event trees, the resulting models will always suffer from significant limitations in giving account of the timing of the events and of the process physics.

3. TIMED GRAPHS

The event tree limitations in giving a proper account of the process physics and event timing hold true for fault trees as well.

The awareness that structures of complex systems are difficult to model by traditional fault tree techniques, especially when they contain cyclic effects such as feedback loops, bidirectional flow or time-dependent effects /3/, has led other authors to use fault graphs as a way to advance beyond fault trees. A fault graph is a failure oriented "digraph" (directed graph with logical connectives that allow cycles). It can trace a piping and instrumentation diagram of a system directly but with the addition of logical operators (AND, OR, NOT).

Digraph matrix analysis (DMA) is used to evaluate cut sets and systems reliability via graph concepts, such as path set (a set of nodes travelled on a path from one node to another) and reachability (the complete set of all possible paths between any two nodes) /3/. The same paper gives examples of graphs able to model conditioned cycle structure and complex networks with bidirectional flow; it also indicates that major demonstrations have been given for nuclear power plant systems involving over 10,000 components. In a subsequent paper, DMA has been shown to be isomorphic to a limited Petri net structure /4/, so that the capabilities of both these models are comparable. Petri nets /5/ were initially used as dynamic modelling tools for timed systems in which certain events may occur concurrently but with constraints on the concurrence, precedence or frequency of these occurrences. Therefore they principally found applications for modelling systems of asynchronous concurrent computation in computer science. They are now being exploited as reliability modelling techniques as well. Stochastic Petri nets have been used for modelling and analysing the performance and the reliability of redundant multiprocessor systems /6/. But far away from computer systems, they have also been used to model and evaluate a drilling procedure /7/. Such a procedure was constituted by a succession of phases each comprising a succession of sequential and/or parallel concurrent operations related to the overall task to be accomplished. However, planning of operations and phases might be disturbed and rearranged by random events because of external circumstances, component failures and human errors (for instance, in the case of strict sequentiality, this might not be respected and an operation "n+i" might have been started before or instead of the operation "n+1" when something perturbs system conditions during the operation "n").

After the modelling, Petri nets can be analysed either by Markov chains or by Monte Carlo routines according to their structure to evaluate the relevant success/failure probabilities.

With respect to event trees and fault trees, such timed graphs appear to undoubtedly present significant advantages in modelling a certain class of systems or situations.

For instance, they appear to better model and evaluate the success of recovery actions when a multiplicty of tasks has to be accomplished by different operators according to a parallel-series sequence of actions. They also might be explored for application to the analysis of emergency plans: where the success also depends on preparedness, concurrence of tools and intervention teams, within certain time limits.

Also they have been shown to deal fairly well with physical cycling effects which in reality may occur on a system because of multiple failures and might be difficult to model via fault trees /8/ or are characteristic for networks with bidirectional flow (like the electrical ones).

According to the objective of the analysis, they can therefore successfully substitute fault trees or event trees, and improve the correctness and the completeness of the model.

The dynamic introduced by such methods is, however, limited to the timing, whereas in certain situations timing of interventions strongly interacts with the evolution of the physics of the process, which must be studied in a separate manner.

The improvement on the description of the physics appears to be limited to quasi steady-state conditions, whereas in incident conditions it is indispensable to describe the physical transient course. Again this must be studied separately by appropriate simulation tools.

4. PROCESS SIMULATION TOOLS

The previously described techniques were aimed at modelling the logical structure of a system by giving account of cause-effect relationships among states of components and events affecting the system and/or to evaluate the frequency of undesired occurrences by taking into account the failure/repair stochastic processes.

Different kinds of plant models are those which attempt to simulate the process physics in steady-state and transient conditions. They do not address the problem of modelling and evaluating random occurrences of failure/repair events; but they attempt to predict the system transient response to assumed component failures or control actions.

In this way it is possible to study the process dynamics of assumed accident sequences in a very detailed way. Therefore they are widely used for plant analysis and for safety evaluations of both deterministic and probabilistic nature. Only the aspects relevant for the purpose of this paper will be further discussed in the following.

For event tree/fault tree analysis such codes are used to identify timing of events, to establish successs conditions for intervening systems, and to evaluate incident consequences.

The states of the systems, the times of the required operator interventions and the occurrence of the initiating events have to be assigned as boundary conditions for each run of the codes. Therefore the system transient behaviour can only be studied for a limited number of conditions.

The availability of a tool for an automated assignment of stochastically changing boundary conditions would certainly result in the possibility of a more complete exploitation of the information implicitly contained in the code. As will be shown, DYLAM can be used in this manner, so that its association with transient codes results in a very powerful tool for modelling both random events and process physics in a self-contained approach. Transient codes also constitute the bulk software of plant simulators used for training of operators, validation of operating procedures and study of man-machine interfaces, as well as design of protections.

Full-scale simulators should also incorporate the DSSs possibly available to the operators. They may also be used to validate DSS concepts and actual design.

5. THE DYLAM APPROACH

Essentially DYLAM (Dynamic Logical Analytical Methodology) is a method for simulation of systems with stochastic configurations. It differs from other techniques because it takes into account the two aspects "process simulation" and "components reliability performance" in a unified procedure. It is based on heuristical bottom-up procedures for identifying the sequences of events leading the system in the undesired conditions. It is characterized by its ability to perform a dynamical analysis, and to consider changes of the system structure due both to control logic and to random events. The principal steps of the DYLAM analysis procedure can be summarized as follows.

5.1 Component Modelling

The analysis proceeds bottom-up. No explicit model needs to be established for the system. Rather, failure models refer to the components constituting the system, and must be assigned by the analyst. These consist of:

- identification of the different failure or degradation states in which a component may occur to be;
- parametrical relations which associate each state with a mathematical equation describing the physics of the component under those conditions; and
- probability parameters characterizing states and possible state transitions.

Modelling is very flexible. The analyst, however, should find a reasonable compromise between two somewhat contrasting needs:

- the completeness of the analysis, which depends on the exhaustiveness of identified and retained component failure models; and
- the increase of the number of retained failure models, which results in an exponential increase of computer times.

Fig.1 shows a possible model of the series of three different valves /9/. To decrease the number of states only one valve (SVR) has been substituted to the series of a regulation valve (V1), a check valve (V2) and a manual isolation valve (V3). Correspondingly, three failure modes have been associated with the valve in such a way that they cover all the cases of interest.

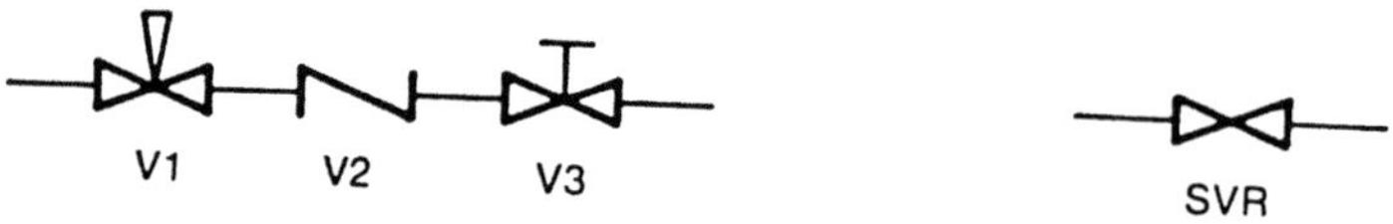

V1 = regulation valve
V2 = check valve
V3 = manual isolation valve

SVR = equivalent valve model

SVR states:	Flow rate relations:	Probability parameters:
0 = nominal	$Q_o = K_1 A(t)\sqrt{\|\Delta P\|}.H(\Delta P,0)$	$A_o;\lambda_{0,1};\lambda_{0,2}$
1 = check valve failed open	$Q_1 = K_1 A(t)\sqrt{\|\Delta P\|}.\Delta P/\|\Delta P\|$	U_1;
2 = blockage of one valve	$Q_2 = K_1 A(t)\sqrt{\|\Delta P\|}.0$	U_2;
3 = manual isolation valve left closed after maintenance	$Q_3 = K_1 A(t)\sqrt{\|\Delta P\|}.H(\Delta P.0)H(t,t_e)$	U_3;
		$\Sigma=1$

Fig.1: Model of a series of valves.

An event variable SVR characterizes the states, it takes the values

0 if all the three valves are in the nominal state;
1 if the check valve V2 fails open;
2 if at least one valve fails blocked;
3 if the manual isolation valve has been left closed after maintenance.

The case SVR = 3 differs from the case SVR = 2, since it is assumed that an operator may open the valve when he recognizes the fault position but he cannot repair the valves if the blockage is due to mechanical causes or obstructions. During job execution, the value of the event variable controls the choice of the appropriate physical model: in the example of Fig.1 the basic relation linking the flowrate Q to the pressure drop ΔP and the valve cross section A(t), variable in time because of the regulation on V1,

$$Q = K\ A(t)\sqrt{|\Delta P|}$$

is multiplied by the factor*

*H(A,B) is the step function: H = 0 if A < B
H = 1 if A $\geq$ B

$H(\Delta P,0)$	which gives account of the unidirectional flow in the nominal case, where V2 is effective;
$\Delta P/\|\Delta P\|$	which expresses the sign of ΔP when V2 fails open;
0	if there is blockage;
$H(\Delta P,0)H(t,t_e)$	which expresses the fact that after a time t_e, depending on the operator response, the system can be restored (in this case the restoration rate of the component is replaced by a time to restore, the model of which is a part of an operator model).

To each state, the initial occupance probability is assigned (A_o,U_1,U_2,U_3) and the transition rate considered among the states. In the examples only $\lambda_{0,1}$ and $\lambda_{0,2}$ are retained, because states 1 and 2 are assumed to be not repairable, transition $3 \rightarrow 1$ is modelled through t_e and the probability of the isolation valve being closed during operation is considered to be negligible.

For other examples and for discussion of all the possibilities offered by the modelling technique, the reader is referred to Refs./2,9-14/.

5.2 System Equation Resolution Algorithms

Once the components have been modelled, the analyst must implement the way in which the resulting equation system has to be solved during the event sequence generation procedure, or decide to use the standard package available in DYLAM where the Reina algorithms /10/ for the resolution of non-linear algebraic equations have been implemented. These are able to solve a large number of problems, however in the most general cases, systems of differential equations or complex networks need to be solved, for which the solution method must be supplied case by case by the analyst. In principle, there are no limitations for applying the DYLAM procedure to very complex simulation tools /15/. However, long computational times are in practice very significant factors in limiting the potential advantages of DYLAM application. Therefore, the physical models to adopt should be as simple as possible, compatible with the requirements of a safety analysis.

5.3 TOP Event Definition

After steps 5.1 and 5.2 the system has been implicitly modelled in all its possible states: all information has been embedded into a computer program in a form ready to be exploited according to the aims of the analysis. Indeed

- information about initiating events and their frequencies, availability of the components and their possible failure modes is included in the component models;
- the way in which the single components interact physically with eachy other is described by the parametric equations and by the way in which the equation system has been structured.

Three component systems:

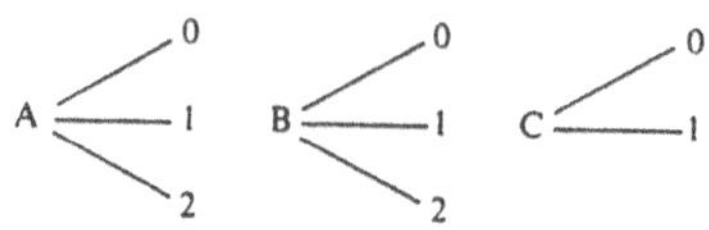

Transition rates of A and B small with respect to transient time C characterized by failure/repair rates:

A0 B0 C0 are the nominal states

Mother sequence:

A1 B2 C1 (at time 0) order of the sequence = 3

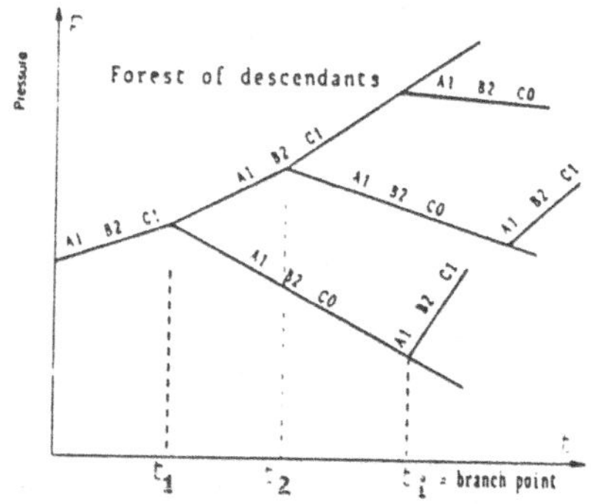

Fig.2: Examples of event sequences.

Limiting rules are applied to the generation of both mother sequences and their possible descendants. First of all the analyst has the possibility either

- to prefix a mother sequence and by a pseudo Monte Carlo procedure let the code generate a chosen number of possible descendants; or
- to prefix a mother sequence and let the code systematically generate all possible descendants compatible with the limiting rules described in the following for the branch points; or, finally
- let the code generate all mother sequences up to a prefixed order with all their possible descendants, compatible with the limiting rules described in the following.

Limiting rules for the mother sequences

- cut-off of all sequences containing two active states at time t = 0; this means that the probability of getting a transient initiated by two different events at the same time point is assumed to be zero when these events are independent of each other;
- cut-off of all sequences containing only passive states; this because the system is in a degraded state which, however, is irrelevant for its safety since no transient occurs which would require the intervention of the degraded protections;
- cut-off of all sequences with an initial probability below a prefixed threshold.

Before starting the event generation procedures, the system conditions (TOP event) which must be searched for are to be assigned. This must be done by programming into a particular DYLAM subroutine the TOP events to be investigated. These can be defined in terms of values of the physical variables of concern (e.g. values which must not be exceeded to avoid melting, explosions or other undesired effects. However, in general, even success or not dangerous conditions might be studied for system identification purposes).

More than one TOP can be studied in a same run. For instance, one might program the FORTRAN equivalent expressions of statements like: "search for all events which provoke the coolant temperature to rise above 150, 180, 200 and 220°C and events which provoke temperature above 150°C and at the same time pressures higher than 80 bars". In this case the occurrence of such TOP conditions is explored versus transient time.

But it would also be possible to ask for an integral or average value such as certain overtemperatures lasting more than a certain duration or the average temperature in the time exceeding a certain value. The possibility of specifying TOP conditions in physical quantitative terms is one of the principal advantages of DYLAM versus other techniques.

5.4 Event Sequence Generation Rules

To exploit the information implicitly contained in the DYLAM simulator when programmed for the specific study case, rules have been developed which allow the analyst to control the generation of the possible event sequences up to the desired resolution. The number of possible event sequences is in reality infinite and, therefore, cannot be explored by any technique unless approximations and cut-off procedures are introduced; these, however, must save the intended useful information.

In the following some basic principles will be shortly recalled. Before doing so, some further definitions need to be introduced (see Fig.2).

- "Mother" sequence is the combination of the states of the components at the beginning of a transient.
- "Active" component is a component able to initiate a transient by changing its state.
- Order of a mother sequence is the number of states in the sequence which are different from the nominal ones. For instance, a transient initiated by the failure of an active component and with another component in a degraded state is a sequence of order two.
- From any mother sequence a forest of "descendant" sequences can be originated by changes of states occurring during the course of the transient because random independent and/or conditional transitions.
- A branch point is the time at which a sequence splits into two or more possible descendants.

Let us consider only one not repairable component, A, with failure rate λ. If A starts safe, then the probability of keeping its state after a time t is $R = e^{-\lambda t}$. When t is such that $R_1 = R(t_1) = W_1$ (assigned in input), a first branch point is generated. A second branch point is generated when $R_2 R_1 = W_1$ and so on. The continuous time process is thus discretized.

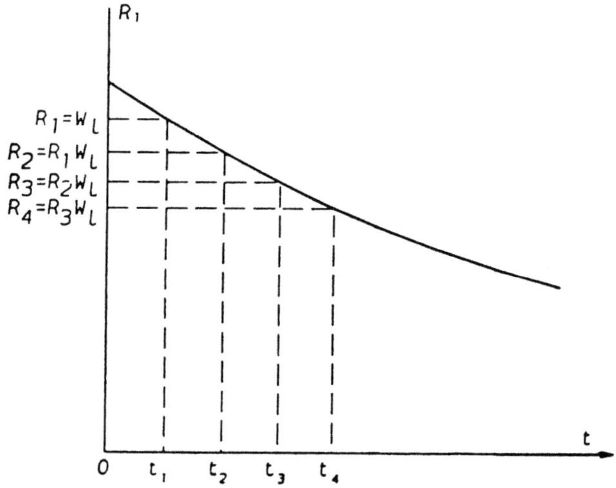

The physics are evaluated for the sequences:

1. A, safe at any t.
2. A, safe at $t < t_1$, failed at $t \geq t_1$.
3. A, safe at $t < t_2$, failed at $t \geq t_2$, and so on.

The associated probabilities vs time are:

	$0 \rightarrow t_1$	$t_1 \rightarrow t_2$	$t_2 \rightarrow t_3$
$P(1) =$	1	R_1	R_2
$P(2) =$	0	$1 - R_1$	$1 - R_1$
$P(3) =$	0	0	$R_1 - R_2$
$\Sigma P(i) =$	1	1	1

The procedure is generalizable whatever the considered number of transitions between states. It concentrates the probability mass of a transition between t_1 and t_2 in only one transition at t_2. It neglects multiple transitions between t_1 and t_2 among states of a same component and assumes that, if a transition at time t_2 results in a TOP event, the same would apply for the same transition at time $t_1 < t < t_2$.

Fig.3: Branch point generation rules.

Limiting rule for a branch point generation

The sequence is assumed to remain in the same state structure unless the probability of remaining in such states becomes less of a prefixed fraction of the initial probabilities. This is introduced because at each simulated change of state a new physical behaviour (e.g. a new transient) has to be evaluated. Then it is necessary for practicality reasons to discretize the continuous time into finite time interval steps, and to evaluate only a finite number of possible branch points. This corresponds to the following simplifying assumptions: if a branch point is selected at a time t_i and the next one at the time t_{i+1}, then the physics is evaluated only for the changes of states occurring at ti and t_{i+1}, and the transient associated with a change of state j at time t_i is given a probability corresponding to the product of the probability of the system being in state j-1 for $t < t_i$, and the probability of the transition $j-1 \rightarrow j$ occurring at a time $t_{i-1} \leq t < t_i$. That is: it is assumed that if this transition would occur in the time interval $t_{i-1} \div t_i$, the transient would have the same course.

By assuming that: the change $j \rightarrow j+1$ corresponds to the first failure of a component K (which has two repairable failure modes λ_1, μ_1 and λ_2, μ_2 $(\lambda_1 > \lambda_2, \mu_1 > \mu_2)$ in the failure mode 1, then the probability $P(K \rightarrow K_1, t_1)$ associated with the transient provoked at time t_1 by this failure is the following

$$P(K \rightarrow K_1, t_1) = (1 - e^{-(\lambda_1 + \lambda_2)t_1}) . (\lambda_1 / \lambda_1 + \lambda_2)$$

probability of a failure having occurred at $t = t_1$ (discretization)	conditional probability of transition $K \rightarrow K_1$ (approximation)

The time t_1 is evaluated according to the resolution desired (see Fig.3 where only a two state not repairable component is considered). The system is considered to remain in the state K_1 until $e^{-\mu_1(t-t_1)}$ (not repair probability) is less than W_ℓ defined as in Fig.3 (the relation $e^{-\mu_1(t_2-t_1)} = W_\ell$ allows us to evaluate t_2, the next branch point).

To save computer time and memory occupation, the forest of descendant sequences is explored by means of the algorithms implemented by Reina /16/ and illustrated in Fig.4, so that physical calculations are not repeated for sequence segments characterized by the same physics. By these rules, the computer time and the resolution degree can be controlled in accordance with the aim of the analysis.

5.5 Results of a DYLAM Analysis

As Fig.5 summarizes, the results consist of:

- time course of possible transients which define also the space where process variables can range according to all possible system conditions;
- time courses of possible TOP event probabilities versus time;

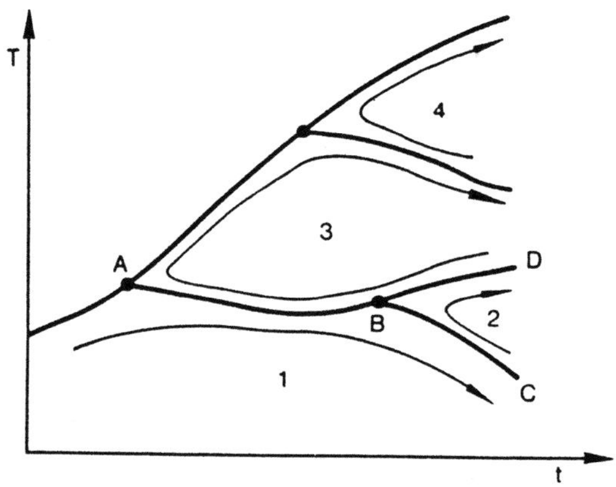

Between A and B the physical behaviour is the same both for descendant ABC and ABD. The process physics are calculated only once by memorizing all values before the branch points, which are explored according to the principles of the procedure indicated.

Fig.4: Descendant sequence exploration.

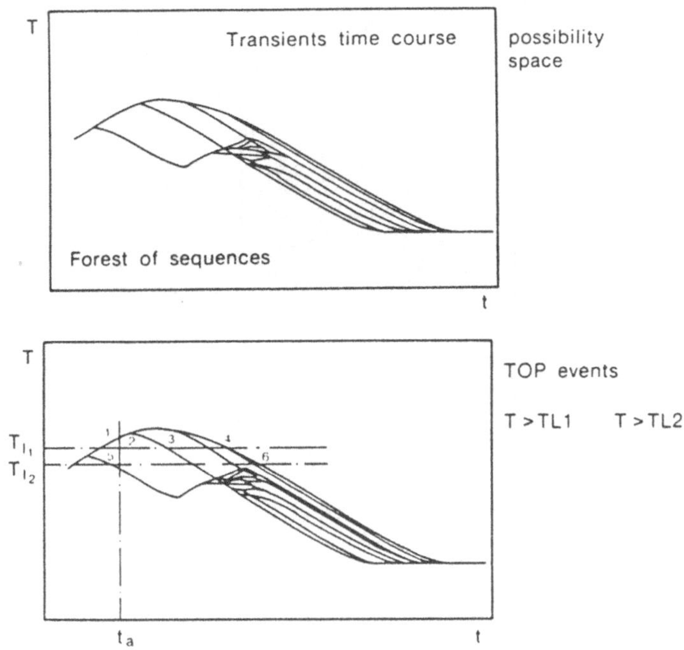

System identification

Sequences leading to TOP event:

$$TL_1 \quad 1,2,3,4 \qquad TL_2 \quad 1,2,3,4,5,6$$

(obviously each sequence would correspond to identified states of components).

Probability of TOP events

$$\text{Prob}(T \geq TL_1) = \sum_{1}^{4} \text{Prob}(i)$$

$$\text{Prob}(T \geq TL_2) = \sum_{1}^{6} \text{Prob}(i)$$

$$\text{Prob}(T \geq TL_1 \text{ at time } t_a) = \text{Prob}(1)$$

Fig.5: Possible results of a DYLAM analysis.

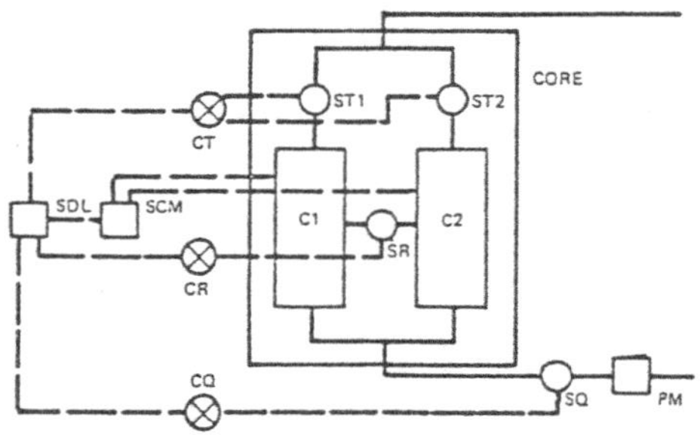

ST1,ST2	= temperature sensor	CR	= on-off reactivity controller
SQ	= flow-rate sensor	SDL	= shutdown logic
SR	= reactivity sensor	SCM	= scram actuator
CT	= on-off temperature controller	PM	= sodium pumps
CQ	= on-off flow-rate controller	C1,C2	= core channels

simplified LMFBR primary circuit

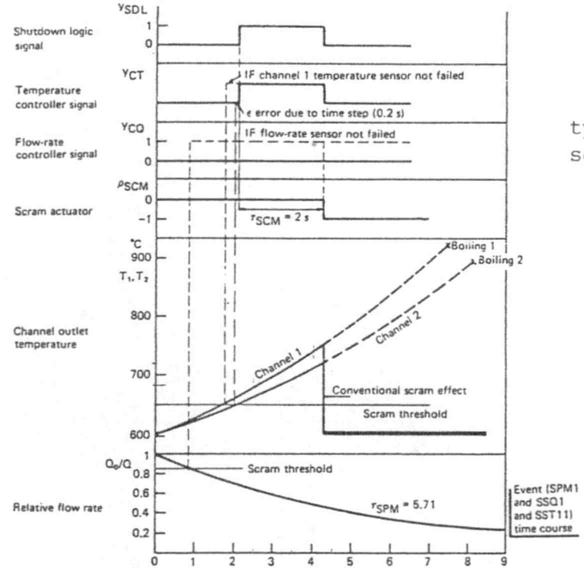

typical sequences

Process physics: approximated by algebraic equations, obtained from precalculations by core accident codes

Equation solution scheme: DYLAM package

Results: sodium boiling probability vs time
initial conditions for core disassembly accidents

Fig.6: LMFBR core accident study case.

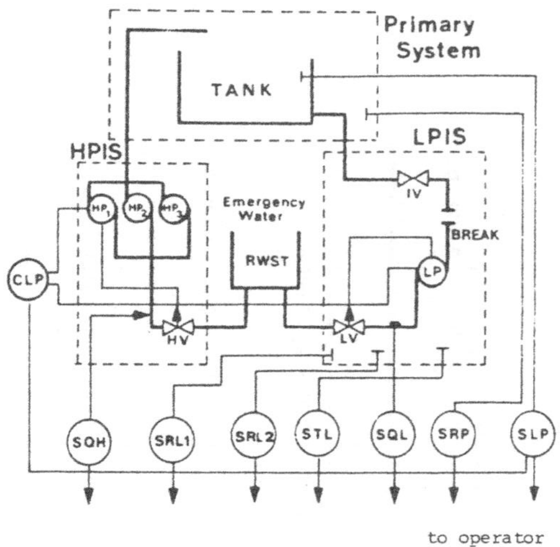

SQL = LPIS flow-rate sensor
SRL1 = LPIS low radioactivity sensor
SRL2 = LPIS high radioactivity sensor
STL = LPIS temperature sensor
SLP = tank level sensor
CLP = tank level controller
SQH = HPIS flow-rate sensor
SRP = primary radioactivity sensor

Tank system and instrumentation assumed for controlling LOCA.

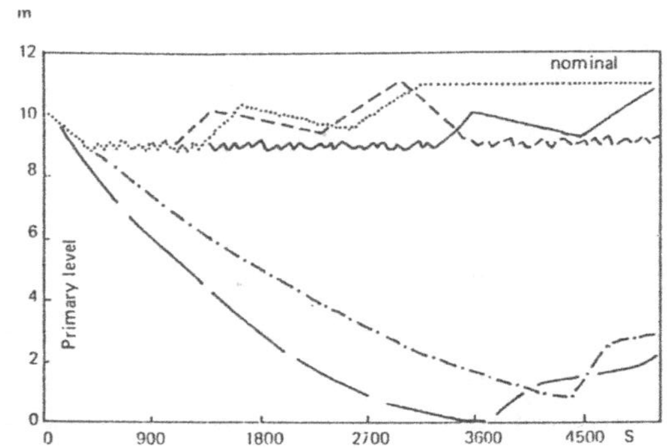

A sample set of simulated possible transients, including operator interventions.

-Process physics simulated by a differential equation expressing tank voiding with times corresponding to fuel melting times.
-Equation solution scheme: DYLAM package.
-Results obtained: demonstration of feasibility to simulate the operator process of perception-diagnosis-action-recovery after errors (see Fig.13) and of usefulness of DYLAM for studying man-machine interfaces.

Fig.7: Man-machine interaction study case.

The nitric acid controlled cooler

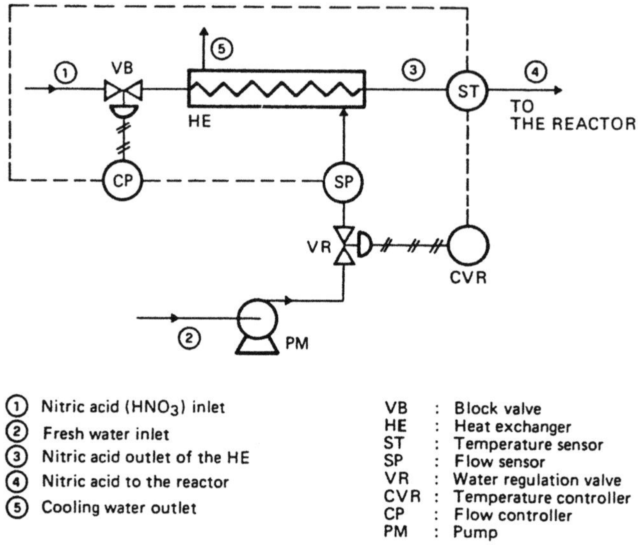

A typical sequence

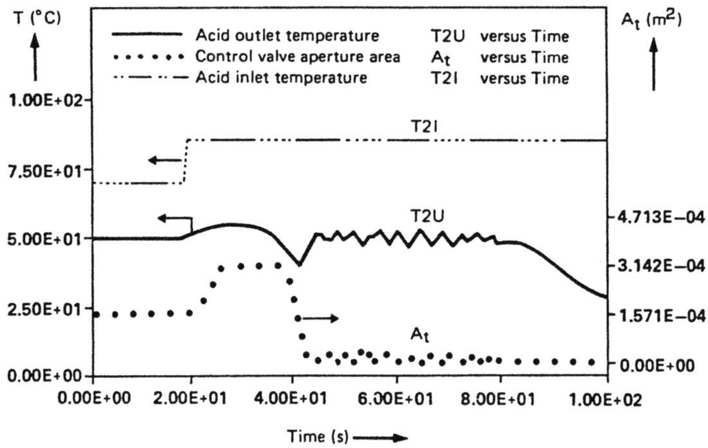

- Process simulated by simple differential equations solved by the DYLAM package.
- Results: probability of high temperature to the reactor inlet ④ (possibility of a run away reaction) as a result of internal malfunctions and/or external initiators (high temperature at inlet ①).

Fig.8: Chemical unit control loop.

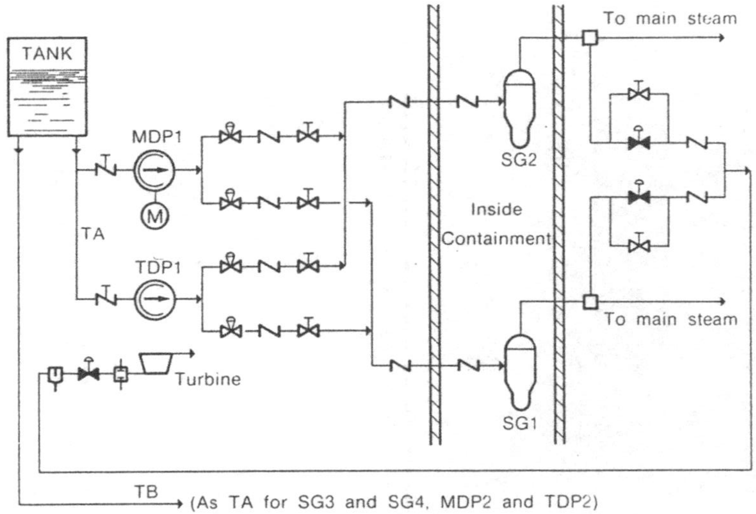

Auxiliary feedwater system

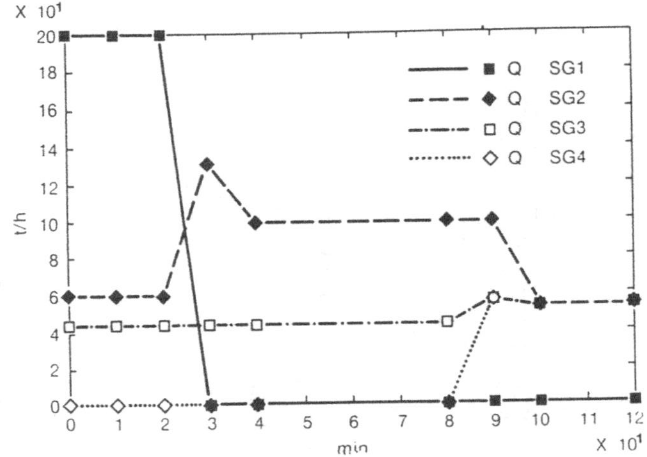

Flow in the steam generator as depending on multiple failures and operator regulation actions

- Top event: insufficient flow to steam generators in the case of a feedline break of SG1 (DYLAM used as a fault tree equivalent tool)
- Hydraulic equations solved by ad-hoc algorithms
- Results: top event probabilities, including operator actions with distributed delays and/or errors. Furthermore, system identification in terms of time available to the operator to avoid mission failures and tank voiding

Fig.9: The auxiliary feedwater system study case.

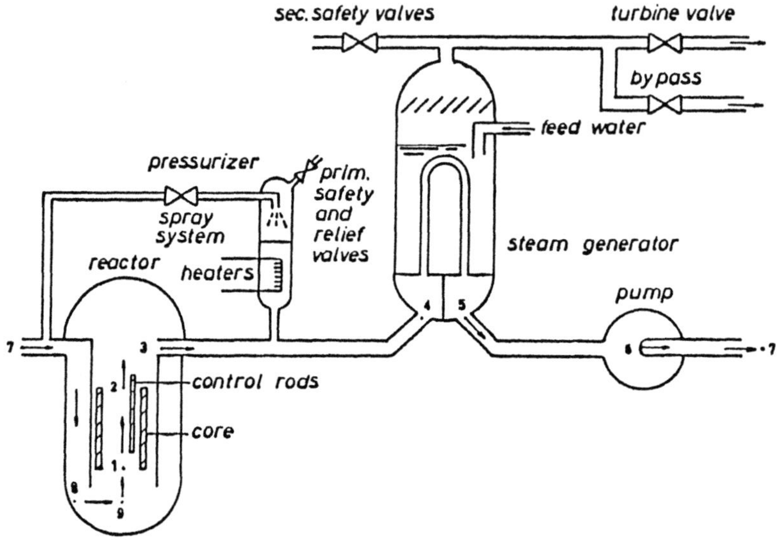

PWR core-primary circuit and steam generator

- Study of pressure/temperature transients by coupling the system code ALMOD-JRC (a modified JRC version of ALMOD produced by GRS) with DYLAM as a driver.

- Investigations of valves failures after random number of opening/ closure cycles.

- Identification of problems connected with accident simulators, when based on the DYLAM concept and of feasibility of the approach.

Fig.10: PWR case study.

- identification of sequences leading to TOP conditions and of the associated probability.

6. REVIEW OF STUDY CASES AND APPLICATIONS

When dealing with complex processes, the aspects connected with the achievement of reasonable running times must be attentively considered; otherwise the DYLAM capability of analyzing in an automated way a very large number of possible transients would be a purely theoretical one. Therefore, particular consideration must be devoted to the physical models assumed, the number of component states included and to the system equation solution algorithm.

In an early LMFBR case study (Fig.6, /17/) the process physics was described by algebraic equations derived from the outputs of accident simulation codes through fitting procedures and the algorithms for solving non-linear algebraic equations implemented into DYLAM could be used easily. Subsequently, some simplified systems were analyzed by introducing simple differential equations which could still be solved by the standard DYLAM algorithms, and, namely: a man-machine interaction study case (Fig.7, /11/) and a chemical process control loop (Fig.8, /18/).

A more significant application with respect both to the complexity and to the concreteness of the study case was performed by Cacciabue et al. (Fig.9, /9/) who analyzed a PWR Auxiliary Feedwater System: in addition to reliability estimations, other very interesting results could be obtained concerning operator actions and allowable intervention times.

Further steps towards analysis of more complex systems have been taken by:

- coupling the DYLAM algorithm with a PWR system code (Fig.10, /15/) to demonstrate feasibility of DYLAM acting as a driver of new simulators, in which it performs like a generator of stochastically changing boundary conditions for the process to be analyzed;
- coupling DYLAM with simulators of chemical batch reactors (Fig.11, /19/) to demonstrate its advantages for safety assessments of chemical processes and protection design; and to a simulator of "Domino" effect accident scenarios in storages of explosive/flammable substances (Fig.12, /14/).

7. CONCLUSIONS

As a result of the previous discussions, with respect to other system modelling techniques, DYLAM is particularly advantageous for the description of dynamic incident scenarios where it can substitute or complement event tree techniques; and for reliability assessment of systems whose mission is defined in terms of values of process variables to be kept within certain acceptable limits where it can substitute or complement fault tree techniques.

It is not a purely probabilistic approach and its use in the

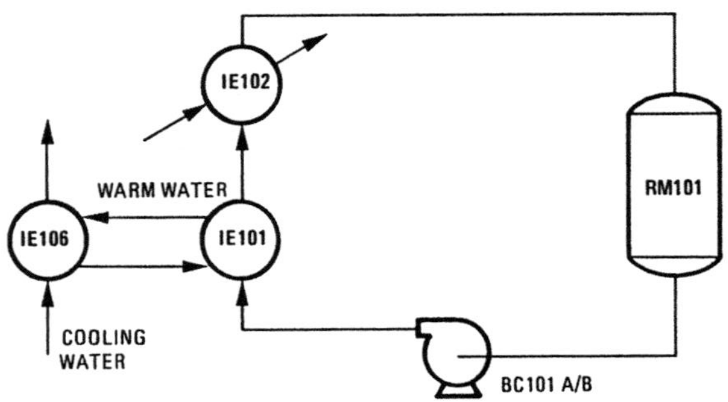

Chemical batch reactor

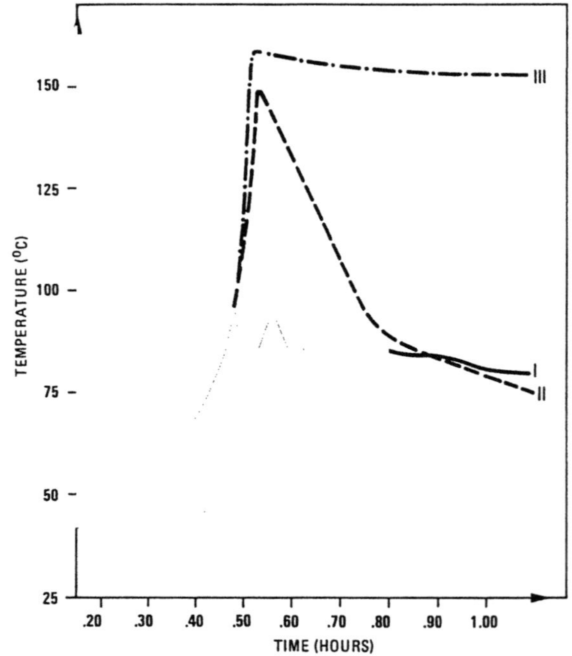

I : NORMAL OPERATION
II : FAILURE OF HEAT EXCHANGER IE 102
III : PUMPING SYSTEM FAILURE BC 101 A/B

NOTE: The other failure sequences produce transients included between curves I and III

Typical transients

Fig.11a: The chemical batch reactor study case.

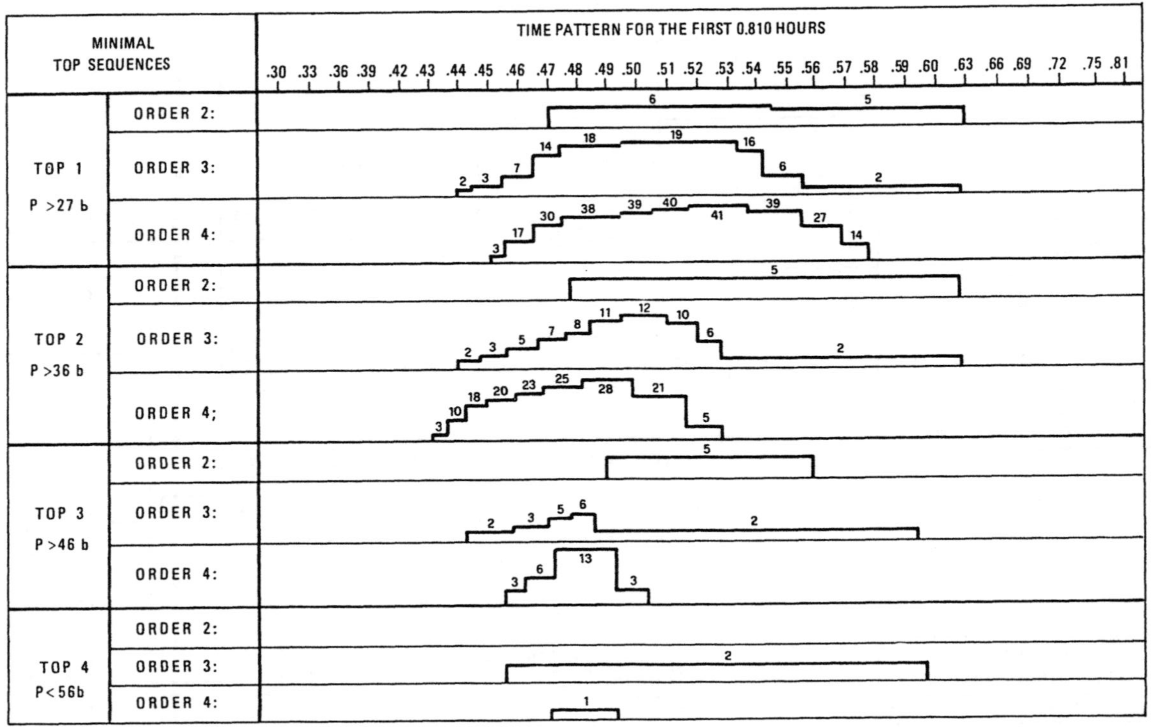

Typical results: number of sequences vs time duration of top conditions

Fig.11b: The chemical batch reactor study case.

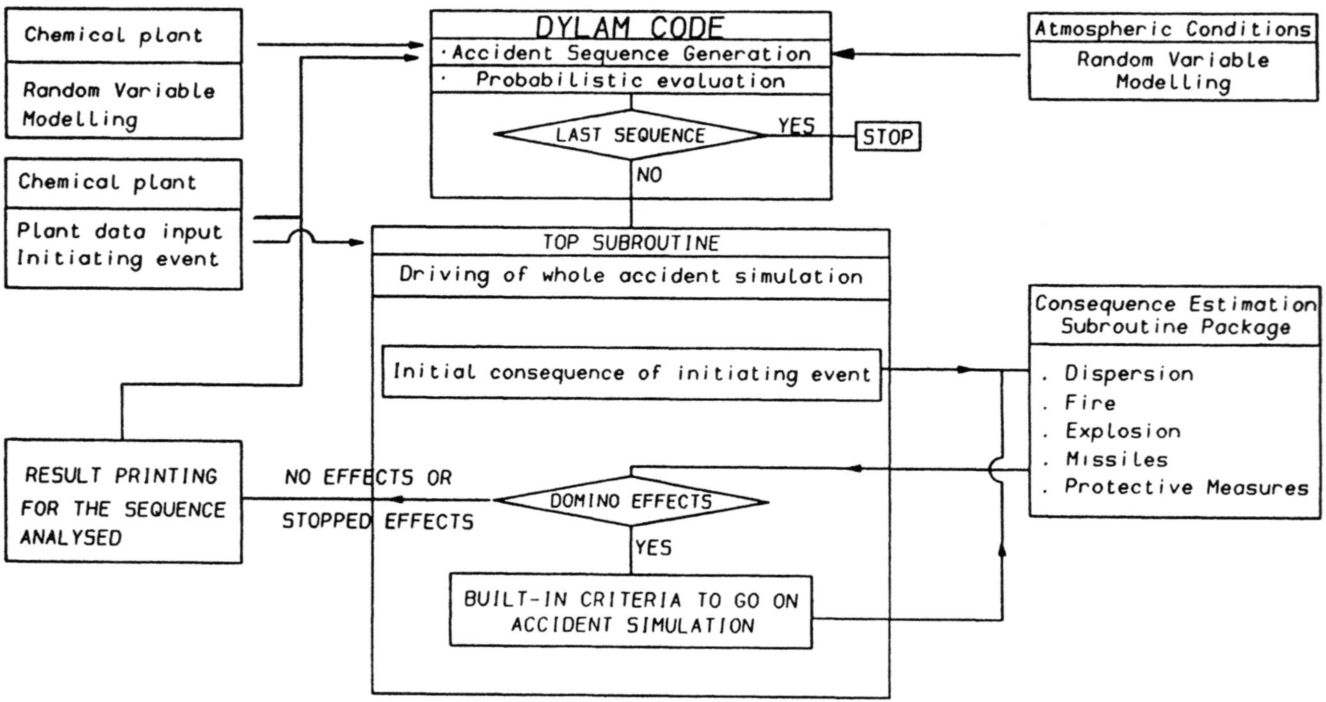

Fig.12: Flow chart of the DOMINO software package.

cases where physical calculations are not relevant would be too costly and less accurate than other approaches.

It is worthwhile to better specify what is intended by complementing other techniques: because of the combinational explosion of possible states, it is useful to model by DYLAM only those components which really interact with the process physics (e.g. the availability of the signal for actuating a protection can be conveniently derived by FTs and its value is an input to the DYLAM model of the protection and so on).

DYLAM can also be used for identification of system behaviour and, therefore, as a design tool especially as far as needed protections and operator procedures are concerned.

Indeed, as shown in /11/, it offers a correct frame for studying man-machine systems and, therefore, for designing decision aids for operators in emergency situations. This is achieved by the possibility of generating numerous incidental scenarios including process malfunctions, instrument failures, operator actions and errors.

The advantageous use of DYLAM for modelling human reliability in incident control can be easily understood when thinking that it determines the time course of the process variables and, therefore, of the states of signals and alarms available to the operator for performing diagnosis, actuating the prescribed procedures and recovering wrong interventions /11/.

Finally, the procedures indicated for the generation of multiple branches might be implemented in process accident simulation codes, to control the choice among different models when the uncertainties connected with the phenomena do not allow chains of events to be deterministically described in a unique manner.

The above indications have been confirmed by the study cases analyzed and by the interest with which the methodology is being considered by other organizations (see, for instance, /1/).

ACKNOWLEDGEMENTS

The extensive collaboration with Dr. G. Reina (Studio Mesa, Milano) in the development of the DYLAM technique is gratefully acknowledged.

REFERENCES

1. Izquierdo, J.M. and Sanchez-Perea M. (1990) 'DYLAM-TRETA. An approach to protection systems software analysis', next paper of these Proceedings.
2. Amendola, A. (1988) 'Accident sequence dynamic simulation versus event trees', Reliability Engineering and System Safety 22, 3-25.
3. Alesso, H.P., Prassinos, P. and Smith, C.F. (1985) 'Beyond fault trees to fault graphs', Reliab. Engng. 12(2), 79-92.
4. Alesso, H.P. (1985) 'On the relationship of digraph matrix

analysis to Petri net theory and fault trees', Reliab. Engng. 10(2), 93-104.

5. Peterson, J.L. (1977) 'Petri nets', Computing Surveys 9(3), 223-52.
6. Bobbio, A., Cumani, A. and Del Bello, R. (1986) 'Reduced Markovian representation of stochastic Petri net models', Systems Sci. 10(2), 5-23.
7. Signoret, J.P., Gaboriand, M. and Leroy, A. (1988) 'Study cases of petroleum facilities as comparison bases for different methods', in: Reliability Engineering, Proc. Ispra Course, Madrid University, September 1986, Eds. A. Amendola and A. Saiz de Bustamante, A.P. Kluwer, Dordrecht (NL).
8. Amendola, A. (1986) 'The systems reliability benchmark exercise', final report EUR 10696, CEC-JRC, Ispra (Italy).
9. Cacciabue, P.C., Amendola, A. and Cojazzi, G. (1986) 'DYLAM vs fault-tree: the AFWS case study', Nucl. Technol. 74, 195-208.
10. Amendola, A. and Reina, G. (1984) 'DYLAM-1: a software package for event sequence and consequences spectrum methodology', EUR 9224 EN, CEC-JRC, Ispra (Italy).
11. Amendola, A., Reina, G. and Ciceri, F. (1985) 'Dynamic simulation of man-machine interaction in incident control', Proc. of the 2nd IFAC/IFIP/IFORS/IEA Conf. on Analysis Design and Evaluation of Man-Machine Systems, Varese (Italy), September 10-12, 1985, Pergamon Press, Oxford.
12. Amendola, A. (1988) 'The DYLAM approach to systems safety and reliability assessment', EUR 11361 EN.
13. Amendola, A., Labath, N.A., Nivolianitou Z. and Reina, G. (1988) 'Application of DYLAM to the safety analysis of chemical plants', Int. J. of Quality and Reliability Managemenent 5,2, 48-59.
14. Labath, N.A. and Amendola, A. (1989) 'Analysis of "Domino" effect incident scenarios by the DYLAM approach', 6th Int. EFChE Symp. on Loss Prevention and Safety Promotion, Oslo, June 19-22, 1989.
15. Cacciabue, P.C., Colombo, D. and Tozzi, A. (1986) 'Coupling the system code ALMOD-JRC and the DYLAM driver for studying the consequences of reactor accidents in the framework of system response analyser', Technical Note N°I.05.C1.86.19, CEC-JRC, Ispra (Italy).
16. Reina, G. and Amendola, A. (1987) 'DYLAM-2: description and how to use', Technical Note N°I.87.128, CEC-JRC, Ispra (Italy).
17. Amendola, A. and Reina, G. (1981) 'Event sequences and consequence spectrum: a methodology for probabilistic transient analysis', Nucl. Sci. Eng. 77, 297-315.
18. Nivolianitou, Z., Amendola, A. and Reina, G. (1986) 'Reliability analysis of chemical processes by the DYLAM approach', Reliability Engineering 14, 163-182.
19. Labath, N. and Nivolianitou, Z. (1986) 'Application of DYLAM methodology to the reliability analysis of a batch chemical plant', 1st Eur. Workshop on Fault Diagnostic, Reliability and Related Approaches, Rhodes (Greece), August 31 - September 3, 1986.

DYLAM-TRETA. AN APPROACH TO PROTECTION SYSTEMS SOFTWARE ANALYSIS

J.M. IZQUIERDO & M. SANCHEZ-PEREA
Development Unit. Nuclear Safety Council
C/ Justo Dorado, 11
Madrid 28040
Spain

ABSTRACT. A methodology for the safety assessment of the software of protection systems using dynamic reliability is presented.

It includes a conceptual framework with appropriate definitions and acceptance criteria together with a description of the tools available to implement the technique. The approach is illustrated via an example.

1. Introduction

1.1. CONTROL THEORY VERSUS PROTECTION THEORY

Control theory is a well established discipline that has provided during years a systematic approach for the design of control systems. Indeed, its related literature production is overwhelming and a set of mature methodologies have been established for the build-up and assessment of an optimum control system.

An important distinction should be made between the software aspects, associated with the definition of system functional requirements (i.e. control programs for set points of comparators, optimization of values for controller time constants or, more generally, of the overall performance, and generation of normal operating procedures for manual system operations) and the hardware aspects associated with the implementation of functional requirements (i.e. the development of technology to meet them and the definition of performance requirements -reliability of operation; redundance, diversity, robustness, etc-). Most of the so called "control theory" aims to software rather than hardware optimization.

Whenever a control system operates, it is inevitable the design of a protection to cope with its potential failures. Protection systems (i.e. the system and/or operation team that generate the signals to

A. G. Colombo and A. Saiz de Bustamante (eds.), Systems Reliability Assessment, 183–203.

actuate protections) appear then to parallel control systems and it would be expected a similar attention to a systematic approach for their conception. Surprisingly, when considering the equivalent <u>software</u> aspects of their performance and design, a scarcity of published contributions become evident and no equivalent mature "protection theory" has been established. This is not the case of the hardware where reliability techniques have filled the gap and today reached a high level of maturity.

This situation is partly due to the fact that many protection systems can be rather simple, particularly if the protection intervention is not too inconvenient. However, the complexity of high risk installations where the protective measures are aggressive and there is a strong need for meeting the criteria of "not to intervene if unnecessary", and "very important to do it if necessary", provides incentive toward optimization and a systematic approach for software aspects is being called for.

The problem is particularly sensitive to regulatory bodies in order to base their decisions on a sound "corpus doctrinae" and prevent arbitrariness in judgements for accepting proposals and issuing licences. This is further supported by the fact that experimental verification of the assumptions made is very limited, analytical aspects being thus more important for protection than for control.

1.2. SCOPE AND OBJECTIVES

This lecture aims to show some principles and methodologies which can be applied for the design and assessment of software aspects of protection systems, focusing more in the assessment side where there is a more evident lack of systematic approaches.

A simple but representative example is consistently considered throughout for illustration purposes.

1.3. DISCIPLINES INVOLVED

Software aspects of protection systems (i.e, determination of appropriate initiation signals with associated alarms, precautions, protected system operation limitations and emergency procedures) have in the past been designed by defining a set of "design basis events", simulating their transient history under certain assumptions and trying to demostrate that for these scenarios the facility will not violate prespecified damage limits.

Failure probabilities are indirectly considered, often only in a qualitative sense, by imposing different damage limits to design basis events with probability of occurrence lying in different ranges.

Although the incorporation of reliability techniques allows a more detailed evaluation of these failure rates, the rest of the analysis is deterministic in nature, and reflects an extension of control theory methods where it is also usual to define "design basis events" like step and ramp perturbations, in spite of the essential differences in role between both types of systems.

It can be easily identified:

a) The control system (level control), the main feeding line, a switch-over valve to the alternate feeding line and a normal discharge line for disposal of the fluid to a controlled area of applications.
b) The automatic protection consits of i) a trip system which trips the pump and/or closes valves A, B, C and ii) a limitation/safeguard system formed by: a) Pump restart and automatic opening/close of valve C in case of uncontrolled blowdown without tank or piping ruptures b) Automatic close/opening of valves A, B, C and D in case of ruptures, to promote a controlled blowdown.

The software aspects of the control system include optimization of the normal operations of charge and discharge, automatically and/or manually, through appropriate combination of optimum values of the PI controller constants, the reference level control program and a set of operating instructions.

The software aspects of the protection system include the determination of:

i) The initiation signals for stop/starting the pump and for close/opening of the protection valves.
ii) Operation precautions and limitations to prevent failures to occur and assure assumptions made in the risk analyses, including systems operability and limiting conditions for operation.
iii) Definition of an alarm system to initiate potential operator intervention.
iv) Design of a set of emergency procedures.

We are looking for a procedure to verify that, once completely defined, this software maintains an appropriate risk level without unduly penalizing the facility operation. As mentioned before items iii) and iv) are not within the scope of this lecture.

TABLE 1

PHYSICAL SYSTEM DATA AND SPECIFICATIONS

Geometrical data:

. Heights:	. Valve cross sections:	. Pipe volumes:
H1=15m	$Si=0.1m^2$ (i=1,...,6)	$V1=2.1m^3$
H2=12m	$S7=0.05m^2$	$V2=2.0m^3$
Hw=0-2.2m	$Stank=1m^2$	$V3=0.25m^3$

Hydraulic data (IS units):

. Valve friction constants:	. Pump characteristic:
($Q=K_v$*A*sqr(ΔP)*sign(ΔP)	ΔP=-300*Q+179000
K3=68.91	
K5=KA=460.94	
K2,4,6,B=777.35	
K7,C=490.5	

Dynamic data:

. Opening/closing time constants:	. Pump decay time constant: 30s
Isolation valve: 10s	. PI controller: P=10
Control and protection valves: 5s	$I=75s^{-1}$
Check valve: 0.5s	. Switch from pipe 1 to pipe 2:
	$Q_1<0.01$ kg/s

On the other hand, through implicit assumptions about the expected dynamic behavior of considered scenarios, (particularly concerning whether the intervention of protective measures is a requirement*) event tree techniques try to incorporate the evolution of process variables (temperatures, flows, etc) in an equally qualitative and incomplete sense which do not allow software aspects to be fitted into the analysis.

Both, probability (reliability techniques) and damage (transient analysis) are intrinsic to any coherent treatment of risks and the maturity reached by both disciplines makes the way towards a better treatment by removing some of these deficiencies originated by attempts to decouple both aspects in order to simplify the problem.

Finally, manual actions due to human intervention are a part of the protection system, subject however to a decision making process much more complex than the initiation criteria of automatic devices. The area of man-machine interaction is equally required to be integrated (or rough assumptions will inevitably be made as well). This aspect however will be only peripherally touched here.

1.4. EXAMPLE

As an ilustration, consider the simple hazardous fluid storage tank facility depicted in Fig. 1 and Table 1.

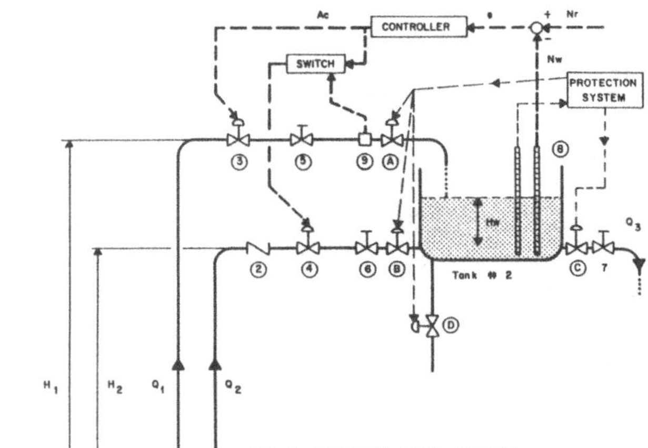

FIG. 1 STORAGE TANK FACILITY

* i.e, the protection is expected to actuate even if the dynamic behaviour does not meet initiation signal conditions. -if it does not, it is atributed to a failure in the protection hardware-

2. Basic principles and methodology

In the following, a set of useful definitions and principles are outlined focusing only on the verification of the adequacy of a given protection system (assessment).

For more detailed explanations and its extension to design and optimization see ref. 1 for the particular case of nuclear power stations, but easily extrapolable to other facilities.

2.1. DEFINITIONS *

2.1.1. Installation, I:
Set of interrelated systems to obtain a benefit through the accomplishment of one or several installation functions.

2.1.2. System, S:
Set of i) components, ii) articulated in a certain way, iii) with characteristics and iv) specifications to meet a certain system function as a part-task of the installation functions.

2.1.3. Protection associated to I, P_I:
Set of systems and/or system features (i), ii), iii), iv) in the system definition) that prevent the facility to produce undue risks.

2.1.4. Automatic Protection System:
Set of devices that generate the signals for automatic actuation of protections. These will include some or all of the following

. Limitation System: Function, to generate signals to reduce damage without stopping I, i.e drive it into a stable state in some operating mode.
. Trip System: Function, to generate signals to quickly stop I, i.e. drive it into a stable state in some non operating mode.
. Safeguard System: Function, to generate signals to reduce damage once I has been stopped. In conjunction with trip and limitation systems, drive it into a safe state in some non operating mode.

2.1.5. Logical Variable:
Integer variable describing state of components.

2.1.6. Process Variable:
Variable describing a process evolution (temperatures, flows, etc.).

. Safety Variable: Variable in which terms initiation criteria of protections are formulated.
. Damage Variable: Variable in which terms damage limits are formulated.

* only for the purposes of this work

2.1.7. *Combined (logical and process) state of a system (or Installation) of N components:*

Set of combined states of system components

$$[\; n_1, n_2, \ldots, n_N > \quad \& \quad [\; x_1, x_2, \ldots, x_N >$$

Logical state — Process state

COMBINED STATE

where $n_j(t)$, $x_j(t)$ define each individual component combined state

$n_j(t)$ logical state (i.e. nominal, non operating, failed, derated etc) of component j.

$x_j(t)$ process variable state vector for component j, enough to evaluate system damage evolution.

. Steady State: Time independent state. Note that there may be incompatibility of certain logical states with steady process states.
. Stable State: A process state such that, if subject to a small perturbation, the system or installation comes back to that state once the perturbation dies out.
. Safe state: A combined steady state from which any initiating event or driver transient will produce no top sequence in the associated dynamic event tree. (See Section 2.2.3. below).

2.1.8. *System Transient:*

Time evolution of system process states.
. Driver Transient: Evolution from an initial steady state.

2.1.9. *Event:*

Change in logical states
. Active (Passive) Event: Event triggering (not triggering) a transient.
. Initiation Event: Active event triggering a driver transient.
. Single Event: Event involving changes in a single component.
. Sequence of events: Event resulting from single and/or common events occurring in succession within finite time intervals.

2.1.10. *Fault:*

Event implying degradation of system functions
. Common Failure (or Common Event): Non-single simultaneous fault (or event).

2.1.11. *Perturbation:*

A deviation from the current evolution in the process states.
. Initiating Perturbation: A perturbation of steady states.

2.2. THE DBEP CONCEPT AND ASSESSMENT METHODOLOGY

2.2.1. Probability Densities.

In general both, logical and physical states have a probability density that depend on time and on the combined state. Here however, only the logical states will be statistically considered, but their probability density*, $P(n_1, n_2, \ldots, n_N)$, may be a function of the process state i.e.:

$$P(n_1, n_2, \ldots, n_N) = F(n_1(t), n_2(t), \ldots, n_N(t), x_1(t), x_2(t), \ldots, x_N(t))$$

a distinctive feature from the usual reliability techniques that implicitly assume independence. The term dynamic reliability versus static reliability will be used to discriminate both approaches.

This means that in the course of a transient, the evolution of the process variables may change the probability of occurrence of an event; for instance, if the process variable conditions for protection intervention are met, some protection components will change from the non-operating to the operating state in a deterministic way.

2.2.2. The DBEP risk acceptance criterion.

An acceptance criterion should be established based on a design basis envelope (DBEP) defined on a risk diagram of probability of occurrence of a sequence of events versus its generated damage as the boundary within which the points representative of the identified sequences should be in order to maintain an acceptable risk level. We will call a "top sequence", any sequence that generates at least one point outside this boundary. The damage should be defined in terms of process variables (damage variables) and will therefore be specific of the type of system considered. This criterion reduces to the standard method of classifying events by probability ranges and using different damage limits for different classes if the DBEP becomes a step-wise curve (see Fig. 2). Note that a DBEP curve should be generated for each damage criterion.

FIG. 2 DBEP'S USED IN THE EXAMPLE

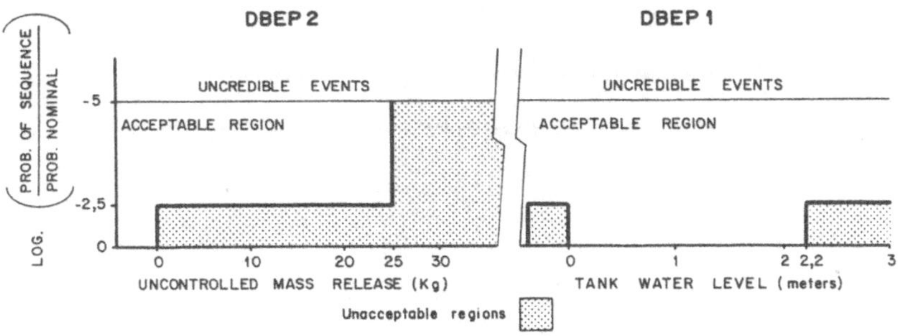

* (i.e. $P(n_1, n_2, \ldots, n_N)\,dt$ is the probability of remaining in the logical state $[n_1, n_2, \ldots, n_N]$ between t and $t + dt$)

2.2.3. Protection system assessment methodology. Dynamic event trees.

The assessment of the adequacy of a given protection system lies on the verification of the risk acceptance criterion through the generation of a high number of points on the risk diagram by collecting pairs (probability, damage) as a function of time from dynamic event trees. The dynamic event trees are obtained through simulation of a comprehensive set of driver transients triggered from initiating events and/or perturbations of steady states* and taking into account all of the transient histories arising from the dynamic evolution of the combined states i.e. including events result of the evolution itself.

2.3. EXAMPLE

Referring to the storage tank facility of section 1.4, Table 2 describe the logical and process states of all components numbered according to Fig. 1 and Fig. 2 shows the DBEP risk acceptance criterion curves selected. Here the potential damage considered is the uncontrolled release of the tank fluid (damage limits of zero release for frequent events and limited amount of mass release for low probability events) and the operation inconveniencies created by tank blowdown for frequent events.

Thus,unacceptable tank level and excessive mass release are the damage criteria, and process variables required to determine them are damage variables.

TABLE 2

LOGICAL AND PROCESS STATES OF THE STORAGE TANK COMPONENTS

LOGICAL STATES

	PUMP	CHECK VALVE	CONTROL VALVE	ISOLATION VALVE	LEVEL SENSOR	FLOW-RATE SENSOR
0:	Normal	Normal	Normal	Normal	Normal	Normal
1:	no flow	open	open	open	min. output	min. output
2:			closed	closed		max. output

PROCESS STATES

Components	Sensors
PUMPS: Pressure gain, flow and speed.	FLOW: Measured flow.
VALVES: Pressure drop, flow and stroke position.	LEVEL: Measured level.

* These states may be obtained with static reliability fault tree analysis

3. Tools for implementation of the methodology

3.1. THE SRA PROJECT

At this time, the EC Joint Research Center at ISPRA is involved in the development of the so called System Response Analyzer (SRA)(ref.2) with sufficient capability to provide a tool to be used for this purpose. The System incorporates a machine model module able to deal with the process state evolution, a reliability module to deal with the logical states and a human module still under development. CSN is cooperating in this effort (ref.4) by using its own machine model, the TRETA package (ref.3), which is modular and flexible enough for this application. For simplicity the software aspects of man-machine interaction will be excluded from the scope of this contribution.

SRA development extends the field of application of simulators to the area of risk and reliability analysis, in particular to the identification of relevant sequences, to the modelling of human behaviour, and to the validation of operation procedures, always considering the effect of the physical evolution of the plant.

The machine and operator models have to be able to simulate the diverse configurations possibly encountered as a consequence of the physical evolution of complex system transients and of the component stochastic behaviour. The DYLAM driver ensures the dynamic flow of events, generating the time dependent configurations according to a given logic as well as to causality principles.

3.2. THE DYLAM-TRETA SIMULATION PACKAGE

3.2.1. Dylam.

DYLAM has been designed in the following way:
The system to be analyzed is modelled not only in its nominal behaviour, but also in every derated/failed one. This is done by identifying the components that may fail and modelling them with the machine module for any possible logical state. Together with the definition of the component logical states, probabilities for each one and the corresponding transition rates ought to be given. Finally, one or more damage criteria (i.e., undesired conditions) are defined as the occurrence of a certain temperature, pressure, stress, level, etc. or a combination of them.

Once the system is modelled from the physical and probabilistic point of view and the damage criterion to be studied is chosen, the driver has to generate sequences covering all possible events according to the implemented cut-off rules. All sequences leading to the damage criterion can be calculated.

The driver generates a set of different combinations of initial logical states for the components. This initial set is limited by the user-selected maximum number of components in non-nominal state and its compatibility with a steady process state.

The next step consists of considering an initiating perturbation and/or event and simulating the system thereafter. This event will trigger, in its temporal progression, new events that, in turn, can act on the physical evolution of the system in a continuous dynamic interactive process.

As it was mentioned previously, to every component state we associate a probability density and to every state change a transition rate. There are five different ways to define the probabilistic data of the components, representing effects such as aging, dependency on physical variables, or even dependency on the state of another component. Therefore, the probability for the system to remain in a given logical state can be calculated based on the probabilistic data for the components state.

This probability will eventually decay in time, and when its value drops below a given percentage (provided in the data), a branch point is generated. At this branch point, new single events will come up, as many as components with the possibility of changing their state. Nevertheless, the study of the current transient will continue with no state changes, but as soon as it is finished, the driver will go back to this branch point to continue the study with one of those new sequences.

How the components state probabilities are defined in DYLAM and how the overall probability of sequences of events is evaluated are topics which do not lie within the scope of this report. See references 5-8 for the details of these procedures.

3.2.2. Treta.

3.2.2.1. Structure:

A.- Main segments

There can be distinguished three parts in TRETA:

a) A general driver called CAPTA, able to solve block diagram networks of numerical models of ordinary non linear differential equations, in the same way as simulation languages operate. This feature allows progressive enlargements of the simulated system. CAPTA is dealing with:
  - input data management,
  - during simulation, time point renewal,
  - for each time point, sequential computation of each diagram defined at input,
  - feedbacks management, by using an iterative scheme,
  - storage management of partial outputs.

  CAPTA also uses an auxiliary code, called COLAPSO, which allows to identify "linear loops and paths", and so, to generate a "collapsed topology" of linear parts by using only one transfer function for each linear input-output relation. The collective transfer function is then transferred into the time domain with Laguèrre polynomial techniques, and can be coupled with the non-linear portions.

b) Block catalogue, also divided in three subcatalogues:
  - a component library,
  - a module library, and

- an algorithm library.

Today these libraries correspond to a "Pressurized Water Reactor", but CAPTA versatility allows any kind of systems. In fact a hybrid BWR-PWR system is within scope.

B.- Problem description

In order to show the basic structure of the program, it is necessary to explain how the problem is set up. This is done by using a BLOCK DIAGRAM, where the following must be specified:

- input signals to the system,
- component interconnexion (topology of the diagram),
- components, and
- output signals of the system.

A block diagram is composed of BRANCHES, and these include:

- an ADDER which obtains the MAIN INPUT to the block,
- a BLOCK, which represents the physics of the branch,
- a set of COMPLEMENTARY INPUTS, also needed by the block, and
- a set of OUTPUTS of every branch.

C.- Data handling

For each branch of the block diagram, the following data must be assigned:

1) Topological data, in order to define the sequentiality of the computation. These are:
   - a branch number, which indicates the computation sequence;
   - the numbers of the branches contributing to the main input of the branch; and
   - when a feedback appears, the number of the branch where computation must continue, and the parameters needed to test the convergence of the iterative process used to solve it.
2) Physical data. These include:
   - the type of block, included in the catalogue;
   - the numbers of the branches corresponding to the complementary inputs; and
   - all those data needed for each specific type of block.
3) Data for the output display.

D.- Structure of the library of models

Every library routine (component and module routine) has two parameters:

- IF, number of the branch making use of this routine,
- IR, operational calling mode,

and it is structured according to the IR parameter as follows:

a) if the call is done with IR equal to 0, the physical data of the IF-th branch are read;
b) if the call is done with IR equal to 1, the input data of the IF-th branch are printed with their explanation;

c) if the call is done with IR greater than 1, the output or outputs of the IF-th branch are calculated for the current instant. IR equal to 2 corresponds to the nominal state of the component, and higher values to other possible states;
d) if the call is done with IR equal to-1, the physical variables needed for restarting the transient from this time point are stored in the physical variable buffer;
e) if the call is done with IR equal to-2, the physical variables stored in the buffer are read in order to restart a transient from a given time point.

E.- Component/module/algorithm concepts

The organization of the library corresponds to these three concepts.

A part of the system block diagram is said COMPONENT if it may have several operation states or modes, each of them with a value of IR and a fixed topology associated. These would include:
- nominal, failed on mode A, failed on mode B, etc. working states;
- linear, non linear, quasistationary, etc. models used for different working conditions, or even for different simulation requirements.

Components generally will have a physical meaning. Examples are: valves, pumps, sensors, etc.; reactor kinetics, steam generator level and/or pressure, etc. PID controllers, and so on.

The concept of MODULE refers to those parts of the system block diagram, or of any component, which can not change its state and will perform a kind of function often needed. Modules generally will not have a specific physical meaning, in fact the same module can be used by two different components, or in two different contexts. Examples are: momentum equation solver module, integrators, non linear functions, first order delays, pure delays, etc.

The ALGORITHMS are general computation tools used by the component or module subroutines. Examples are: differential equation solvers, algebraic equation solvers, convolution algorithms, etc.

A component has a fixed "external topology" for each state formed by interconnected modules, while a module has its "internal topology" built by algorithms and/or other modules. That is, component routines can use module routine calls, while module routines can use algorithm routines and even other module routines. This feature allows optimization between flexibility and efficient numerical solutions.

Component and module routines have the IR structure, although modules do not have IR values corresponding to different states.

More details about TRETA can be found in ref. 4.

3.2.3. TRETA-DYLAM linkage:

In order to accomplish the tasks outlined in section 2.2. and implement the methodology explained in chapters 1 and 2, TRETA and DYLAM have been linked.

These two programs together apply the basic principles intended to be developed (see ref.1) in the sense that they are able to simulate all the possible transients that take place in the evolution of the combined states of an installation. Thus, a damage variables-versus-probability plot can be made to compare with the design basis envelope.

More details related to the TRETA-DYLAM linkage can also be found in ref.4.

3.3. EXAMPLE

Figure 3 shows the block diagram used in the physical modelling of the example considered, without any protection system. The branches are represented with numbers indicating the order in which the calculation proceeds. Box-shaped branches are specific blocks (component or modules) modelling various parts of the physical system (i.e. pipe nr. 1 stands for the module solving momentum equation in that pipe, valve 4 is the block modelling control valve A in all its possible states). Physical data and component probabilities plus transition rates are provided in Tables 1 and 3.

Figure 4 represents the additional protection systems and the way that specific protections against damage conditions is implemented.

TABLE 3

EVENT PROBABILITIES AND TRANSITION RATES

.Check valve(2): Independent state probability

p0(0)=0.96
p0(1)=0.04

.Control valves(3,4): Independent state probability with transient rates

		0	1	2
p0(0)=0.90	0	-	5.3E-4	5.3E-4
p0(1)=0.05	TR= 1	1.2E-4	-	0.0
p0(2)=0.05	2	5.3E-4	0.0	-

.Isolation valve(5): State-dependent state probability

It depend on EV 6 (isolation valve) as follows:

		0	1	2 (EV 6)
p(st EV 5 | st EV 6):	0	0.9	0.9	0.7
	1	0.1	0.1	0.0
	2	0.0	0.0	0.3
	(EV 5)			

.Isolation valve(6): Independent state probability with transient rates

		0	1	2
p0(0)=0.80	0	-	0.0	0.0
p0(1)=0.05	TR= 1	0.0	-	0.0
p0(2)=0.15	2	1.18E-4	0.0	-

.Level sensor(8): Independent state probability with transient rates

p0(0)=1.0 tr01=5.24E-4
p0(1)=0.0 tr10=0.0

.Flow-rate sensor(9): Independent state probability with transient rates

		0	1	2
p0(0)=1.0	0	-	5.27E-4	5.27E-4
p0(1)=0.0	TR= 1	0.0	-	0.0
p0(2)=0.0	2	0.0	0.0	-

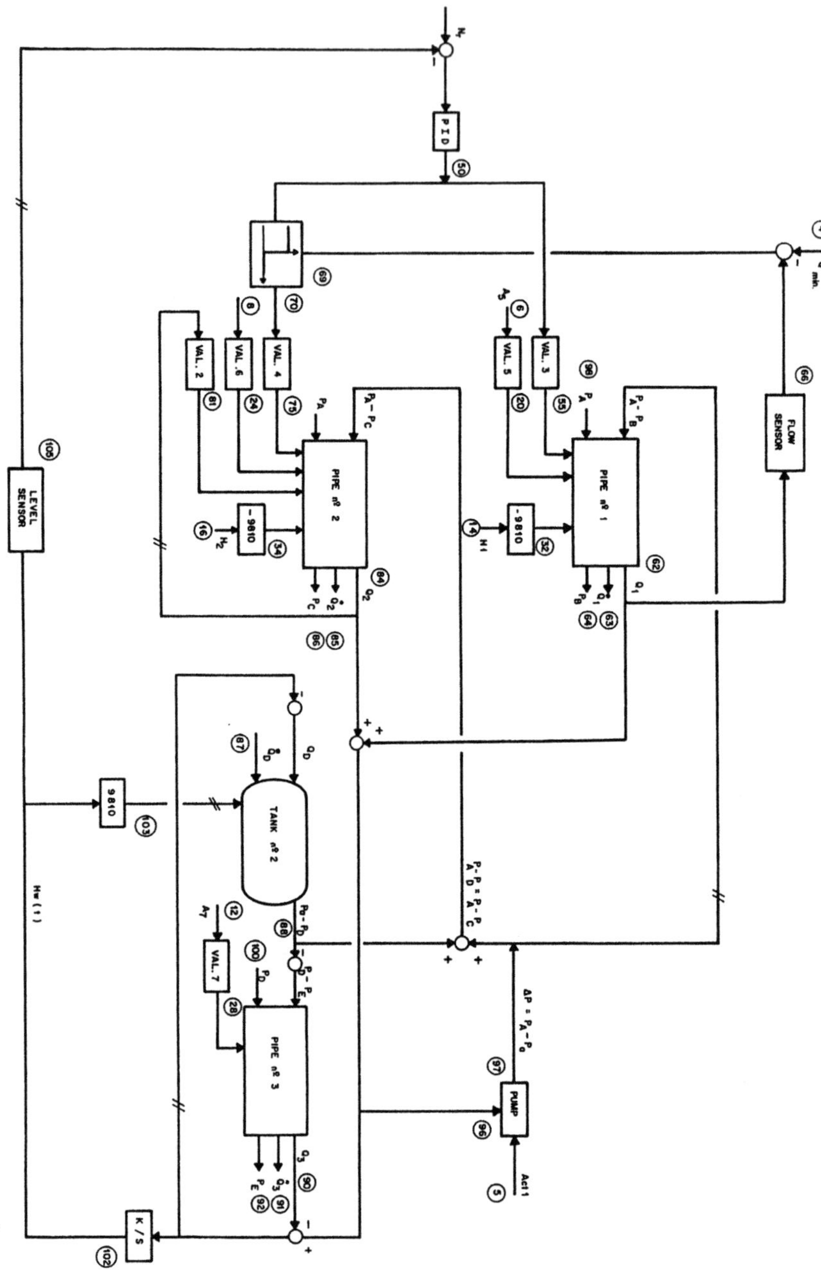

FIG.3 BLOCK DIAGRAM OF THE EXAMPLE OF FIG. 1

4. Results and discussion*

Once the joint package DYLAM-TRETA has been set up let us proceed to the assessment of the protection of the sample case in section 1.3.

4.1. SIMULATION WITHOUT PROTECTION

A first simulation is made without any protection, and for only six minutes. The system is driven to follow a normal operation, i.e. to rise water level from an initial steady state of 1.0 meter to 1.1. meters in tank nr. 2. The number of sequences analyzed was 37. Nine damage conditions were detected during these 360 sec, some of them falling outside the DBEP, clearly indicating the need for protection and the dangerous paths. In fact, only 4 sequences would not violate some DBEP in the long term.

From this first simulation four sequences have been selected. Three of them result in damage condition 1 (=overflow in tank nr. 2), while the other is representative of a passive event that may become significant. Fig. 5 a, b, c, d show water height, as given by TRETA and probabilities as given by DYLAM for these sequences.

The need for a protection against these events is now apparent. So we add a system that stops the installation and operates three new valves (A, B, C, D) in pipes nº 1, 2, 3 and tank nº 2. The new TRETA block-diagram is obtained by adding the portion of fig. 4.

FIG.4 BLOCK DIAGRAM OF THE PROTECTION SYSTEM

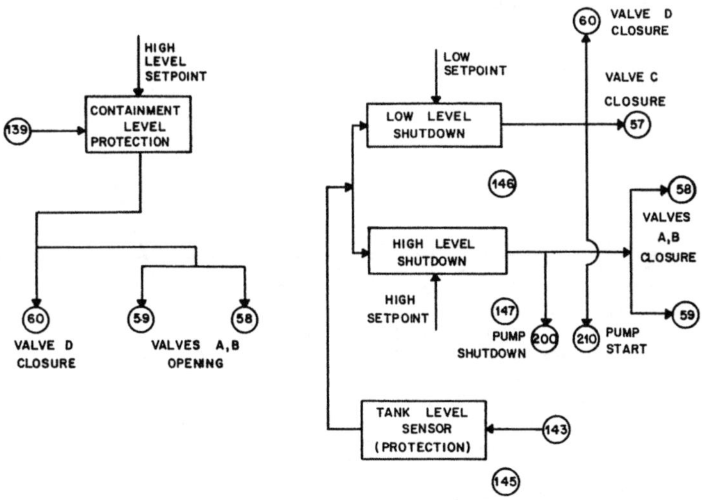

* These results were obtained with the cooperation of Ricardo Bolado and Enrique Meléndez whose contribution is greatly appreciated. Also, as shown in ref. 4, we are grateful to the "Instituto de Investigación Tecnológica del ICAI. Madrid" for help in the DYLAM-TRETA linkage.

FIG. 5A FAILURE OF LEVEL SENSOR AND CONTROL VALVE

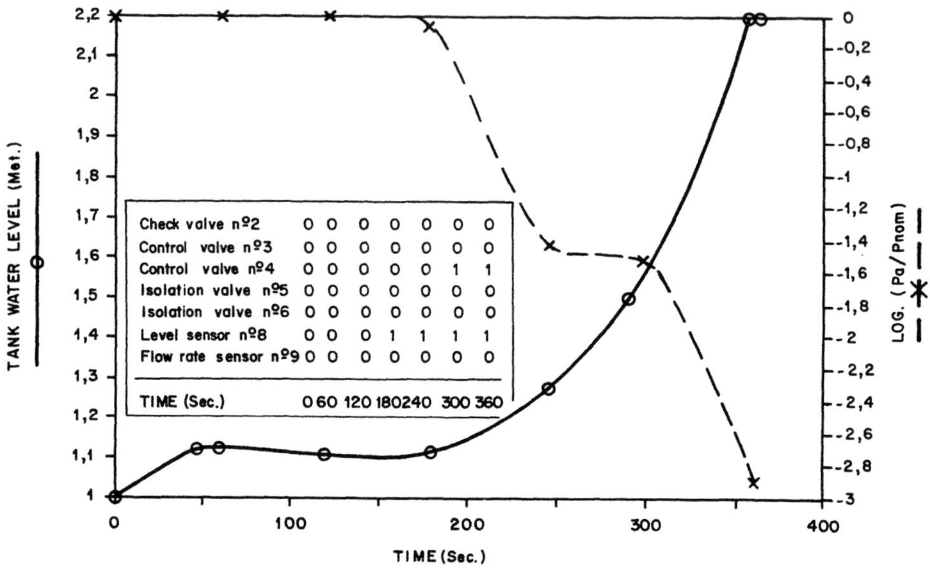

FIG. 5B FAILURE OF LEVEL AND FLOW RATE SENSORS

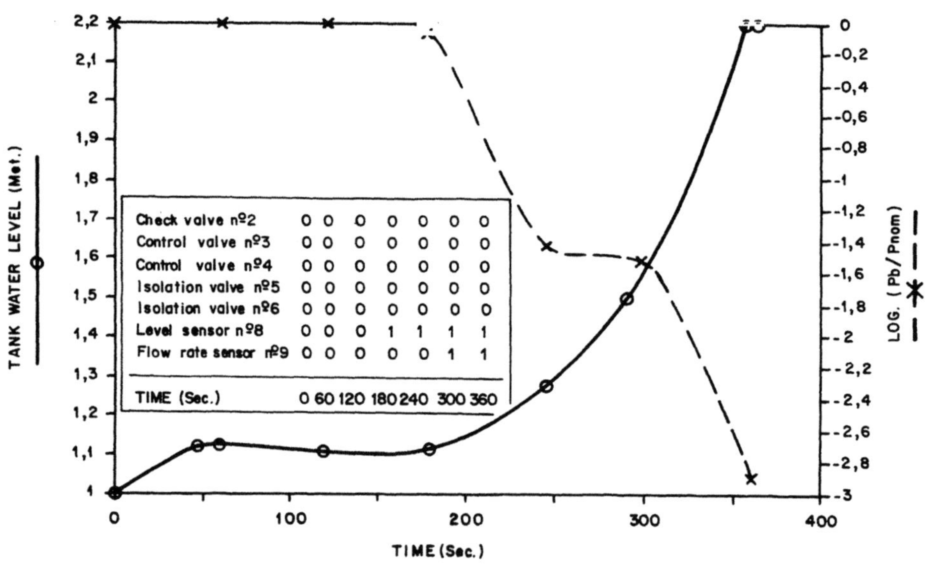

FIG. 5C FAILURE OF CONTROL VALVE Ner2

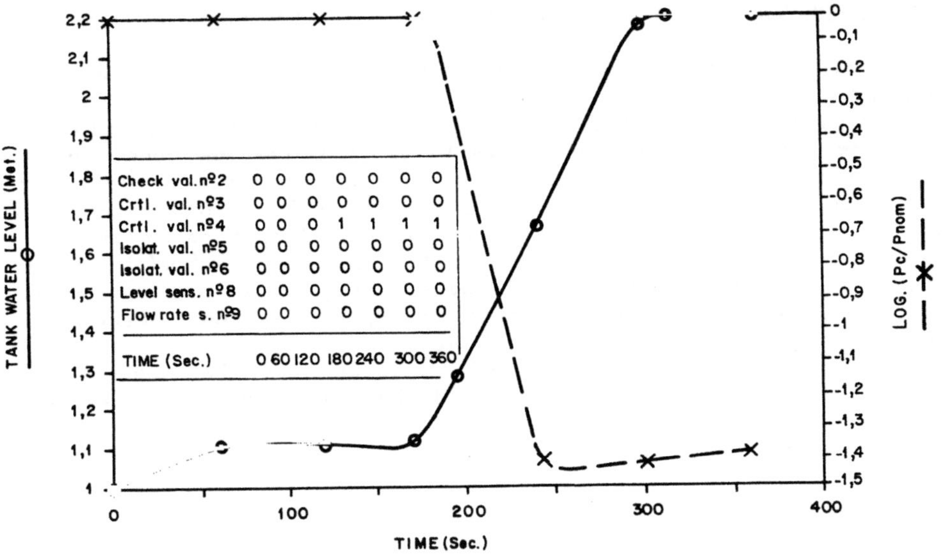

FIG. 5D CLOSURE OF VALVES Ner4 AND 3

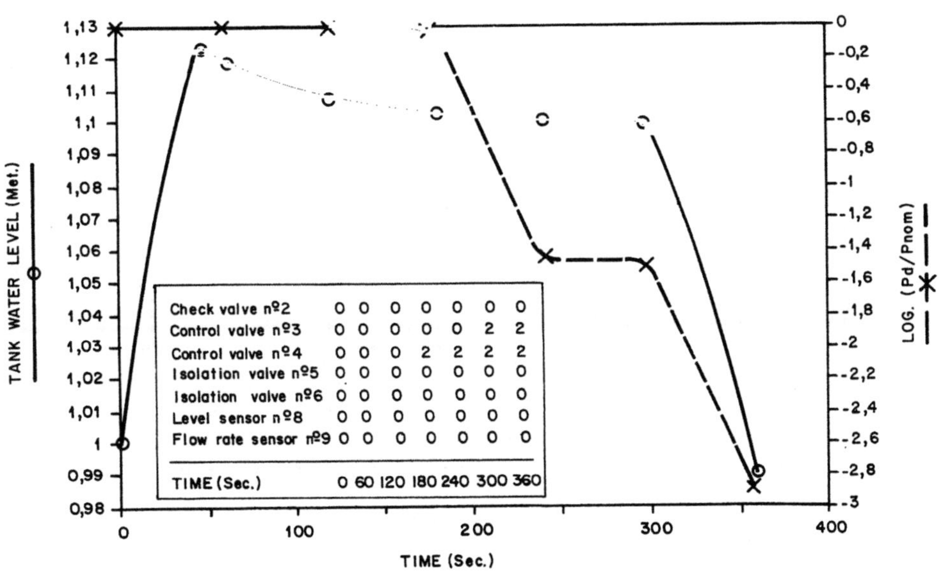

4.2. ASSESSMENT OF THE TRIP SYSTEM. NEED FOR A SAFEGUARD SYSTEM

Three different trip systems are evaluated to assess their relative quality.

In type 1 the two safety valves A, B, are simultaneously closed on a 2.1 meter level trip set point while valve C remains open. In type 2 valve B is delayed in actuation through an "integral of level error" criterion and valve C is closed, delayed by 10 seconds from valve B closure, both in order to prevent the installation stop if unnecessary. In type 3 the safety valves closing time is increased with respect to type 2 for economic reasons. Only overfilling events will be considered and it is also ascertained wether safeguard system intervention is required for this type of faults.

In Fig. 6 relevant portions of the DBEP1 for the three cases are shown, as obtained with TRETA-DYLAM dynamic event tree results. 91 sequences were obtained from which 21 resulted in significant points and are identified in the figure with the corresponding sequence of events.

FIG. 6 SIGNIFICANT SEQUENCES OF THE DYNAMIC EVENT TREE (Treta-Dylam results)

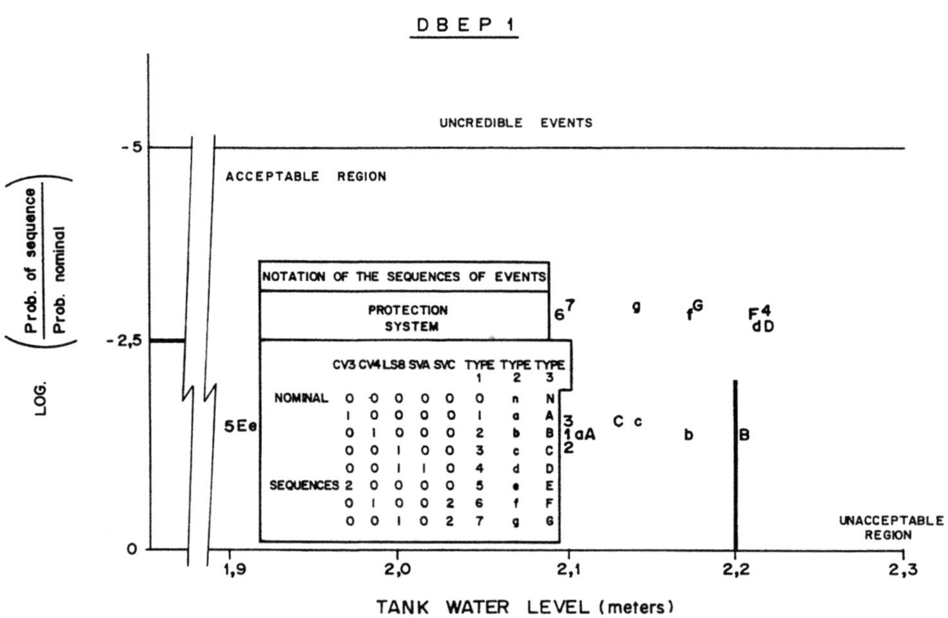

It is concluded that type 1 is safe but stops the installation too often, type 2 is excellent but more expensive than type 3 that also satisfies overfilling DBEP requirements except for one sequence (seq.B). Three sequences were identified that require safeguard intervention in any one of the three systems. In all cases the level span for normal operation is from 0.2 to 1.9 meters, resulting in little differences of limiting conditions for operation, while systems of type 2 and 3 are more sensible to details of the control system set points.

In Fig. 7 a few selected sequences are detailed in its time evolution.

FIG.7 TIME EVOLUTION OF SOME SEQUENCES

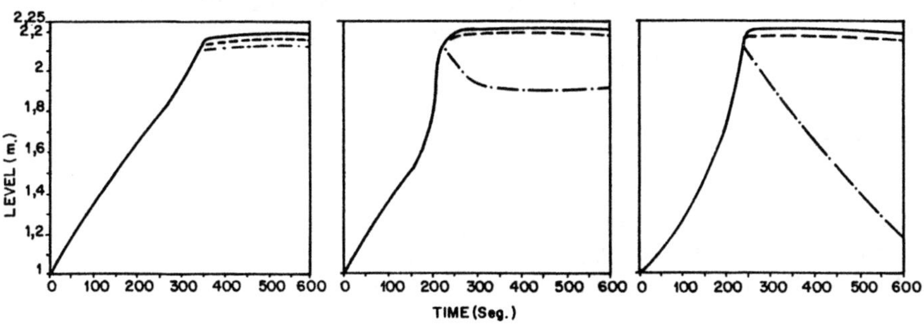

PROTECTION SYSTEM Nº 1 —·—
" " Nº 2 – – –
" " Nº 3 ——

5. Conclusions

a) The TRETA-DYLAM package is able to generate dynamic event trees. They allow the incorporation into the risk analyses of software aspects.
b) The application of this metholology to the assessment of the software of protection systems overcomes deficiencies of current techniques concerning coupling between dynamics and reliability.
c) TRETA is a good machine module for SRA purposes.

6. Acknowledgments

We are grateful to Ricardo Bolado and Enrique Meléndez for their contribution, and to Fernando Dominguez and Alfonso Fernández for their careful typing and artwork.

7. References

1.- Izquierdo, J.M. & Villadóniga, J.I.
"BASIC PHILOSOPHY OF THE NUCLEAR REGULATION"
ANS-ENS Brussels Meeting. October 1978

2.- Cacciabue, P.C., Amendola, A. & Mancini G.
"SHORT DESCRIPTION OF THE SYSTEM RESPONSE ANALISER"
Technical note NO. I.05.C1.86.25. Commission of the European Communities JRC-ISPRA. December, 1986.

3.- Unidad de Programas de Desarrollo del C.S.N.
"TRETA: UN PROGRAMA DE SIMULACION GENERAL CON APLICACION A TRANSITORIOS EN CENTRALES NUCLEARES"
Revista de la Sociedad Nuclear Española. Octubre 1987.

4.- Villa Navarro, J.R. & Sánchez Perea, M.
"FEASIBILITY STUDY OF THE LINKAGE OF TRETA COMPUTER PROGRAM WITH SRA MODELS"
Research Contract NO 3220-87-08 ED ISP E (JRC ISPRA; 1988).

5.- Amendola & Reina.
"DYLAM-1, a Safety Package for Event Sequence and Consequence Spectrum Methodology"
EUR 9224 EN. JRC-ISPRA, 1984.

6.- Amendola, Reina & Ciceri.
"Dynamic Simulation of Man-Machine Interaction in Incident Control"
JRC-ISPRA, Feb. 1985.

7.- Amendola & Reina.
"DYLAM-2: Description & How to Use"
EUR Report. JRC-ISPRA, 1986.

8.- Amendola.
"The DYLAM Approach to System Safety and Reliability Assessment"
EUR Report. JRC-ISPRA, 1987.

SECTION 3

HUMAN RELIABILITY AND UNCERTAINTY ANALYSIS

HUMAN RELIABILITY MODELS

A. AMENDOLA
Commission of the European Communities
Joint Research Centre - Institute for Systems Engineering
21020 Ispra (Va) - Italy

ABSTRACT. The lecture reviews methods for human reliability assessment. These include merely qualitative methods for the analysis of procedures and man-machine interfaces; probabilistic models based either on decomposition approaches or on statistical distributions of operator time responses or on expert judgement; and models simulating the cognitive behaviour of the operator in his decision making and actions. The cognitive models are still in a development stage. On the other hand, the results of a comparative exercise on human reliability assessment demonstrates that consistent results are very difficult to achieve by adopting current probabilistic models.

1. INTRODUCTION

Human factors are well recognized to have a dominant impact on the safe and reliable operation of technical systems: aviation, process and power plants, drilling and mining, informatic systems, etc. Human errors may affect design and procedures, as well as decisions or actions for controlling transients or in reacting to perturbations and hardware failures. Even any maintenance policy must be opportunely studied in order that test and repair actions, when erroneously executed, do not leave systems in dangerous conditions which, when not detected and restored, reveal their adverse effects during operation. By provoking common cause failures /1/ they may have the capacity to nullify redundancies and multiple defences built-in in a system, and, therefore, may be responsible of major disasters. Therefore, human factors deserve an accurate analysis in availability and reliability assessment.

70% of aviation accidents are provoked by human malfunctions, similar figures apply to the shipping and chemical industries /2/. In the first nuclear power reactor risk study (the well known WASH-1400 report) the accident risk imputable to human errors was estimated to range between 50 and 60%. The patterns of major accidents occurred afterwards confirmed the dominant impact of human malfunctions, but they revealed new aspects, not sufficiently considered in the past.

A. G. Colombo and A. Saiz de Bustamante (eds.), Systems Reliability Assessment, 207–230.

Indeed, in addition to human fallibility against which defences can be built in a system by appropriate man-machine interfaces, operator procedures, personnel training and motivation, etc., there are other factors which heavily affect safety and reliability, and which are imputable to humans being voluntary actors rather than executors of more or less complicated functions. These factors are not covered by the human reliability models which are the subject of the present lecture. For instance, sabotage is not included normally in a risk and reliability assessment. Defences against sabotage might be the subject of other kinds of analysis (safeguards, security, etc.). Also the constraints imposed by the safety regulations might be voluntarily violated because of the will of achieving a contrasting objective (to perform at any cost an experiment: Chernobyl; to decrease operational costs: Bhopal; to reach the prefixed destination even if a driver is aware of some failures in his vehicle, etc.). Even violations are normally not included into risk and reliability assessments. They depend heavily on general management and organizational factors /2/, whose assessment is becoming an urgent issue to which, after the Bhopal and Chernobyl accidents, several international bodies are giving an increased attention /3/.

Apart from violations of specific safety regulations, a management not sufficiently attentive to the overall safety problem is the primary cause of a general relaxation with respect to the safety rules which results in ignoring "possible precursors" of major incidents; in omitting communication of significant information; in misdocumentation of changes and the reasons for which; in allowing working on the site without instructing on possible hazards, etc. In other words, mismanagement can be considered to be the most important "common cause failure"; on the other hand, it "is the most problematic and least developed issue from a risk and reliability viewpoint" /2/.

The above described limitations must be kept in mind when judging about the numerical results of any system reliability assessment. But even after the exclusion of these topics, assessing human reliability or better the reliability of the complex man-machine system is a very troublesome task; human failure rate is highly depending on situations and, therefore, useful statistical data are hard to be found, if not for very simple tasks (i.e. response to well defined stimuli like the actions performed at an assembling chain); on the other hand, predictive models are hard to be consistently used and doubts exist on their validity /4,5/.

In reality, controversial discussions continue to exist between human factor professional psychologists and reliability engineers about the possibility to get useful data: "Programs that have attempted to obtain Human Error Probabilities (HEPs) have met with only limited success. Improvement of this deficiency would be tremendously expensive, and validation of the resulting HEPs is, while theoretically possible, practically infeasible" /6/. Only recently some models to overcome the gap between cognitive psychology and reliability modelling have been proposed: they attempt, however, to simulate operator's decision making and actions in his response to a changing work environment, and, at the moment, do not allow probability assignments.

This paper briefly reviews proposed models and open problems and, therefore, can be a useful introduction to the various topics implied in human reliability assessment. For the same purposes other useful papers are, among others, those quoted at references /2 and 5-10/.

2. DECISION MAKING, ACTIONS AND ERRORS

Human errors may occur in any phase of the design, construction and operation of a plant. Design and construction errors are also responsible for many kinds of "dependent failures" /1/. They may be discovered in the commissioning period or during system operation. Normally, quality assurance programs are designed and implemented to minimize the occurrence of such errors.

This lecture is especially devoted to the analysis of human reliability during plant/system operation, where humans are requested to maintain, supervise and control the technical systems. Two main aspects must be distinguished /11/. On the one hand, operators must perform test and maintenance actions at prefixed time intervals (in this case they have normally to follow well defined procedures), or identify failures and thereafter repair the failed system. In the second case, operators must perform a diagnosis before starting actions. In some cases they must act under time constraints. In general, either the concerned items are disconnected from the process or the tasks do not affect the process.

On the other hand, operators are demanded to supervise the correct behaviour of the plant during normal steady-state operation, to supervise or to manually control start-up or shut-down transients, to diagnose the occurrence of dangerous incidents, to supervise the behaviour of automatic control systems, and/or to intervene to control and to mitigate the accident. In this case the operator is a part of a dynamical control chain /12/ (see Fig.1) and the accident dynamic

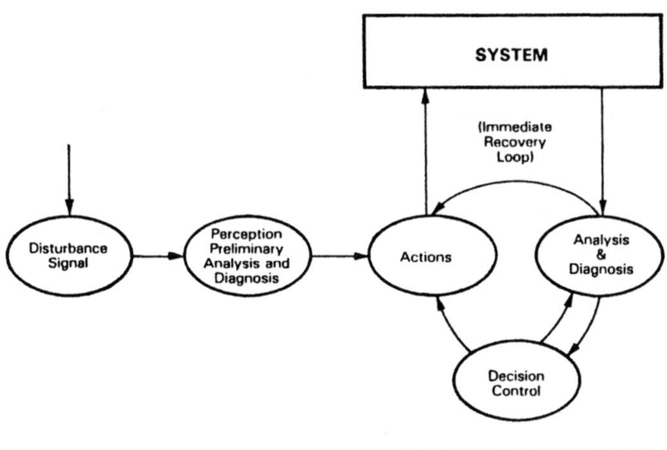

Figure 1. Dynamic man-machine interaction model

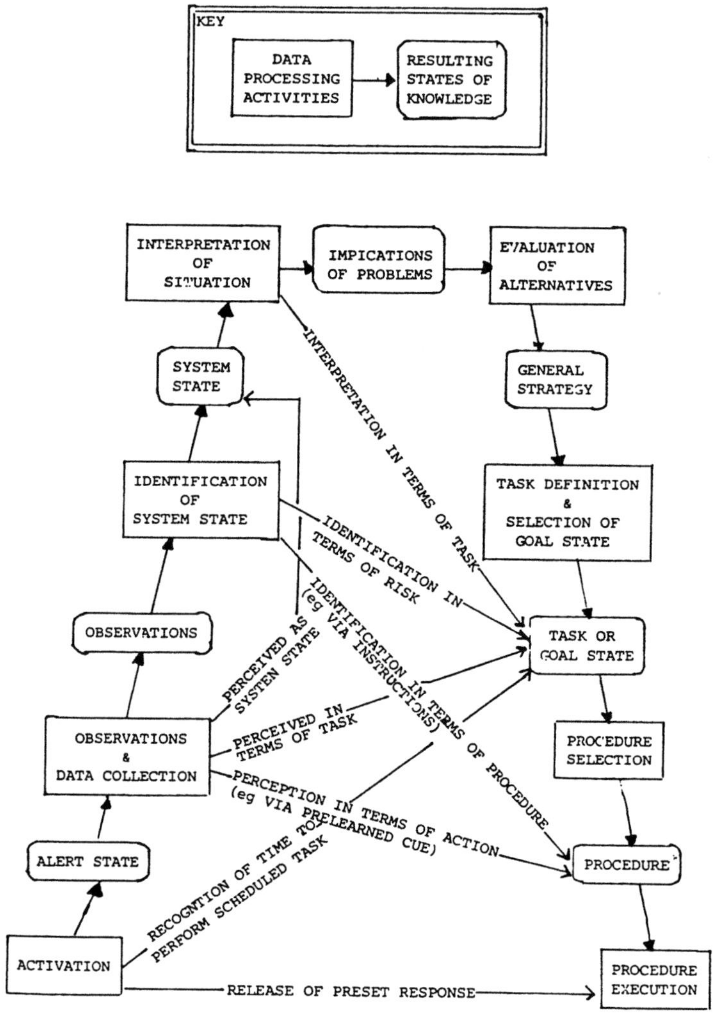

Figure 2. Main decisional model from Rasmussen.

determines the timing of the events. Time becomes a stringent constraint. Timing changes are also provoked by the actions of the operator himself. Man cannot be studied separately from the system and vice-versa. The "metasystem" of concern is the very man-machine system.

Fundamental contributions for understanding the role of the operator in controlling a process have been derived from cognitive psychology /13/. Cognitive psychology "emphasizes three major characteristics which distinguish it from behaviourism (dominating approach before the 1950s):

- It emphasizes "knowing" rather than merely responding as in the behaviouristics stimulus-response bonds;
- It emphasizes mental structure/organization which is inherent in all living creatures and provides an important impetus to cognitive development;
- the individual is viewed as being active, constructive and planful rather than a passive recipient of environmental stimulation" /2/.

In this direction, after the establishment of a multifacetted taxonomy for description and analysis of events involving human malfunction, models have been proposed which are very useful to better understand the complex man-machine interaction process.

The well known model proposed by J. Rasmussen is depicted in Fig.2. The lowest level represents the "skill-based behaviour" on which immediate responses to stimuli can be situated. The intermediate level represents the "rule-based" behaviour, that is the application of series of codified procedures. The highest level is the cognitive behaviour: the operator, before taking any action, must diagnose the situation and this occurs according to the problem solving ability of human mind and is based on the deep knowledge that man has on the system under his control its multiple failure modes and available recovery possibilities, etc. It is because of this ability that man is supposed to act in unforeseen situations. These cannot be resolved by automatic systems which, in principle, might substitute man for the lower level decisional tasks. Expert systems might be designed as decision aids but man is always asked to supervise even the most sophisticated computer control systems; in this sense any new advanced control tool shifts the role of the operator towards "knowledge-based" behaviour.

Associated to his decisional model Rasmussen identifies possible malfunctions as in Fig.3. In this scheme, other definitions of human error types /16,17,18/ such as slips and mistakes, can be easily situated. Reason defines slips "as errors which result from some failure in the execution stage of an action sequence, regardless of whether the plan which should have guided them was adequate to achieve its objective"; whereas a mistake "is defined as a deficiency failure in the judgemental and/or inferential processes involved in the selection of an objective or in the specification of the means to achieve it, irrespective of whether or not the actions directed by this decision-scheme run according to plan". Thus, the mistakes arise at a knowledge level, whereas slips occur during rule or skill-based

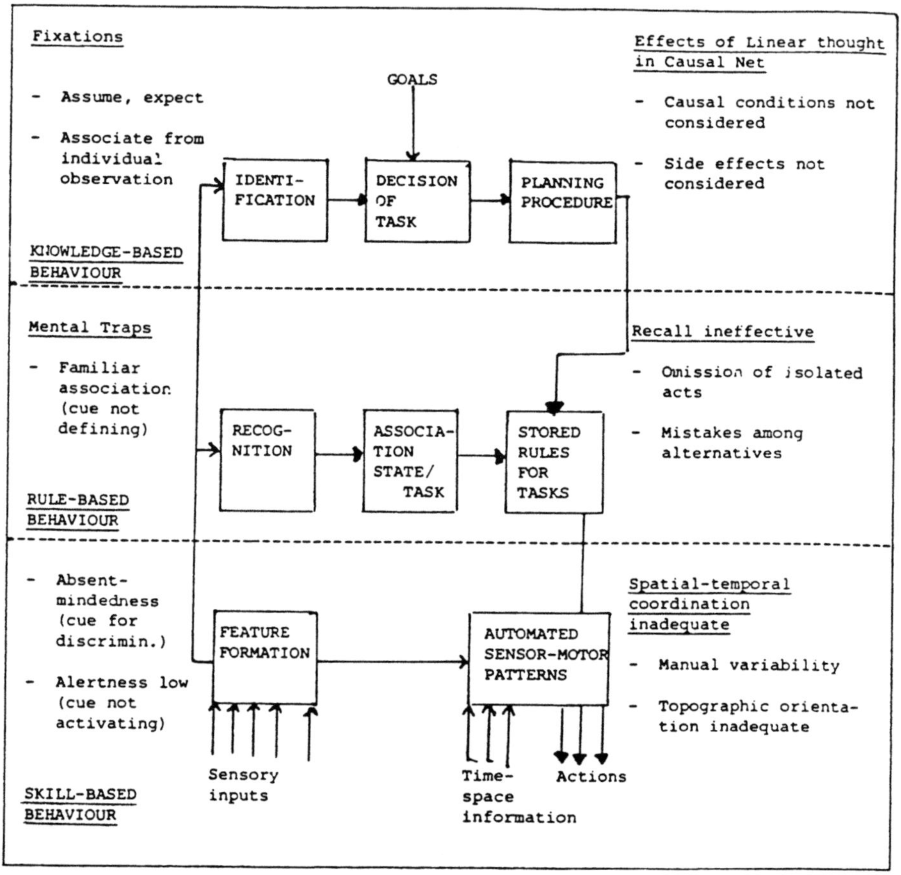

Figure 3. Typical human malfunctions (Rasmussen, 1980).

execution tasks.

One of the most important mechanisms for a mistake is the "fixation". An operator is familiar with certain patterns of transients; if alarms signal that something is wrong, the operator tends to select the most familiar situations according to a frequency gambling, and afterwards to choose among the symptoms those which confirm his expectation. This mechanism can impede recovery from an early diagnosis error. Putting the operator in very unfamiliar situations, simulation of rare events by means of training simulators is the way to educate the operator as a problem solver. Decision support systems are being more and more introduced to aid the operator in his most difficult tasks, giving him a rapid access to knowledge bodies, which should avoid fixation-provoked mistakes /19/.

3. STRUCTURED APPROACHES FOR THE QUALITATIVE ASSESSMENT OF MAN-MACHINE SYSTEMS

For nuclear power plants, aeronautics and, to a lesser extent, in the process industry, full-scale simulators are being employed to train operators but also to check the adequacy of "hard" and "soft" man-machine interfaces. There are, however, several proposed systematic procedures that can be used to assess adequacy of design and to identify weak points and possible human error events. The careful application of such procedures before any reliability modelling and assessment has been recommended by the results of international comparative projects on risk assessment (Benchmark exercises) /4,5,20/. Fig.4 shows a modified FMEA format suggested from the exercise at ref./20/ in order to qualitatively analyse man-machine systems in incident situations. Examples of task analysis for chemical processes can be found in ref./2/. In performing such kinds of qualitative analysis, care must be given to the identifications of the actions which might lead to "common cause failures".

4. RELIABILITY MODELLING

The usual methods of fault trees/event trees have been adapted to model human malfunctions in logical diagrams in which they appear together with the other usual component failures. These methods are useful to assess the sensitivity of the system to certain errors; they suffer, however, from significant limitations since, after the decomposition of a complex system into single parts, it is very difficult to take adequately into account the dependency structures which exist especially in the different actions of man (see considerations on planning and fixation at item 2). Furthermore, the timing of the dynamic interaction in Fig.1 can hardly be modelled by static diagrams like fault trees or event trees. Fig.5 shows an example /11/ of a fault tree model for an undesired top event, such as the unavailability of a component as a result of operations performed on the component during testing or maintenance or during operation tasks; the task is thus subdivided into steps related to actions on components. Failures leading to the undesired event can be divided

System	State (etc.)							
Event requiring intervention	Event detection	Demand required to the operator	Failure mode	Failure cause	Components concerned by the error	Effects on the system	Potential for self-correction	Recovery actions
	(signals demanding operator to act)	(It must be resolved into the single tasks and prescribed timings)					(Changes in the signals from which operator may be aware of his error)	(How to recover and time available for recovery)

Figure 4. Format for qualitative analysis of operator tasks.

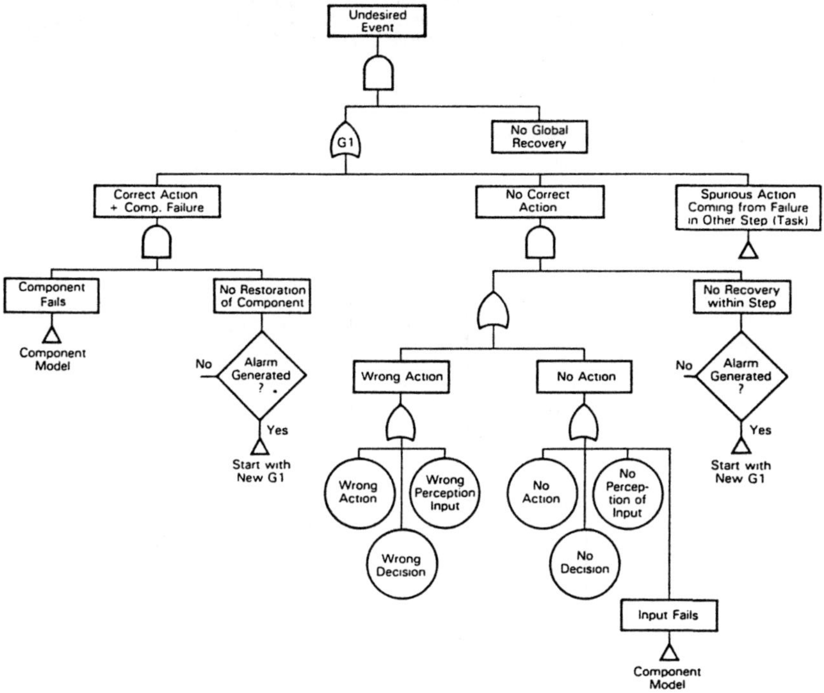

Figure 5. Static model fault tree.

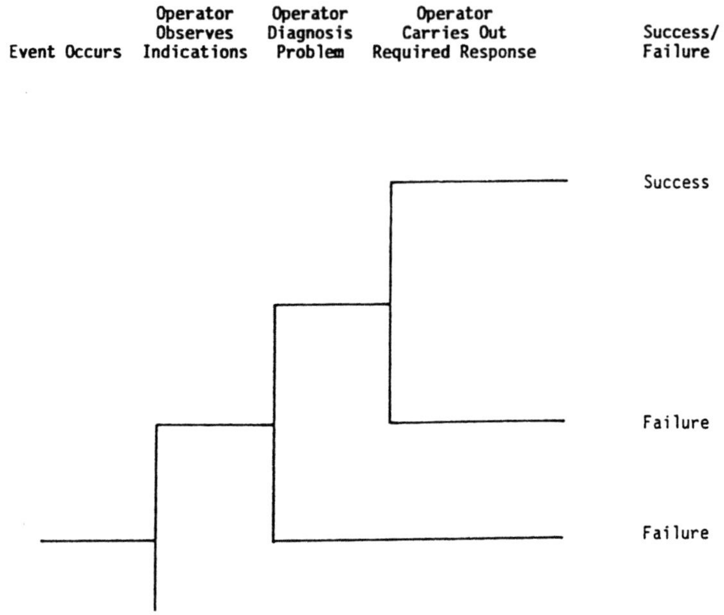

Figure 6. Operator action tree.

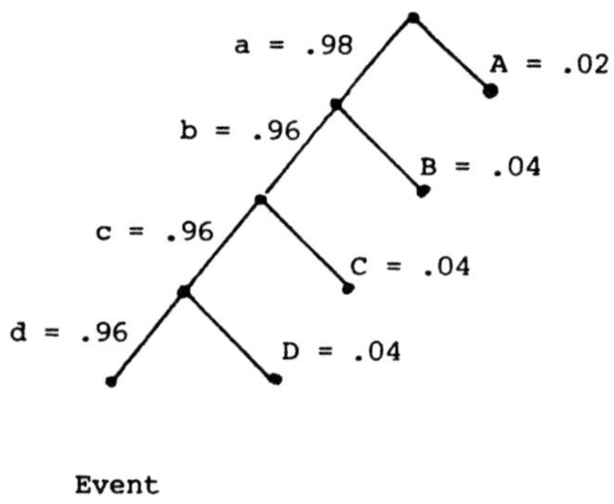

A = Control room operator omits ordering the following tasks

B = Operator omits verifying the position of MU-13

C = Operator omits verifying/opening the DH valves

D = Operator omits isolating the DH rooms

Figure 7. HRA event tree.

into three types:

1) The combination of a correct action on the component with a failure of the latter (e.g. valve correctly commanded to open by remote control, failing because of a fault in the actuator). If the failed state can be observed by an indicator, a new action (new fault tree) is initiated by the operator; he must perceive the signal, make a decision and restore the component.
2) An incorrect action or an omission. In this case also an indicator may signal the fault and lead to recovery.
3) A spurious action on the component originated by a correct action on a "wrong" component in the same or in a different task. This event represents a link between actions performed on different components.

Recovery actions are also of three types:

1) Originated by the existence of indicators and already described above.
2) Global. These recovery actions are still performed at the component level and may be triggered by the carry-over of another action on the same component, but in a different step of the procedure and by the carry-over of an action on another component which reveals the incorrect state of the component under consideration. In the first case the incorrect state of the component will be described by an OR gate of a series of AND gates, each of these describing the failure of the component in one step of the procedure and the lack of recovery in the other relevant steps.
3) General. This recovery refers to actions at the task level by which the full operation of the system(s) or of the subsystem(s) is tested.

In general, it has been suggested /21,22/ to identify explicitly the human induced causes of failures (e.g. maintenance error), even when the failure rate attached to the component failure mode already includes such failure causes. In this way an adequate attention can be focused on the human-induced weak points, whereas in the probability assignment double-accounting of maintenance errors can be avoided.

A similarly constructed logical tree has been proposed to model and assess management explicitly from the risk viewpoint: the Management and Oversight and Risk Tree (MORT) /23/.

Models based on event trees are also widely used, such as the operator action tree /24/, which focuses on decision making (see Fig. 6) and the HRA trees (Human Reliability Analysis, also known as THERP) /25/ (see Fig. 7), which provide logical diagrams for analysing expected and faulty operator actions mainly based on procedures.

5. RELIABILITY ASSESSMENT

After having performed qualitative analysis and modelling, steps which give already a strong insight into the quality of the man-machine system, many times a success probability figure is wished for the human actions to be integrated into the overall reliability/risk assessment procedure. The methods used for estimating human

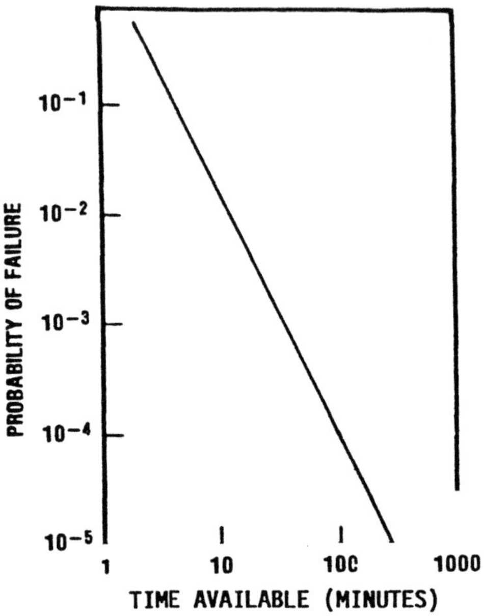

Figure 8. Failure probability versus time for operators /26/.

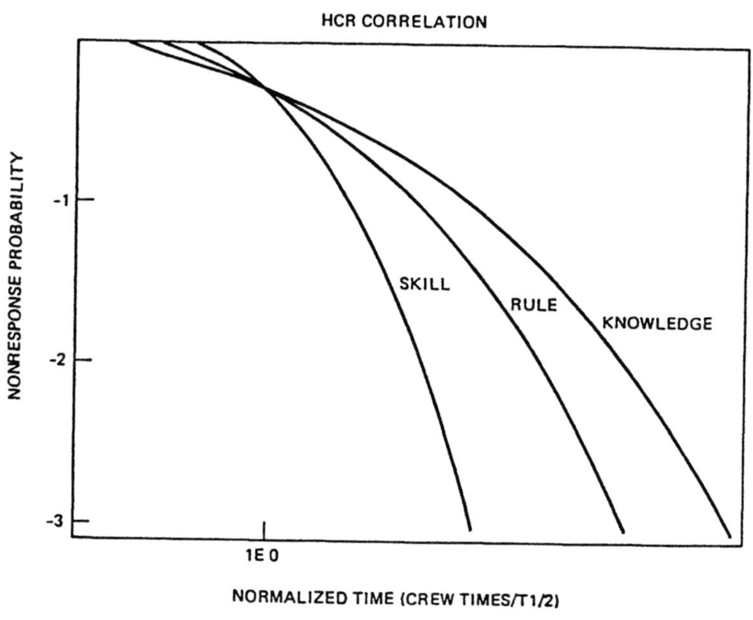

Figure 9. Normalized crew non-response curves for skill, rule and knowledge-based cognitive processing.

probabilities are basically different according to whether the tasks are classified as manipulative/procedural or cognitive/diagnostic.

In the first case, where time is not a main constraint, the data from the Swain and Guttmann's handbook /25/ are normally used to assign probbilities in HRA trees. The data are given for a very wide class of single tasks. The HEPs (Human Error Probabilities) given in the handbook should then be adjusted by so-called PSF (Performance Shaping Factors) in order to take into account differences between the standard case and actual situations (e.g. stress, poor training, bad management, etc.). Since the very detailed contents of the handbook is pratically the only source of data, these are widely used. They were derived from different sources, even from military records.

Error probability as a function of time has also been introduced in the handbook after recognition that the time constraint is one of the dominant factors for the success of diagnosis-planning-actions during incident control. In such cases the operator trees have been preferred /21/. To quantify the events in OAT (like that of Fig.6), at first simple failure probability curves against time available have been proposed /26/ (see Fig.8). Afterwards a more sophisticated model has been proposed by distinguishing between the decision making levels proposed by J. Rasmussen (Fig.2). The probability/time curve has been linked to the kind of involved behaviour (HCR - Human Cognitive Reliability Model - see Fig.9 /27/).

One distribution is given for each of Rasmussen's three modes of behaviour, i.e. knowledge, rule and skill-based. Analytically, the curves are expressed by a Weibull distribution:

$$P(t) = e^{-\left\{\frac{\frac{t}{T_{\frac{1}{2}}} - c_{\gamma j}}{c_{\mu j}}\right\}^{\beta_j}} \tag{1}$$

where:

$T_{\frac{1}{2}}$ = median time taken by the crew to complete the task (this time can be corrected by performance shaping factors);

$\beta_j, c_{\gamma j}, c_{\mu j}$ = coefficients associated with the mental processing of type j, which can be calibrated by using simulator data.

In Fig.9, time is normalized by dividing the actual time by the median time.

Other models are based not on data but on structuring expert judgement. In this procedure three main steps can be distinguished:

a) The identification of the relevant PSFs (in the Seabrook Nuclear Plant Risk Assessment /28/ the two principal performance-shaping factors were judged to be the time constraint and the potential for misdiagnosis. This last is a very important aspect which can be assessed via the so-called confusion matrices: these identify symptoms in common to two or more possible incidents, which therefore might lead the operator to confuse one incident with another one).

IDEAL POINT ELICITATION

Thinking only about the performance shaping factor below , what position on the scale would be IDEAL for a course of action in the present context?

LITTLE TIME

1

2

3

4

5 to

6

7

8

9

MUCH TIME

Your best possible action value is : 7

Is this rating OK? Y

Figure 10. SLIM rating of tasks (example).

b) The development of a numerical rating of the influence of each PSF on each action.
c) The conversion of these ratings in probability figures.

At this aim a well structured approach has been proposed by D. Embrey /29,30/: in the SLIM (Success Likelihood Index Methodology) experts are elicitated to identify the relevant PSFs, to name end points for each factor (e.g. a scale between 1 and 9), and to define the "ideal" points in this scale; i.e. the points at which human performance is judged to be optimal. An example of a possible result from such elicitation process is presented in Fig.10.

An overall SLI is given by summing up the R_{ij} rating of task j on the i-th PSF multiplied by a weight W_i attributed to the i-th PSF, according to the formula:

$$SLI_j = \sum_i W_i R_{ij} \qquad (2)$$

Conversion from each index to (P_f) error probability is made by:

$$\log(P_f) = a\ SLI + b \qquad (3)$$

where a and b might be determined by known anchor points, i.e. events for which reliable data are available.

Recently, a "logistic model" /31/ has been proposed for transfering this SLIM index into probabilities in order to avoid some possible numerical inconsistencies (like probabilities less than 0, etc.) and giving a more reasonable behaviour of probabilities vs SLI:

$$\log(1/P_s - 1) = \log(a) + b\ SLI \qquad (4)$$

by indicating this time with P_s the success probability.

Eq.(4) just changes the shape of the probability curve in a reasonable way, but basically does not change the philosophy of the approach.

6. LIMITATIONS OF THE PROBABILISTIC EVALUATIONS

The results obtainable by the previously described methods have been submitted to several kinds of criticism. Psychologists tend, in general, to stress the cognitive aspects of human behaviour and, by criticizing the engineering models, refuse to accept the idea that statistical data can be obtained and used /6/. Engineers find the use of PSFs, but also of the data and models to be difficult because of non univocal results obtainable. Criticisms have been also expressed with regard to the correctness of the probabilistic approach implied /32/.

In reality, a recent "experiment", carried out under CEC-JRC coordination, has demonstrated how much results obtained by different analysts for a same study case may differ, even when based on same procedures and data /4,5/. (The following summary follows the corresponding content of ref./33/). A Human Factors Reliability

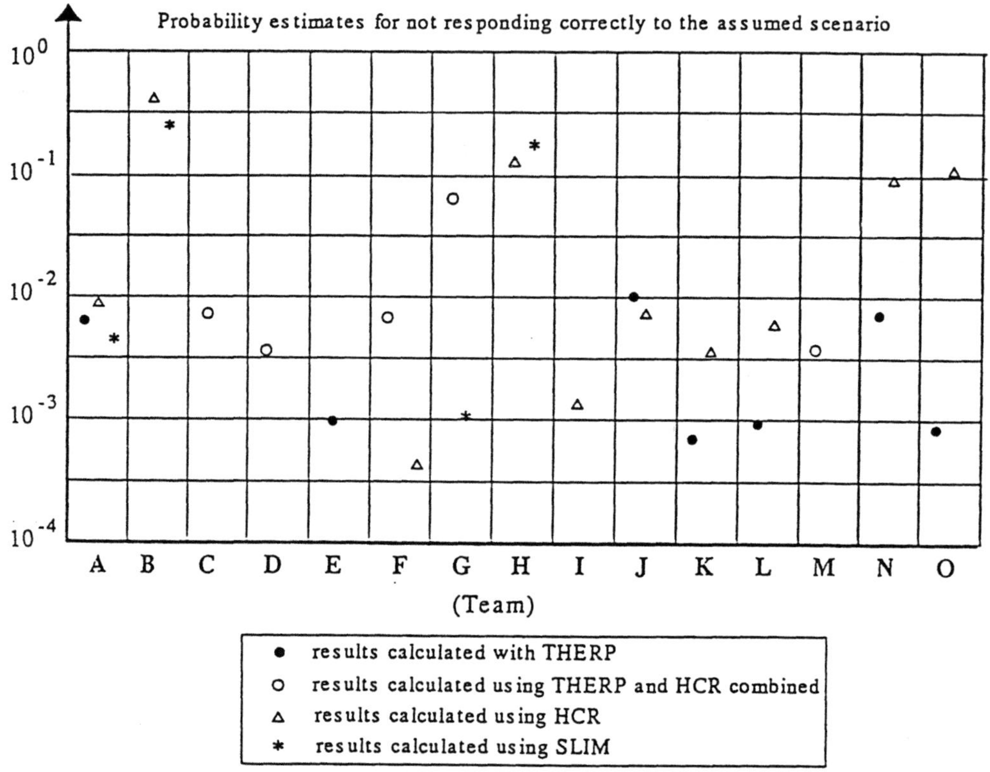

Figure 11. Summary of probabilistic results for the transient study case.

Benchmark Exercise (HF-RBE) has been organized into which 14 expert teams from European Organizations plus a USA team (NUS sponsored by EPRI and NRC) have participated with the aim to:

1) compare the various procedures and approaches used to identify human error mechanisms;
2) compare the modelling techniques and data used;
3) compare the probabilistic techniques and assess the degree of consistency in the results.

The HF-RBE basically addressed two different problems:

1) The analysis of routine functional test and maintenance procedures with the aim to assess the likelihood of test induced system unavailability, the likelihood of failures to remain undetected during the test and the possibility of test induced transients.
2) The analysis of human actions during interventions in case of an operational transient with the aim of assessing the likelihood that the operators will fail to correctly diagnose the situation and take the proper corrective measures. This study involved analysis of cognitive errors (diagnosis and selection of reponse strategy) and of actions to be performed under given time constraint.

For the first problem a study case was designed referring to the test procedures for the emergency feedwater system of the Grohnde Nuclear Power Station. The workplan consisted of: a phase during which the key human interactions leading to test caused unavailability,' transient initiation or test inefficiency were identified, modelled and quantified; and a phase during which the models and quantification techniques were compared on the basis of reference human interactions. For the second problem, the operational transient study case was designed. It consisted of the analysis of the operator response to a particularly rare incident (a loss of power combined with a partial loss of the emergency feedwater system). The spread of the results for the operational transient study case are summarized in Fig.11. As Fig.11 shows, very significant spread in the human error probability estimates was observed. The reasons of this spread can again be mostly attributed to modelling and assumptions. The application of a same data base (THERP /25/) for a well defined HRA tree, including selection of data and adaptation to the case under analysis (PSFs), has been shown to lead to a spread of about a factor of 10. The following important conclusions could be formulated:

1) The major contributions to the analyst to analyst variability are linked with differences in breakdown and modelling: different assumptions used implicitly during the modelling, different level of decomposition during action sequence breakdown and subsequent lumping of actions on the model, different ways of including recovery and dependency structures in the model.
2) A detailed decomposition may appear to be more complete but it may fail to catch the real important error and error recovery mechanisms. Especially when using a large, detailed human error data base (such as in THERP) there is a risk of modelling only those mechanisms which appear in the data base and miss other important mechanisms, whichever detailed and complex the decomposition is made. It was interesting to note that the PSFs

which were given the highest importance during the application of the SLIM approach were not considered in the application of THERP because there were no provisions made in THERP for them!

3) Within a team, results obtained with different methods agree rather well. In most cases THERP gives the most optimistic error probability estimates. The agreement of the results can be due to the fact that in many cases data from the THERP approach are used, implicitly or explicitly, in a calibration procedure inherent to the SLIM approach and to which the SLIM results are very sensitive.
4) Action sequences involving cognitive as well as more procedural actions were shown to be difficult to model and to assess. For more cognitive tasks, the HCR method is applied. Combining HCR with other methods like HRA trees for typical routine intervention tasks presents difficulties:
   a) if the sequence is decomposed into actions to be assessed with HCR and others to be assessed with other methods, there is a problem in the allocation of time windows to different actions of a sequence subject to a shared time constraint;
   b) if the sequence is assessed in a holistic way or parts of the sequence are lumped together, there is a problem in estimating the median times needed by HCR, for the lumped sets of actions and to attribute them the appropriate decision model behaviour (skill, knowledge, rule-based).

After the results of the HF Benchmark project it appears that the traditional system decomposition approach is not adequate to get consistent probability assignments for man-machine systems. Of course, such kind of modelling keeps its validity when directed to identify weak points in the process interface with the operator and/or in the procedures.

As far as probability assignments are concerned, the SLIM approach might be the most consistent one especially when directed to judge a whole macroevent (i.e. when no task decomposition is performed). Indeed, it does not pretend to predict "objective" estimates. On the contrary, it attempts to structure expert judgement which is well in line with Bayesian inference theory /34/.

7. DYNAMIC MODELLING OF COGNITIVE PROCESSES IN MAN-MACHINE INTERACTIONS

The difficulties already discussed in breaking down operator behaviour in a sequence of different tasks can be better understood when seen from a psychological viewpoint: the sequence of actions is the result of a previously developed planning process which, in turn, is the consequence of a diagnosis of the situation. This implies that the operator's primary intentions and different internal representations of the system, in terms of structural and functional schemes, have to be taken into consideration in order to evaluate the overall process of decision-action performed by an operator of any plant or system.

The analysis of operator errors during a man-machine interaction event must be concerned with the existence of these schemes, accounting not only for the executional aspect of errors (slip) but also and substantially for their activation (mistakes) at cognitive level. The tendency should be to evaluate cognitive processes which may introduce a "dependence structure" among the single actions as well as quantifying the external manifestations, or slips, made in performing the intended plan. Consequently, human reliability analysis in man-machine systems can be improved by complementing classical reliability models with such detailed cognitive models /9/.

This promising line of research started at the JRC with some pioneer papers /11,12/ and is being now developed by a multidisciplinary team involving system analysts, social scientists and artificial intelligence experts /35-39/. A short description of the basic model /35/ is given in the following.

The human model under development is framed within a structure that includes:

- a dynamic stochastic process or machine model achieved by adopting the DYLAM technique /40/;
- an operator model as interactive counterpart (Fig.12);
- a "driver" or controller of the generation of stochastic incidental sequences according to the DYLAM technique /40/.

The human model is constructed in such a way that the complementary aspects of human behaviour are considered:

- high-level decision making (HLDM), such as situation assessment, planning, plan modification, etc.;
- low-level decision making (LLDM), where the actual relationship amongst intention-execution-plant response is implemented;
- errors, which account for slips or mistakes being made at any level of the human activity, depending on a variety of factors, including environmental conditions, internal beliefs, fixations, stress, etc.

In practice the human simulative model really interacts with the system according to the recognition of a fault situation as detectable from the process variables to be supervised and controlled (pressure, temperature, flow rate, etc.). The time change of the process variables are calculated in a self-contained approach by the DYLAM technique /40/.

A preliminary model of formulation of diagnosis has been created which is based on a tree like structure, whereby a first diagnosis is initially performed by the operator in an inductive way (bottom-up dynamics) on the basis of the data coming from the machine. As soon as a minimum number of data has been gathered, the model is capable to work its way up the tree to the formulation of a first diagnosis. This activation of a set of hypothetical schemes relies essentially on some fundamental cognitive primitives such as: similarity-matching and frequency gambling /41/. Then the tree structure is re-examined in a deductive way (top-down), with the goal to verify the first diagnosis, which was based on a limited and possibly incomplete set of data. The successive planning leads to the identification and selection of actions or procedures to be carried out on the plant itself. Starting with a main intention or "top-goal", the result of this process is the construction of a hierarchical network where the different sub-goals and sub-tasks to be performed by the operator are schematically

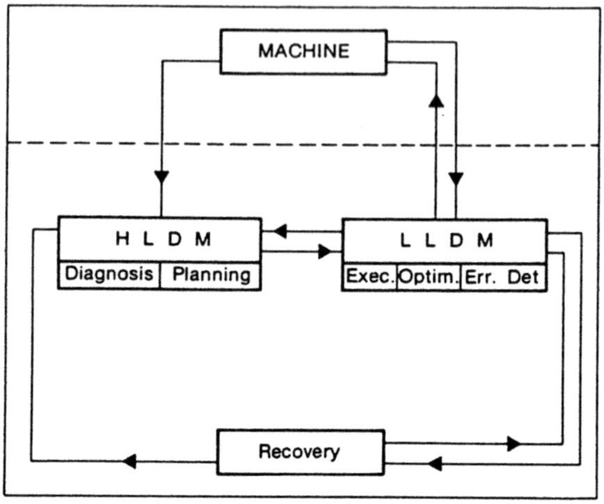

Figure 12. Conceptual flow chart of the human model.

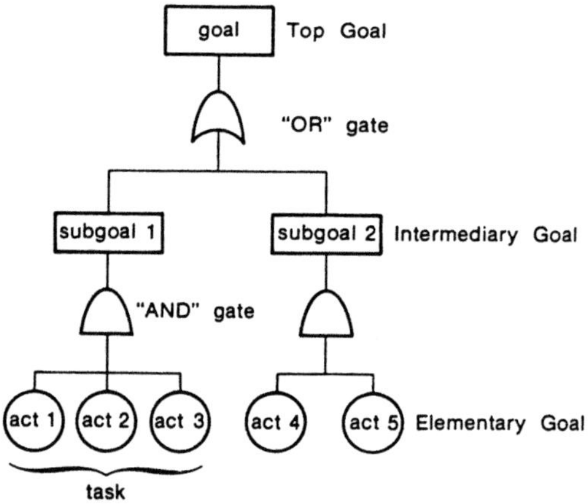

Figure 13. Goal, tasks and act concepts.

arranged in a "tree" type structure and are linked to each other by different gates "AND" and "OR" (Fig.13). A "goal" is an element of the network at any level. An "act" is the last elementary goal of the network, i.e. the action that the operator will perform such as, for instance, "close a valve", "stop a pump", etc. A "task" is the set of acts the operator has to perform in order to attain a certain goal at any level. Uncertainties in the operator's understanding of the situation, the achievement of a goal, etc., are expressed using fuzzy measurement of uncertainties /42/.

According to this integrated approach a man-machine system can be investigated in its dynamic interactions giving an adequate account of the dependency structures that cognitive psychology has identified to dominate human error mechanisms.

8. CONCLUSIONS

The paper has critically reviewed human reliability models directed essentially to assign success/failure probabilities to human actions and has shown the large uncertainties involved in such procedures. It has also indicated a new promising line of research which is able to produce useful results for qualitative identification of dynamic man-machine systems. These new developments need still validation by comparison with outcomes from simulators interfacing with real operators. This means that human reliability models are still far from achieving the maturity needed for reliable prediction of human behaviour and for assigning consistent probabilities to success/failure events.

REFERENCES

1. Advanced Seminar on Common Cause Failures Analysis in Probabilistic Safety Assessment (1989) Proc. of the Ispra Course held in November 1987, A. Amendola (ed.), Kluwer Academic Publishers, Dordrecht (NL).
2. Watson, I.A. (1988) 'Human factors in reliability and risk assessment', in Reliability Engineering, Proc. of the Ispra Course held in Madrid, September 22-26, 1986, A. Amendola and A. Saiz de Bustamante (eds.), Kluwer Academic Publishers, Dordrecht (NL).
3. Rasmussen, J. (1989) Int. Workshops to develop a multidisciplinary research program based on a holistic system approach to safety and management of risk in large-scale technological operations, Karlstad, November 6-9.
4. Poucet, A. (1988) 'Survey of methods used to assess human reliability in the Human Factors Reliability Benchmark exercise', in Reliability Engineering and System Safety, 22, 257-268.
5. Poucet, A. (1989) 'Final report of the Human Factors Reliability Benchmark Exercise', CEC-JRC, Ispra, EUR 12222 EN.
6. Hopkins et al. (1982) NUREG/CR-2833.
7. See the lectures by Bagnara, S. and Embrey, D.E. (1986)on Human Reliability in Theory of Reliability, Proc. of the Int. School of

Physics, A. Serra and R.E. Barlow (eds.), North Holland.
8. Dougherty, E.M. Jr. and Fragola, J.R. (1988) 'Human reliability analysis', J. Wiley & Sons, N.Y.
9. Mancini, G. (1986) 'Modelling humans and machines', in NATO ASI Series, Vol.F21, Intelligent Decision Support in Process Environments, E. Hollnager, G. Mancini and D. Woods (eds.), Springer-Verlag, Berlin-Heidelberg.
10. De Keyser, V. (1987) 'Human error, experience and the structure of knowledge in the reliability of process systems', in Reliability Modelling and Applications, Proc. of the Ispra Course held at JRC, November 25-29, A.G. Colombo and A.E. Keller (eds.), D. Reidel Publishing Company, Dordrecht (NL).
11. Amendola, A., Mancini, G., Poucet, A. and Reina, G. (1982) 'Dynamic and static models for nuclear reactor operators - needs and applications examples', IFAC Conf. on Analysis, Design and Evaluation of Man-Machine Systems, Baden-Baden, September 27-29.
12. Amendola, A., Reina, G. and Ciceri, F. (1985) 'Dynamic simulation of man-machine interaction in incident control', IFAC/IFIP/IFORS/IEA Conf. on Analysis, Design and Evaluation of Man-Machine Systems, Varese (I), September 10-12.
13. Howard, Darlene V. (1983) 'Cognitive psychology', Macmillen Publishing Co., Inc., New York, Collier Macmillen Publishers, London.
14. Mancini, G., Rasmussen, J. et al. (1981) 'Classification systems for reporting events involving human malfunctions', EUR 7444 EN.
15. Rasmussen, J. (1981) 'Models of mental strategies in process plant diagnosis', in Human Detection of System Failures, J. Rasmussen and W.B. Rouse (eds.), Plenum Press, New York.
16. Norman, D.A. (1981) 'Categorization of action slips', Psychology Review, 88, 1-15.
17. Reason, J.T. and Mycielska, K. (1982) 'Absent-minded. The psychology of mental lapses and everyday errors', Prentice-Hall, Englewood Cliffs, New-Jersey.
18. Reason, J.T. (1984) 'Of the nature of mistakes', Int. Workshop on New Technology and Effects on Human Errors, Bad Homburg, February 13-15.
19. Hollnagel, E., Mancini, G. and Woods, D. (eds.) (1986) 'Intelligent decision support in process environments', NATO ASI Series, Vol.F21, Springer-Verlag, Berlin-Heidelberg.
20. Amendola, A. (1985) 'Systems Reliability Benchmark exercise, final report', EUR-10696.
21. Hannaman, G.W. et al. (1984) 'Systematic human action reliability procedure (SHARP)', EPRI NP-3583, Interim report.
22. White, R.F. (1986) 'A suggested method for the treatment of human error in the assessment of major hazards', Reliability Engineering, 15, 171-199.
23. MORT - Management and Oversight Risk Tree, Developed by USA DOE (Dept. of Energy), International Risk Management Institute, Vol.VI, N°2, October 1983.
24. Wreathall, J.W. (1982) 'Operator action trees, an approach to quantifying operator error probability during accident

sequences', NUS report 4159.
25. Swain, A.D. and Guttmann, H.E. (1983) 'Handbook of human reliability analysis with emphasis on nuclear power plant applications', NUREG/CR-1278.
26. Hall, R.E., Fragola, J.R. and Wreathall, J.W. (1982) 'Post event human decision errors: operator action tree/time reliability correlation', NUREG/CR-3010, BNL-NUREG 51601.
27. Hannaman, G.W., Joksimovich, V., Worledge, D.H. and Spurgin, A.J. (1986) 'The role of human reliability analysis for enhancing crew performance', Int. ANS/ENS Top. Meeting on Advances in Human Factors in Nuclear Power Systems, Knoxville (Te).
28. Pickard, Cowe and Garwick, Inc. (1983) 'Seabrook station probabilistic safety assessment'.
29. Embrey, D.E. et al. (1984) 'SLIM-MAUD: an approach to assessing human error probabilities using structured expert judgement', NUREG/CR-3518, Washington (D.C.).
30. Embrey, D.E. (1985) 'SLIM-MAUD: a computer-based technique for human reliability assessment', Int. ANS/ENS Top. Meeting on Probabilistic Safety Methods and Applications, San Francisco (CA).
31. Vestrucci, P. (1988) 'The logistic model for assessing human error probabilities using the SLIM method', Rel. Eng. and Syst. Safety, 21, 189-196.
32. Apostolakis, G., Bier, V.M. and Moslek, A. (1987) 'Expert opinions and uncertainty modelling in human risk assessment: a critical evaluation', Post-SMiRT 9 seminar on Accident Sequence Modelling: Human Actions, System Response, Intelligent Decision Support, Munich (FRG), August 24-25, Rel. Eng. and System Safety, Vol.22, N°1-4, 201.
33. Poucet, A. and Amendola, A. (1988) 'State of the art in PSA reliability modelling as resulting from the international Benchmark exercises project', NUCSAFE 88 Conf., Avignon, October 2-7.
34. Clarotti, C.A. (1988) 'Inference, a Bayesian approach', in Reliability Engineering, A. Amendola and A. Saiz de Bustamante (eds.), Kluwer Academic Publishers, Dordrecht (NL).
35. Amendola, A., Bersini, U., Cacciabue, P.C. and Mancini, G. (1987) 'Modelling operators in accident conditions: advances and perspectives on a cognitive model', Int. J: Man-Machine Studies, 27, 599-612.
36. Bersini, U., Cacciabue, P.C. and Mancini, G. (1987) 'Cognitive modelling: a basic complement of human reliability analysis', 9th SMiRT post-conference seminar on Accident Sequence Modelling: Human Actions, System Response, Intelligent Decision Support, München, August 24-25.
37. Cacciabue, P.C. and Bersini, U. (1988) 'Modelling human behaviour in the context of a simulation of man-machine systems', in J. Patrick and K. Duncan (eds.), Human Decision Making and Control, North Holland, Elsevier, Amsterdam.
38. Decortis, F. and De Keyser, V. (1988) 'Time: the Cinderella of man-machine interaction', 3rd IFAC/IFIP/IEA/IFORS Conf. on Man-

Machine Systems, Oulu (Finland), June 14-16.
39. Bersini, U., Cacciabue, P.C. and Mancini, G. (1988) 'A model of operator behaviour for man-machine system simulation', 3rd IFAC/IFIP/IEA/IFORS Conf. on Man-Machine Systems, Oulu (Finland) June 14-16.
40. See the papers by Amendola, A. and Izquierdo, J.M., Sanchez-Perea, M. on the DYLAM approach in these Proceedings.
41. Reason, J. (1986) 'Recurrent errors in process environments: some implications for the design of intelligent decision support system', in the book quoted at ref./19/.
42. Zadeh, L.A. (1978) 'Fuzzy sets as a basis for a theory of possibility', Fuzzy Sets and Systems, Elsevier, Amsterdam, 1, 7-28.

FIXATION ERRORS : FAILURES TO REVISE SITUATION ASSESSMENT IN DYNAMIC AND RISKY SYSTEMS

V. De Keyser
Université de Liège,
bd du Rectorat, 5 - Bât. B 32
4000 Liège, BELGIUM

D. D. Woods .
The Ohio State University
Columbus, Ohio 43210
U.S.A.

ABSTRACT. Many critical real world human problem solving situations take place in dynamic event-drivers environments, where the evidence comes over time and situations can change rapidly. In these situations people must amass and integrate, uncertain, incomplete and changing evidence. A major source of human error in dynamic domains seems to be a failure to revise situation assessment as new evidence comes in. This paper will be concerned with the identification of the main descriptive patterns of fixation errors and with how to build new sytems to reduce this type of error. It will also begin the process of building a theory of fixation errors.

1. Introduction

Many critical real-world human problem solving situations take place in dynamic, event driven environments where the evidence comes over time and situations can change rapidly. In these situations people must amass and integrate uncertain, incomplete and changing evidence.

The event-driven nature of these incidents adds a level of complexity to problem solving (i.e., diagnosis, situation assesments, planning) that is not observed in more static environments. In dynamic worlds, there is no single well-formulated diagnosis of the situation. Rather, individuals make provisional assessments based upon partial and uncertain data. These assesments are incrementally updated and revised as more evidence comes in. Furthermore, situation assesments and plan formulation are not separate sequential stages, but rather are closely interwoven processes with partial and provisional plan development and feedback leading to revised situation assesments (Kleinmuntz, 1985; Woods & Roth, 1986; Klein, in press).

The results of several studies (Woods, 1984; De Keyser, 1981; De Keyser at al., 1987; Johnson & Thompson, 1981; Johnson et al., in press, Cook & McDonald, 1988; Cook & McDonald,1989) strongly suggest that a major source of human error in dynamic domains is a *failure to revise* situation assesment as new evidence comes in. Initial situation assesment tends to be accurate, in the sense of being consistent with the partial information available early in the event. Errors become manifest later, in the evolution of the event, as people fail to revise their assesments in response to new evidence which indicates a deviation of the event from expected path (e.g., due to multiple failures). These results suggest that a major source of human error in dynamic

A. G. Colombo and A. Saiz de Bustamante (eds.), Systems Reliability Assessment, 231–251.

domains is fixation, or perseverance: failure to revise situation assesments and planned actions when the situation is changing.

Informally, analyses of major accidents have recognized fixation errors as significant contributors and the recommendations can be interpreted as classic folk treatments to reduce fixation errors. For example, The President's Commission report on the Three Mile Island accident (Kemeny et al., 1979) recommendations included: add a new person to the team who has a different background and viewpoint (the Shift Technical Advisor, give the problem solvers more time (related to incubation ideas), give the problem solvers a new kind of representation of the state of the plant (displays and procedures organized around safety functions).

It is easy to call fixations "diabolic" -practitioner behavior we cannot explain, but this reveals more about gaps in the "experts" understanding than about the practitioner himself. There may be many different cognitive processing mechanisms which contribute to the kinds of fixation effects observed in dynamic problem-solving situations. If decisions aids are to be designed to reduce fixation errors, it is important to clarify the processes behind these errors and to establish, if possible, a psychological typology of the processes involved.

2. Definition of fixation

A fixation occurs when a problem-solver fails to revise his/her/its situation assesment or course of action when

1. the situation assesment or course of action has become inappropriate given the actual situation, and
2. the inappropriate judgement or action *persists in the face of opportunities to revise.*

With respect to the first criterion, the initial judgement or course of action under consideration may or may not have been correct at the time it was made, given the evidence available at that point in the evolving incident. It is from hindsight or the position of an omniscient observer that the judgement is seen to be erroneous or to become erroneous as the situation develops. Thus, the judgement or action can be incorrect at the time it was formed or because the situation changed after it was formed.

The second part of the definition contains some of the critical features of a fixation that differenciate it from other error forms. One critical feature is that there is some form of *persistance* over time in the behavior of the fixated person or team. Second, opportunities to revise are cues, available or potentially available to the problem-solver, which could have started the revision process if observed and interpreted properly. In part, this feature distinguished fixations from simple cases of inexperience or lack of knowledge (Cook et al., 1989).

One difficulty is to define the standard to use to determine what cue or when a cue *should* alert the problem-solver to the discrepancy between the perceived state of the world and the actual state of the world. There is a great danger of falling into hindsight bias when evaluating after the fact whether a cue "should" have alerted the problem-solver to the discrepancy.

One condidate for this standard is the "fresh or neutral observer" criterion (Figure 1). In this technique the investigator presents evidence available at one point in the evolving incident (i.e., the information that was available at an hypothesized opportunity to revise) to observers who have not undergone the incident evolution. The neutral observer then makes a judgement or interpretation about the state of the world, relevant possible future trajectories and relevant courses of action. The question is whe-

Fig. 1 : Neutral Observer Test

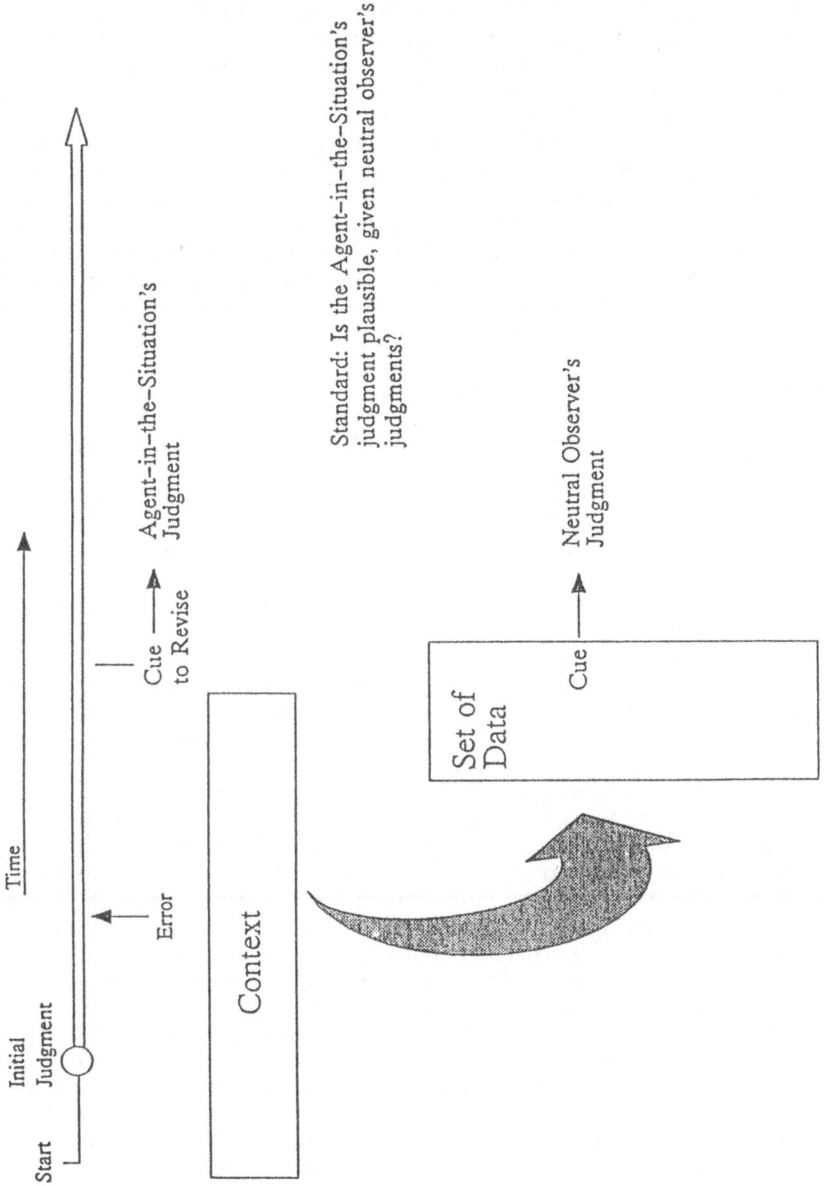

ther the path taken by the actual problem-solver is one that is plausible to the neutral observers (i.e., they entertain that path as a serious candidate).

The basic defining characteristic of fixations is that the immediate problem-solving context has biased the problem-solver in some direction. In naturally occuring problems, the context in which the incident occurs and the way the incident evolves activates certain kinds of knowledge as relevant to the evolving situation (Kahnement & Miller, 1986; Gettys et al., 1987) which affects how new incoming information is interpreted. After the fact or after the correct diagnosis has been pointed out, it seems obvious, even to the fixated person or team (Cook et al., 1989). In line with this notion, laboratory studies establish fixation (or facilitation) effects by comparing problem-solving performance on a target case following some experimenter-defined context to performance on the same case following a "neutral" context. The parallel to this standard in field cases of possible fixations is to compare the performance of the actual problem-solver to the performance of other problem-solvers who do not share the same context. Note that the neutral observer criterion is related to one standard technique for correcting fixation errors - provide a fresh observer by institutionalizing a person with a different perspective.

3. Descriptive patterns of fixation

We have observed in field studies, three main patterns of behavior during fixation errors. The first one could be called "everything but that". The operator seems to have many hypothesis in mind, but never entertains the correct one (see for instance Kouabenan, 1982; Cockerill-Sambre, 1984). His external behavior looks incoherent because he is often jumping from one action to another one without any success.

The second one is the opposite: "this and nothing else". The operator is stuck on one strategy, one goal, and he definitely refuses to change. We can observe a great persistence in his behavior, repeating the same action sometimes eight or ten times (see Daniellou, 1982; Fichet-Clairfontaine, 1985; De Keyser, 1981). For an observer this behavior is the easiest to detect, because of the unusual level of repetitions despite an absence of results. The operator often detects the absence of results himself but without changing his strategy. In this case, the operator has a hypothesis about the cause of the disturbance or about the appropriate response strategy.

The third one is "everything is OK". In this case, the operator does not reacts to the change in his environment. Even if there are a lot of cues and evidence that something is going wrong, he does not pay attention to them. He prefers his own internal representation of the world to the one indicated in the displays. He may not be absolutely wrong: false alarms can be numerous in some systems, and to be blind to them is an excellent strategy. But once more the problem lies in the context: an excellent strategy in one context can be the worst in another (De Keyser, 1981).

Other descriptive features of fixations episodes include the form of the persistence: for example, repeating the same ineffective action over and over again; checking, rechecking, checking again the discrepant evidence as occured in Oyster Creek nuclear power plant incident (cf., Pew et al., 1981); trying a diverse set of unmotivated responses in a sort of random exploration almost as if the person is thinking : "Maybe if I do something/anything the discrepency will be resolved". This pattern also occurred during the Oyster Creek incident (it may be related to the concept of diagnostic interventions - actions taken to produce information that will aid in building an accurate situation assesment).

4. Psychological categorization

Examining several cases of fixation errors in continuous processes brings to mind several possible mechanisms that can alone or in combination produce fixation.

- Inefficiencies or breakdowns in attentional checking.
- Limits in the form and content of the practitioner's knowledge of the world.
- Factors of knowledge organization and access that produce inert knowledge - is relevant knowledge activated and utilized in the actual problem-solving context ?
- Difficulties that arise during problem formulation and hypothesis generation.
- Strategies for building a situation assesment (diagnosis) given the probability of multiple factors or faults and for revising situation assesment as new evidence comes in over time.

Each of these mechanisms may contribute to fixations in different ways. Let us examine each.

4.1. ATTENTIONAL CHECKING

In complex and dynamic situations, the operator, due to limited resources, has to focus his attention on some features of the situation. Observations are compared to some internal representation of what the situation should or should not be. A mismatch between the perceived current state of the process and the 'normal" state drives processing and leads to practitioner interventions. Because of limited resources and the evolving nature of the situation, he or she cannot take into account all of the huge amount of information that the world provides. Based on a variety of factors, the practitioner must decide where to look when (Moray, 1986).

There are three dimensions to the process of attentional checking.

- Sensitivity. This is the ability of the problem-solver to detect differences between actual and reference states. Note that meeting this cognitive demand is complicated because reference states often depend on the system context and on expectations about system behavior in different circumstances. It appears, from several studies (De Keyser et al., 1987; Van daele, 1988; Decortis, 1987) that experienced practitioners are better able to discriminate fine details in changes of the state of the process.
- Degree of Extension. This refers to the size of the attentional field. It can depend on many factors, for example, as mental workload increases, the amount of information the practitioner takes into account can decrease even as the level of risk increases (Sperandio, 1972 ; De Keyser, 1986). Organizational factors can affect the size of the attentional field as well. For example, in rigid work organizations it has been found that practitioners tend to focus their attention on a restricted field - those items that he feels he can control (Assenheim, 1970).
- Depth of Checking. This corresponds to the level in an abstraction hierarchy at which differences are observed. For example, if an action has been taken, the operator can be focused on low level differences in the state of the process such as did the actuator device respond in the "normal" way or did an associated display indicate that the command was received. But he also could focus on higher level consequences of this action - did inflow increase or decrease, is the disturbance which motivated the action subsiding. The success

of attentional checks can depend on the level at which the person is focused. If he is focused at one level, but the observability of the mismatch is low at that level, then the detection opportunity will be missed. Note that this can occur at any particular level of abstraction.

These aspects of attentional checking provide an economical way to reason and plan in complex environments, where, obviously, the operator cannot process all of the potentially available information. But in dynamic worlds, attentional checking has to go on periodically since the system is changing over time. This reassessment should often lead the practitioner to restructure his course of action. Attentional checking and action planning are tightly connected (Woods, Roth & Pople, 1987).

Fixations could arise from attentional processes because critical data that would activate relevant knowledge (a possible hypothesis or the fact that an intervention is not successful) is not seen. This could be an important element in Type 1 (eveything but that) and in Type 3 (everything is OK) fixations.

4.2. LIMITS IN THE FORM AND CONTENT OF THE PRACTITIONER'S KNOWLEDGE OF THE WORLD

It may happen that the operator doesn't possess the knowledge to recognize or respond to a situation. This lack of knowledge, leading him to mistakes and stereotyped action schemata may take different forms:

- The functional behavior of the system is not well understood. The operator does not understand all the links between the variables and how they can interact. This is often a problem of training.
- The relationships between the environment (climate, men) and the process are not clearly apprehended. This is more a question of experience than a question of classical training, although system simulations can provide some assistance. The operator does not understand the variable shapes that a process can adopt in different environments. It introduces new atypical patterns of symptoms which have a high risk of error especially for novices.
- The process of action is not well controlled, because he knows what has happened, but he does not know what to do to mitigate the incident. This often involves missing the side effects of actions. It also frequently involves misunderstanding the dynamics of the system due to lack of knowledge of the time behavior of the system and to inadequate dynamic information in the available displays (see De Keyser et al., 1987).

Actually, the process of knowledge acquisition has to be continuous throughout the working life. Indeed, gaps in knowledge do not lead necessarily to fixation errors in most cases. If an operator has a good feedback on the effects of his or her actions and if there are visible traces of the evolution of the system, the external conditions to structure his experience are present. And in this way - feedback on his actions, checking traces on the evolution of system state - some operators are able to discover or rediscover the laws of the system (De Keyser and Piette, 1970; De Keyser, 1974). But external factors such as work organization, social climate, shift work can produce stress, fatigue, mental workload, factors that if combined with limited knowledge increase the vulnerability to fixation errors.

Gaps in knowledge may be related to Johnson's (Johnson & Thompson, 1981) concept of chasm or missing bridge difficulties.

4.3. INERT KNOWLEDGE

Lack of knowledge may be one part of the puzzle, but more the critical question may be factors that affect whether relevant knowledge is activated and utilized in the actual problem-solving context, factors such as problem formulation and attentional checking (e.g., Cheng et al., 1986 ; Bransford, Sherwood, Vye & Rieser, 1986). The question is not just does the problem-solver know some particular piece of domain knowledge such as what is the relationship between two entities, but does he/she know that it is relevant to the problem at hand and does he/she know how to utilize this knowledge in problem-solving? We tend to assume that, if a person can be shown to possess a piece of knowledge in any circumstance, then this knowledge should be accessible under all conditions where it might be useful. In contrast, a variety of research has revealed dissociation effects where knowledge accessed in one context remains inert in another (Gentner & Stevens, 1983; Perkins & Martin, 1986; Kotovsky et al., 1985). Thus, the fact that people possess relevant knowledge does not guarantee that this knowledge will be activated when needed. The critical question is not to show that the problem-solver possesses domain knowledge, but rather the more stringent criterion that situation relevant knowledge is accessible under the conditions in which the task is performed. This has been called the problem of *inert knowledge* - knowledge that is accessed only in a restricted set of contexts.

Inert knowledge may be related to cases that are difficult, not because the problem-solver doesn't know the individual pieces of knowledge needed to build a solution, but because they have not put the pieces together in this way before.

4.4. PROBLEM FORMULATION

An operator can be stuck in his reasoning because he cannot find a suitable way to formulate the problem he faces. One way that this occurs is when the situation poses a paradox in terms of goals that place conflicting constraints on the operator (Watzlawick et al., 1967). Work organizations are very good at creating such kinds of conflicting injunctions: production versus safety, following the rules versus handling an unusual situation, quality versus time, etc. Usually, to be able to overcome the competition between the different goals, the operator has to reformulate the problem in a more general form. Sometimes, he has to find a compromise path by revising his hierarchy of goals or fixing new priorities among them.

Part of problem formulation is deciding what set of hypothesis are plausible or worth pursuing - hypothesis generation (Gettys & Fisher, 1979; Manning & Gettys, 1981; Gettys, Mehle & Fisher, 1986). Hypothesis generation is a particularly important part of problem-solving given limited resources and time pressure. Fixation errors may be seen as failures to sample the space of potential hypotheses which could account for the currently perceived pattern of evidence. For example, Johnson (Johnson, in press; Johnson et al., 1981) studied experienced medical diagnostician's performance on a problem prone to fixation. One class of errors occurred in hypothesis generation and included failures to call to mind the correct alternative.

When problems unfold over time, hypothesis generation and hypothesis evaluation activities are not separate sequential stages, but rather intermixed and interacting activities. This suggests another source of fixation errors where hypothesis generation is terminated prematurely leading to failures to revise. For example, a highly plausible hypothesis could block access to parts of the hypothesis space. Manning & Gettys (1981) found exactly this result in a study of the effects of providing an initial hypothesis as a retrieval cue on hypothesis generation performance. Johnson et al. (1981) also found revision errors in the performance of experienced diagnosticians where they were unable to shift from a hightly plausible but incorrect initial hypothesis to the actually correct hypothesis. Similarly, initially cued parts of the hypothesis space may block retrieval of other parts (cf., Arkes & Harkness, 1980).

4.5. STRATEGIES FOR BUILDING AND REVISING SITUATION ASSESSMENTS

Fixation errors may also arise depending on the strategy used to build a situation assessment (diagnosis). Because there is the possibility of multiple factors or faults in an incident (studies of actual disasters and near disasters show that there is a high probability of multiple faults), fixation resistant situation assessment requires building and reasoning about multiple views of the situation (Coombs & Hartley, 1987; Woods et al., 1987). Fixations could arise because possible explanations are not thought of ans pursued when the operator builds an initial situation assessment. For example, this could be an important element in Type 1 fixation ("everything but that").

Because provisional situation assessments are necessary in dynamic worlds, there is a need to revise situation assessment a new evidence comes in over time. Fixation prone behavior could result from a failure to consider how new findings should force revision of the earlier situation assessment (see Woods et al., 1987 for an extensive discussion of fixation prone revision of hypotheses). Fixation prone diagnostic strategies could be an important element in Type 2 fixations ("this and nothing else"). One path to revision failures is based on the strategies used by the problem solver to process evidence about possible diagnostic hypotheses. One problem in evidence processing that can lead to revision failures is the confirmation bias where the person only looks for evidence that would confirm the current hypothesis and therefore never adequately considers viable alternatives. Another problem in evidence processing is what Rasmussen et al. (1987) call over-reliance on familiar processing shortcuts.

Another path to revision failures occurs because there are often multiple explanations each of which accounts for a part of the total set of interesting findings. This means that a new piece of evidence can be rationalized as due to a different factor (e.g., sensor failure) and preserve the integrity of the current hypothesis. Furthermore, because there are often multiple factors and faults which produce the total pattern of evidence, it is often correct to preserve the current hypothesis and look for other factors that may explain the new findings. This raises the fundamental question of, given the possibility of multiple faults, what is good explanation (see Coombs & Hartley, 1987).

In highly coupled worlds, incidents often consist of multiple failures or factors that interact to produce the observed pattern of findings. When confronted by this type of problem-solving situation, fixations take the form of mistaking on factor related to the state of the world as the single explanation for that state of the world (cf., Bechtel, 1982, who describes several examples of this error in the history of science). This becomes an error - *premature localization* to a single factor - when attribution to a single factor delays or prevents identification of the set of factors that actually contribute to and produce the observed situation.

For example, consider a portion of world where multiple parts are simultaneously capable of effecting a single goal. Thus, any one is capable of being the source if the goal is disturbed. Now if more than one of these processes are simultaneously active, then the effect of one contributor can partially occlude the effects

Fig. 2 : Fixation Incident Analysis

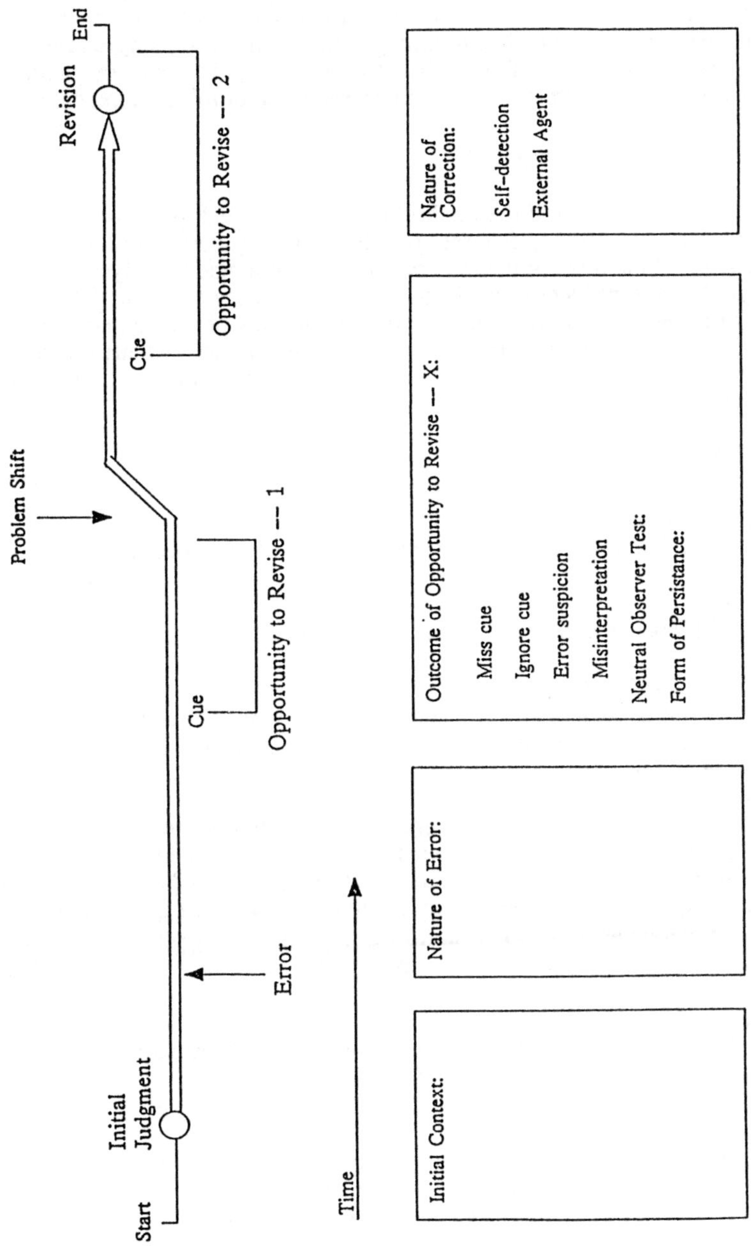

of other contributors. In this error form the first contributor that is identified (e.g., the more visible or more salient) is then thought to provide a complete explanation for the observed data with the result that diagnostic search is terminated prematurely.

Finally, note that fixation prone and fixation resistant situation assessment may involve a tradeoff between an efficient strategy for simpler (i.e., single fault cases) or more frequent cases and a more thorough but more costly (in terms of mental effort) strategy to handle rare but more complicated (interacting multiple factors) cases.

5. Fixation Incident Analysis

This section contains a sketch of the method that we are developing for the analysis of potential cases of fixation across different problem-solving worlds. Figure 2 provides a graphic illustration of the form of the method for fixation incident analysis.

The goal is to build a corpus of fixation cases. Analysis of the data in the error corpus will aid in the development and testing of theories of mechanisms underlying fixation errors and in the development and testing of countermeasures to prevent or mitigate fixation errors. An illustration of the application of the method to an actual incident is contained in the appendix.

5.1. ANALYSIS CATEGORIES

* *Initial judgment and background.*
  . What is original judgment and the context in which it is made?

* *The error.*
  . When did it become erroneous and how was it incorrect ?

* *Opportunities to Revise.*
  . What prompts/cues to revise occured over what time intervals (salience, strengh, direct evidence of mismatch or triggers to reconsider the topic ?
  . Outcome of opportunity to revise:
  (a) missed cue;
  (b) observed - misinterpreted, observed - discounted or ignored;
  (c) evokes - error suspicion, evokes - prompt to reconsider a topic, evokes - recognition of specific unexpected finding, evokes - pointer to alternative hypothesis, evokes - pointer to why the unexpected finding occurred.
  . Form of persistence - repeat action many times, repeated checking of the discrepant evidence, diverse irrelevant actions.

* *Neutral Observer Tests.*
  . Was the opportunity to revise a reasonable one? Compare evolved thinking versus neutral observer's interpretation of the situation.

* *Incident Evolution.*
  . Changes in reversibility of courses of action; when do negative consequences accrue, i.e., the time available in which to successfully revise without a step change in negative consequences on some dimension (note that this is one way that problems change and evolve: step changes in outcomes mean that the practitioner is confronted with a new problem).

* *Revision and Correction.*
  . How was fixation corrected if it was corrected? self-detection via incidental cues, reasoning through evidence/hypotheses, hints or fresh viewpoint.

This approach to error analysis recognizes that errors are often corrected before negative consequences accrue, that they are difficulties in defining errors (specifically confounding outcome and process measures), and that errors are part of a dynamic process so that they must be analyzed in terms of how they emerge and develop over time.

Analysis of an incident always involves stages of interpretation. In collecting data about fixation errors one has to be very careful to maintain the data and evidence on which the interpretations in the fixation incident analysis are based (Leplat & Cuny, 1970).

Some aspects of a fixation incident can be established based on records of plant behavior during the incident, e.g., operator actions, plant parameter behavior, machine breakdowns. Others are based on general knowledge of the function and structure of the domain in question, e.g., goals, temporal factors, procedures, displays, communication channels). Others are based on the context established by work practices, environment, training, operator culture (e.g., shift schedule). Some interpretations (mindsets, missed cues, intentions) may be based on verbal protocols, interviews with participants, analysis of behavioral protocols. Finally, some may be based on analysis/empirical inquiry with other members of related populations (e.g., investigations to uncover what practitioners understood about how a system functioned before the incident occured).

6. How to Prevent and Mitigate Fixation Errors

It is important to understand the mechanisms that contribute to this type of information processing breakdown so that strategies to reduce their occurrence or to help people recover from them can be devised. Our earlier categorization of contributors points to different cognitive processes that may produce or contribute to fixation prone behavior : missing or buggy knowledge, inert knowledge, limits on attentional checking, difficulties in problem formulation, strategies for situation assessment and re-assessment that are biased towards the initial hypothesis.

Designers can affect these cognitive processes to increase or reduce fixation prone behavior through a variety of means. One approach is to change the world itself (the difficulty of the problems posed) and the environment that tasks are performed in to reduce factors that encourage fixations. Another is to improve the nature of problem-solving so that correct situation assessment is formed and the correct response strategies are adopted. A third approach recognizes that initial missteps are always likely to occur in dynamic event-driven worlds and tries to improve recovery from fixations so that they do not persist long enough to have serious consequences (an error recovery approach). We will briefly suggest some possible aiding strategies within the categories of changing the work environment, training, and cognitive ergonomics.

6.1. ENVIRONMENTAL DESIGN

One possibility is to focus on the environment in which tasks are carried out. Work organization, mental workload, may have a negative influence on the degree of extension in attentional checking, reducing the number and altering the set of variables the operator takes into account and leading him to stereotypical actions. The variability of the environment also increases the risk of errors. A good design of the systems or process itself (Turner, 1984) and an efficient flow of information prevents many errors.

6.2. TRAINING

Training should increase one's basic knowledge of the functioning of the system (theoretical laws, interactions between variables, etc.). It should also help the operator to discover the variability of the world and to establish links between different types of knowledge about systems to be controlled: how it works, how the information is displayed, what it means, and how to act on it (Montmollin & De Keyser, 1985; Moray, 1987). Training experiences can help the student explore and learn about system variability in different ways:

- Interactions between the process and its environment, which produces system variability variability within certain fuzzy limits (see Decortis, 1987). This corresponds to the common sense reasoning of an operator which tells him that "every day is different".
- Similarity and dissimilarity in patterns of cues in incidents which are the pertinent details that distinguish one pattern from another.
- Problem reformulation. Training exercises can be set up to force an operator to reformulate problems, to use analogical reasoning, to discover under the superficial features of a problem its deep structure.
- Training in dynamic simulation. The purpose of this training is to force the operator to reassess periodically the situation, changing his attentional focus, leading him to discover the dynamics of a disturbance, and their temporal shape.
- Conditionalizing knowledge. Training experiences need to promote an understanding of how concepts and procedures can function as tools for solving relevant problems (Bransford et al., 1986). Training has to be about more than just student knowledge acquisition ; it must also enhance the expression of knowledge by conditionalizing knowledge to its use via information about "triggering conditions" and constraints (Glaser, 1984; Lesh, 1987).

All these measures could improve the ability of an operator to cope with a complex environment. But in fact, they ask critical questions about training. Are the types of information mentioned above available? Is there a database with all the patterns of incidents, with a similarity index (Boy, 1983) and their probability of occurence? Where is the knowledge of the contingencies in the environment and how they may interact with the process? About the different shapes of a disturbance over time? The answer is quite clear: usually these types of information are not available. And the inadequacy of training sessions for developing operational skill reflects ignorance of the real conditions of work in complex environments. The first step is not to train operators, but to train experts in work analysis (Montmollin, 1984) and in the discovery, the gathering of data concerning the facets we have described above.

6.3. COGNITIVE ERGONOMICS IN DESIGN

The absence of feedback of his actions or the allocation of tasks between man and computer (see Moray, 1986) can undermine the operator's supervision of the system and can decrease opportunities to develop sound knowledge of system behavior. Cognitive ergonomics in the design of operator's aids can provide solutions (see

Goodstein, 1981; Woods & Roth, 1989; De Keyser et al., 1987). We will discuss briefly possible support systems to enhance attentional checking, to enhance situation assessment, and to reduce workload.

6.4. MONITORING AND FIELD OF ATTENTION

The interface between the operator and the process can have strong affects on the attentional checking mechanisms. For example, online representations of the world can help or hinder problem-solvers to recognize what information or strategies are relevant to the problem at hand (Wiecha & Henrion, 1987; Woods & Roth, in press). For example, Fischhoff et al. (1978) and Kruglanski et al. (1984) found that judgmental biases (e.g., representativeness) were greatly reduced or eliminated when aspects of the situation cued the relevance of statistical information and reasoning. Thus, one dimension along which representations vary is their ability to provide prompts to the knowledge relevant in a given context.

Poor interface and display system designs can degrade control of attention and monitoring by creating tunnel vision or getting lost effects (e.g., Woods, 1984b; Elm & Woods, 1985).

One technique is to provide higher level integrated and synthetic display of data in contrast to the usual one sensor-one indicator display philosophy (Goodstein, 1981; Woods & Roth, 1988). Synthetic display techniques can help the operator better track the state of the world and get better feedback about the effects of response strategies especially at higher levels of abstraction.

Another technique is to help the operator find the relevant subset of data at the "right" time within the dynamics of the incident (Woods, 1986). However, what is an interesting finding depends heavily on the context. Context sensitive computation and organization of what data are relevant (e.g., Woods, Elm & Easter, 1986; Woods, Roth & Pople, 1987) could be helpful to reduce inert knowledge. For example, this could be important in reducing the Type 1 fixation pattern ("everything but that") if the pattern were due to inert knowledge rather than buggy knowledge.

6.5. SITUATION ASSESSMENT

However, in some fixation cases there have been a large amount of cues which could have signaled that there was a discrepancy between the operator's perception of the state of the world and the actual state.

Some operators may be so confident in their own judgment that they ignore incoming signals because of a lack of reliability in the instrumentation and in the machines. Information may be neglected because they could be false and because of the high cost of obtaining additional evidence to confirm or deny the signal, e.g., sending people out into the plant to see if some information is correct or by collecting redundant or diverse data on the issue if the computer based display system is difficult to navigate and the relevant data is scattered on multiple displays. Sorkin & Woods (1985) showed how high false alarm rates in alerting or diagnostic systems can result in poor performance if one takes into account the role of the human in the loop. Ignoring information due to high false alarm rates could be a contributor to Type 3 fixations (everything is OK).

One solution is to reduce false alarm rates in alerting and diagnostic systems as Sorkin and Woods (1985) point out (shift the response criterion for the same level of sensitivity in terms of signal detection theory). One technique to do this suggested by

Sorkin and Woods is likelihood alarm displays. Another solution is to provide more integrated displays that collect redundant and diverse indications of the state of a part of the process in order to reduce the cost of checking the validity of incoming signals.

Another factor that could account for fixation errors where operators failed to take into account available evidence is in strategies for building and revising situation assessment as was discussed earlier. In this case machine advisors are often suggested as a possible solution. However, to be effective against fixation proneness what is needed is machine advisors who broaden the set of possibilities that are entertained by helping the person to consider multiple views of the situation or different ways to put the available evidence together into a pattern (e.g., Langer & Piper, 1987). Systems that would help the person in this way could counteract Type 1 fixations (everything but that). This means that the typical diagnostic system which ranks hypotheses by strength is not appropriate because it does not necessarily help the user to consider more possiblities and may even focus him or her more narrowly on just the hypotheses with high initial plausibility. Manning & Gettys (1981) found that providing a highly plausible initial hypothesis as a retrieval cue reduced sampling of the space of potential hypotheses (i.e., high value hypotheses which differed from the cued hypothesis were not called to mind). On the other hand, they found that providing a non-salient hypothesis as a retrieval cue resulted in broader sampling of potentially relevant hypotheses.

Another solution path could be machine advisors who increase the person's ability to revise a situation assessment or to reformulate the problem (Woods, 1987; Gadd & Pople, 1987). One approach to this end is critiquing systems where the diagnostic engine is configured to comment on the actions and strategies of the human in the loop (for examples, see Miller, 1983; Langlotz & Shortliffe, 1983). For example, the machine advisor could explicitly generate hypotheses to broaden the range included in hypothesis evaluation. Reminding/critiquing systems can cue practitioners about the implications and side effects of response strategies. This addresses failures to take into account all side effects in highly coupled systems (fixation on one aspect of a decision). Initial studies of human error detection suggest that one can devise ways to build interface and advisory systems that enhance fixation error detection and correction. For example, Perkins & Martin (1986) found that providing a general hint led to the detection of one third of the errors present.

These approaches could be fruitful with respect to Type 2 fixations where there is a fixation on one hypothesis. However, great care is needed in the design of such machine advisors and critiquing systems. The difficulty is that the operator, if convinced that he is on the right track enough to be fixated, would reject any advice from the machine problem solver in the same fashion as he may reject lower level alarm signals (see Roth, Bennet & Woods, 1987; Woods & Roth, 1988). This raises the question of trust between machine agents and human practitioners which Muir is investigating (Muir, 1987). It also raises the questions about the relationship between human practitioners and new forms of automation (both control and decision automation). For example, some of the resulting joint cognitive system architectures lead to problems such as loss of situational awareness (Wiener, 1985) which may exacerbate fixation errors.

All of the processing elements that may contribute to fixation prone behavior which were discussed earlier are exacerbated by workload, i.e., resource/demand mismatches. Mental workload contributes to processing breakdowns by bringing out any deficiencies in the attentional checking process and in situation assessment and revision.

This leads to another approach to machine support systems. The concept is that human performance can be improve if one frees up the processing capacity of the human-in-the-loop. This can be done by offloading overhead information processing tasks to the computer. Some have also suggested that excessive workload can be compensated for through a dynamic allocation of tasks between man and computer so that as the workload increases the computer does more. However, one problem that can arise with this approach is that higher demand (workload) situations are highly correlated with more difficult problems to solve.

7. Brief Sample Fixation Incident Analysis: From De Keyser (1981)

7.1. BACKGROUND

In this case an operator was monitoring and controlling the treatment of recoverable effluents during a maintenance outage. The operator's role was to send the material to one tank and then switch the flow to a second tank when the first was full and so on. The switchover operation is a manual local action. It was an unusual procedure for water treatment which this operator had only done twice before. The procedure had already begun, when he arrived to begin his shift, around ten o'clock in the night. During the time period in question the operator was alone in the control room because staffing had been reduced.

Normally, the flow rate for treatment processes is 35 m^3/hour, and the operator had a very strong expectation of the time it takes to fill a tank at this rate. But on this occasion, the flow had been increased to 55 m^3/hour by an engineer just before the operator came on duty. The change had occurred and was noted in a book, but the operator did not notice that the change had occurred and was not informed of the change. The need to shift flow to another tank is accompanied by warning signals on pressure and liquid level in the tank.

7.2. THE ERROR

When indications of tank overfilling began to appear (signals and alarms), the operator noticed them but did not react (this occured after midnight). Instead, he waited to switch tanks until the normal time interval was up. This process was repeated for a second and third tank (over a time period of about six hours).

7.3. OPPORTUNITIES TO REVISE

Sets of alarms and sensor values accompanied each tank overfilling. The three sets of alarms occured over a six hour interval.

Outcome of opportunities to revise: observed - discounted/ignored.

The post-incident inquiry suggested that the operator did not pay attention to alarms because of several factors. He knew that these alarms did not reflects the actual state of the tank - the full alarm actually meant the tank was about 80% full. As a result, the alarm carried no need for fast response.

He thought that the flow was at 35 m^3/hour and not 55 m^3/hour. The alarms occured before the expected time to fill the tank had elapsed so that it was not necessary to shift flow from one tank to another. His internal expectation of the time needed to fill a tank was stronger than the external signals.

He attributed the alarms to unreliable equipment and instrumentation errors. The alarms were not considered to indicate the actual state of the process and did not trigger deeper investigation, for example, to confirm that they were false signals. This attribution persisted even as more and different alarms occured as the three tanks were overfilled in turn.

7.4. REVISION AND CORRECTION

Fresh viewpoint. The incident ended when alarms triggered by the disturbances alerted other operators. When they arrived, the overfilling was discovered.

7.5. FORM OF PERSISTENCE

Repeated error discounting alarms three times (three tanks). The false alarm explanation persisted even as more and different alarms occured.

7.6. NEUTRAL OBSERVER TEST

How would an operator who had no expectation about filling times or flow rates have evaluated the evidence?

8. Acknowledgements

We would like to thank A. Van Daele, M. Masson from the University of Liège and E. Roth, from the Carnegie-Mellon University, for their collaboration in this study. The work was funded by a Nato Grant for International Collaboration in Research.

REFERENCES

Arkes, H. R. and Harkness (1980) 'A. R. Effect of making a diagnosis on subsequent recognition of symptoms', Journal of Experimental Psychology : Human Learning and Memory , 6, 568-575.

Assenheim, G. (1970) 'Etude d'un système au travers d'un centre privilégié : un poste de contrôle en cristallerie', Le Travail Humain , 1969, 32, 1-12.

Bechtel, W. (1982) 'Functional localization assumptions as impediments to clinical research', in L. R. Troncale (ed.), A General Survey of Systeme methodology, Vol. 2., Intersystems Publications, Monterey, CA..

Boy, G. A. (1983) 'Le système MESSAGE: un premier pas vers l'analyse assistée par ordinateur des interactions hommes-machines', Le Travail Humain, 46, 2.

Bransford, J., Sherwood, R., Vye, N. and Rieser J. (1986) 'Teaching and problem solving : Research foundations', American Psychologist , 41, 1078-1089.

Cheng, P., Holyoak, K., Nisbett, R. and Oliver, L. (1986) 'Pragmatic versus syntactic approaches to training deductive reasoning', Cognitive Psychology, 18, 293-328.

COCKERILL-SAMBRE (1984) 'Prise en compte des facteurs humains dès la conception et l'installation de la coulée continue de la S.A. COCKERILL-SAMBRE-CHERTAL', Action Communautaire Ergonomique, Recherche, 107, 247 - 11 - 026, Liège.

Cook, R.I. and McDonald, J. (1988) 'Cognitive Tunnel Vision in the Operating Room : Analysis of Cases Using a Frame Model', Anesthesiology, 69, A-497.

Cook, R.I., McDonald, J., Smalhout, B. (1989) 'Human Error in the Operating Room : Indentifying Cognitive Lock Up' manuscript submitted for publication.

Coombs, M. J. and Hartley, R.T. (1987) 'The MGR Algorithm and Its Application to the Generation of Explanations for Novel Events', International Journal of Man - Machine Studies, Vol. 27, 679-708. (Also in G. Mancini; D. Woods and E. Hollnagel (eds.) 'Cognitive Engineering in Dynamic Worlds', Academic Press, London.

Daniellou, F. (1982) 'Strategie de résolution d'incidents aux presses automatiques: le poids de la technologie et de l'organisation du travail'. Communication au XVIIIème Congrès de la SELF à Paris.

Decortis, F. (1987) 'Démarrage d'une centrale thermo-électrique, in V. De Keyser and col. (eds.), 'Les Communications Hommes- machines dans les Systèmes complexes', Services de la Programmation de la Politique Scientifique, Bruxelles.

De Keyser, V. (1974) 'Approche psychosociologique de l'expérience ouvrière à travers les systèmes automatisés', thèse de doctorat en Psychologie, Université Libre de Bruxelles, Bruxelles:.

De Keyser, V. (1981) 'La Fiabilité humaine dans les processus continus, les centrales thermo-électriques et nucléaires', rapport CCE DGXII, CEERI, Bruxelles.

De Keyser, V. (1986) 'Technical assistance to the operator: some lines of thought', in E. Hollnagel, G. Mancini and D. D. Woods (eds.), Intelligent Decision Support in Process Environments, Springer Verlag, Berlin-Heidelberg.

De Keyser, V. (1987) 'Structuring of knowledge of Operators in Continuous Processes: Case Study of a Continuous Casting Plant Start-up', in J. Rasmussen, K. Duncan and J. Leplat (eds.), New Technology and Human Error, Wiley & Sons, Chichester.

De Keyser, V. and Piette, A. (1970) 'Analyse des activités des opérateurs au tableau synoptique d'une chaîne d'agglomération en sidérurgie', Le Travail Humain, 33, 2, 341-352.

De Keyser, V. and col. (1987) 'Les Interactions Hommes-machines dans les systèmes complexes', Services de Programmation de la Politique Scientifique, Bruxelles.

Elm, W. C. and Woods, D. D. (1985) 'Getting lost: A case study in interface design', in Proceedings of the Human Factors Society. 29th Annual Meeting.

Fichet-Clairfontaine, P. (1985) 'Etude ergonomique de l'influence de la conception de la salle de commande, de la stabilité du processus et de la diversification de la production sur l'activité des opérateurs dans 4 unités à processus continu', Thèse de doctorat en ergonomie non publiée, Université de Parix XIII, Paris.

Fischhoff, B., Slovic, P. and Lichtenstein, S. (1978) 'Fault trees: Sensitivity of estimated failure probabilities to problem representation', Journal of Experimental Psychology : Human Perception and Performance, 4, 330-344.

Gadd, C. S. and Pople, H. E. (1987) 'An interpretation synthesis model of medical teaching rounds discourse: Implications for expert system interaction', International Journal of Educational Research, Vol.1.

Gentner, D. and Stevens, A. L. (eds.) (1983). Mental Models. Lawrence Erlbaum Associates, Hillsdale, NJ.

Gettys, C.F. and Fisher, S. D. (1979) 'Hypothesis plausibility assessments in hypothesis generation', Organizational Behavior and Human Performance, 24, 93-110.

Gettys, C. F., Mehle, T. and Fisher, S. D. (1986) 'Plausibility assessments in hypothesis generation', Organizational Behavior and Human Decision Processes, 37, 14-33.

Gettys, C. F., Pliske, R. M., Manning, C. and Casey, J. T. (1987) 'An evaluation of human act generation performance', Organizational Behavior and Human Decision Processes, 39, 23-51.

Glaser, R. (1984) 'Education and thinking : The role of knowledge', American Psychologist, 39, 93-104.

Goodstein, L. (1981) 'Discriminative Display Support for Process Operators', in J. Rasmussen and W. B. Rouse (eds.), Human Detection and Diagnosis of System Failures, Plenum Press, New York.

Johnson, P. E. and Thompson, W. B. (1981) 'Strolling down the garden path : Detection and recovery from error in expert problem solving', in Proceedings of the Seventh International Joint Conference on Artificial Intelligence, British Columbia, Vancouver.

Johnson, P. E., Duran, A., Hassenbrock, F., Moller, J., Prietula, M., Feltovitch, P. and Swanson, D. (1981) 'Expertise and error in diagnostic reasoning', Cognitive Science, 5, 235-283.

Johnson, P. E., Moen, J.B. and Thompson, W. B. 'Garden path errors in diagnostic reasoning', in M. J. Coombs and L. Bolec (ed.), Expert Systems, Springer-Verlag, New York, in press.

Kahneman, D. and Miller, D.T. (1986) 'Norm Theory : Comparing reality to its alternatives', Psychological Review , 93, 136-153.
Keinan, G., Freidland, N. and Ben-Porath, Y. (1987) 'Decision making under stress: Scanning of alternatives under physical stress', Acta Psychologica, 64, 219-228.
Kemeny, J.G., et al. (1979) The President's Commission on the Accident at Three Mile Island, National Technical Information Service, Springfield, VA.
Klein, G. A. 'Recognition-primed decisions.', in W. B. Rouse (ed.), Advances in Man-Machine Research, Volume 5., JAI Press, Greenwich, CT, in press.
Kleinmuntz, D. N. (1985) 'Cognitive heuristics and feedback in a dynamic decision environment', Management Science, 31, 680-702.
Kotovsky, K., Hayes, J. R. and Simon, H. A. (1985) 'Why are some problems hard? Evidence from Tower of Hanoi', Cognitive Psychology, 17, 248-294.
Kouabenan, D. R. (1982) 'Representation de la genèse des accidents de travail: déterminants des attributions causales', thèse de troisième cycle, Ecole Pratique des Hautes Etudes, Paris.
Kruglanski, A. and Freund, T. (1983) 'The freezing and unfreezing of lay-interferences : Effects on impressional primacy, ethnic stereotyping, and numerical anchoring', Journal of Experimental Social Psychology, 19, 448-468.
Kruglanski, A., Friedland, N. and Farkash, E. (1984) 'Lay persons' sensivity to statistical information: The case of high perceived applicability', Journal of Personality and Social Psychology, 46, 503-518.
Langer, E. J. and Piper, A. I. (1987) 'The prevention of mindlessness', Journal of Personality and Social Psychology , 53, 280-287.
Langlotz, C. P. and Shortliffe, E. H. (1983) 'Adapting a consultation system to critique user plans', International Journal of Man-Machine Studies , 19, 479-496.
Leplat, J. (1986) La Fiabilité Humaine, A. Colin, Paris.
Leplat, J. and Cuny, X. (1979) Les accidents de travail, Presses Universitaires de France, Paris.
Lesh, R. (1987) 'The evolution of problem representations in the presence of powerful conceptual amplifiers', in C. Janvier (ed.), Problems of Representation in the Teaching and Learning of Mathematics, Erlbaum, Hillsdale, NJ.
Manning, C. A. and Gettys, C. F. (1981) 'The effect of a previously generated hypothesis on hypothesis generation performance', Catalog of Selected Documents in Psychology , Vol. 11.
Mayseless, O. and Kruglanski, A. (1987) 'What makes you so sure? Effects of epistemic motivation on judgmental confidence', Organizational Behavior and Human Decision Processes, 39, 162-183.
Miller, P. L. (1983) 'ATTENDING: Critiquing a physician's management plan', IEEE, Transactions on Pattern Analysis and Machine Intelligence, PAMI-5, 449-461.
Montmollin, M. and De Keyser, V. (1985) 'Expert Logic vs Operator Logic', in G. Johannsen, G. Mancini and L. Martensson (eds.), Analysis, Design, and Evaluation of Man-Machine SYstems, IFAC, CEC-JCR Ispra, Italy:.
Montmollin, M. de (1984) 'L'intelligence de la tâche. Eléments d'ergonomie cognitive', Peter Lang, Berne.

Moray, N. (1986) 'Monitoring behavior and supervisory control', in K. Boff, L. Kaufman and J. Beattie (eds.), Handbook of Human Perception and Performance, Wiley, NY.
Moray, N. (1987) 'Intelligent Aids, Mental Models and the Theory of Machines', International Journal of Man-Machine Studies, Vol. 27, 619-630. (Also in G. Mancini, D. Woods and E. Hollnagel (eds.), Cognitive Engineering in Dynamic Worlds, Academic Press, London.
Muir, B. (1987) 'Trust between Humans and Machines', International Journal of Man-Machine Studies, Vol. 27, 527-540. (Also in G. Mancini, D. Woods and E. Hollnagel (eds.), Cognitive Engineering in Dynamic Worlds, Academic Press , London.
Perkins, D. and Martin, F. (1986) 'Fragile Knowledge and Neglected Strategies in Novice Programmers', in E. Soloway and S. Iyengar (eds.), Empirical Studies of Programmers. Ablex, Norwood, NJ.
Pew, R. W., Miller, D. C. and Feehrer, C. E. (1981) 'Evaluation of Proposed Control Room Improvements Through Analysis of Critical Operator Decisions, Electric Power Research Institute, (NP-1982), Palo Alto, CA.
Rasmussen, J., Duncan, K. and Leplat, J. (1987) 'New Technology and Human Error', John Wiley & Sons, Chichester.
Roth, E. Bennett, K. and Woods, D. D. (1987) 'Human Interaction with an "Intelligent" Machine', International Journal of Man-Machine Studies, Vol. 27; 479-526. (Also in G. Mancini, D. Woods and E. Hollnagel (eds.) Cognitive Engineering in Dynamic Worlds, Academic Press, London.
Sperandio, J. C. (1972) 'Charge de travail et régulation des modes opératoires', Le Travail Humain, 35, 7, 85-98.
Sorkin, R. D. & Woods, D. D. (1985) 'Systems with Human monitors: A signal detection analysis', Human -Computer Interaction, 1, 49-75.
Turner, J. (1984) 'Software Ergonomics: Effects of Computer Application Design Parameters on Operator Task Performance and Health', Ergonomics , 27, 6, 663-690.
Van Daele, A. (1988) 'L'écran de visualisation ou la communication verbale?', Le Travail Humain, 51, 1, 65-79.
Watzlawick, P., Helnick Beavin, J. and Don Jackson, D. (1967) 'Pragmatics of Human Communication. A Study of Interactional Patterns. Pathologies and Paradoxes', W. W. Norton Company, New-York.
Wiecha, C. & Henrion, M. (1987) 'Linking multiple program views using a visual cache', in Proceeding of Interact' 87. Stockholm, Sweden.
Wiener, E. L. (1985) 'Cockpit Automation: A Field Study of Flighterew Transition', (Tech. Rep. NASA CR-177333).
Wiener, E.L. 'A field study of pilot adaptation to highly automated transport aircraft'. In preparation.
Woods, D. D. (1984) (a) 'Visual Momentum: A concept to improve the cognitive coupling of person and computer', International Journal of Man-Machine studies, 21, 229-244.
Woods, D.D. (1984). (b) 'Some results on operator performance in emergency events', in D. Whitfield (ed.), Ergonomic Problems in Process Operations. Inst. Chem. Eng. Symp. Ser. 90.
Woods, D. D., Elm, W. C. and Easter, J. R. (1986) 'The Disturbances Board Concept for Intelligent Support of Fault Management Tasks', in Proceeding of the International Topical Meeting On Advances in Human Factors in Nuclear Power. American Nuclear Society/ European Nuclear Society.

Woods, D. D. (1986) 'Paradigms for Intelligent Decision Support', in Hollnagel, E., Mancini,G. and Woods, D. D. (eds.), Intelligent Decision Support in Process Environments, Springer-Verlag, New York.

Woods, D. D. and Roth, E. (1986) 'Models of Cognitive Behavior in Nuclear Power Plant Personnel. U. S. Nuclear Regulatory Commission, (NUREG-CR-4532), Washington D.C.

Woods, D. D. & Roth, E. M. (1988). Cognitive Systems Engineering. In M. Helander (Ed.), Handbook of Human-Computer Interaction. New-york: North-Holland.

Woods, D.D., Roth, E. & Pople, H. (1987) 'Cognitive Environment SImulation : An Artificial Intelligence System for Human Performance Assessment', U.S. Nuclear Regulatory Commission, (NUREG-CR-4862), Washington D. C.

THEORIES OF MECHANICAL REASONING UNDER UNCERTAINTY

P. GARBOLINO
Scuola Normale Superiore
56100 Pisa
Italy

ABSTRACT. The three most important theories of quantitative reasoning under uncertainty currently implemented in expert systems are reviewed: probability theory, belief functions theory, and possibility theory. They are compared with respect to the requirements of logical consistency they satisfy, and with respect to their adequacy in representing the process of belief updating or learning from new data. It is pointed out the difference between two different kinds of uncertainty, namely, uncertainty due to partial ignorance of the information sufficient to establish the truth-value of a statement, and uncertainty due to the vagueness of the meaning of a statement. The consequences of this difference are stressed. A final warning is given concerning the dangers of trusting too much the mathematical machinery disregarding the appropriateness of the framework in which the machinery has been put at work.

KEY WORDS. Uncertainty, expert systems, degree of belief, probability, Bayes theorem, belief function, Dempster rule, vagueness, possibility, inductive inference, fuzzy logic, combination of uncertain evidence.

1. Introduction

In rule-based expert systems the expert knowledge is represented under the form of "if...then" rules stored in the so called knowledge base. For example, we can have a very simple rule as

"if light X is ON then valve Y is OPEN". (1.1)

The expert system is provided with an inference engine which contains the inference rules to be applied to the knowledge base. An example of such inferences rules is the so called *rule of detachment* or *modus ponendo ponens*, which says:

"if A and (if A then B) then B", (1. 2)

or, in other words, if the sentence "A" is true and, if the sentence "if A then B" is true, then the sentence "B" is true. Storing the rule (1) in the knowledge base amounts to declare it "true" for the inference engine: therefore, if the news that "light X is ON" is added to the knowledge base, the inference engine will fire the rule (1. 2) to infer that the sentence "the valve Y is OPEN" is "true" . Eventually, this news will represent the antecedent of another rule in the knowledge base, and the inference engine will fire again the rule (1. 2) to infer the consequent of this new rule and so on.

A. G. Colombo and A. Saiz de Bustamante (eds.), Systems Reliability Assessment, 253–272.

The above given example is a simple instance of *deductive* inference. Unfortunately, very often the inference cannot be a deductive one, because either our data are uncertain, or the expert knowledge is in itself uncertain, or both. The most important case is when the expert knowledge is in itself uncertain because either it expresses the knowledge of a statistical correlation, as it often happens, or there is some random "noise" in the system as, for example, when the device is not totally reliable so that there exists the possibility that the valve Y is closed even if the light X is on.
In such a case, the rule (1.2) will be taken to represent uncertain knowledge, where its antecedent represents the evidence which gives some support to its consequent or hypothesis. The goal of any theory of inference under uncertainty is to provide consistent rules to calculate the *degree of belief* the truth of the hypothesis is given by the truth of the evidence.
The first requirement that any theory of inference must satisfy is consistency, that can be put this way: there are certain kinds of sentences which cannot be inferred from the knowledge base if they are not already contained in it. For example, in classical logic this kind of forbidden sentences is given by the set of all possible sentences of the form "A and not A". A theory by means of which you are able to infer any sentence at will, and in particular "A and not A", is inconsistent.
The situation is a bit more complicated for the theories of inference under uncertainty, because in the case of partial evidence the truth of the evidence does not imply the truth of the hypothesis. All the theories we are going to review admit that degrees of beliefs are represented by a real-valued function, say φ. Then a plausible minimal consistency requirement for φ is :

(C1) if a sentence A is a theorem in the knowledge base, that is, either it is declared true or it can be inferred from a true sentence, then $\varphi(A)=1$.

Is it possible to assign stronger consistency requirements to our degrees of belief? There is a strong tradition among philosophers and statisticians of the so called Bayesian school supporting the view that a further necessary condition for consistency is the simple additivity of degrees of belief:

(C2) $\varphi(A \vee \neg A))=\varphi(A)+\varphi(\neg A)$.

Here we mention Ramsey (1931), Savage (1954), Shimony (1955), Kemeny (1955), Cox (1961), de Finetti (1974), Lindley (1982), and among A.I. people Cheeseman (1988). According to this view, it's only the mathematical theory of probability which provides consistent rules of inference under uncertainty.
Other, more recent, schools of thought accept only the minimal requirement (C1), and by consequence, develop different rules of inference. In the A.I. literature you will find a lot of *ad hoc* theories of inference under uncertainty. Here we shall concern ourselves only with those furnished with a serious conceptual and mathematical basis, that is, the theory of belief functions: Dempster (1967) and (1968), Shafer (1976) and (1987); and possibility theory: Zadeh (1978), Yager (1980), and Dubois and Prade (1988a).

2. Probabilistic Expert Systems

2.1. THE RULES

The rules of probabilistic expert systems are, of course, some suitable transformations of the known rules of probability theory. A point worthwhile to be noticed is that the rule for belief updating, that is known as Bayes theorem, is an immediate logical consequence of the additivity and the product rule:

$$P(H|E)=P(HE)/P(E)$$
$$=[P(E|H)P(H)]/[P(EH)+P(E\neg H)]$$
$$=[P(E|H)P(H)]/[P(E|H)P(H)+P(E|\neg H)P(\neg H)]. \quad (2.1)$$

Some of the people mentioned above, namely, de Finetti, Cox, Savage and Lindley, have claimed that Bayes theorem is the basic mechanism of consistent inductive inference, because it is the only way to go from the initial probability P to a new probability P', conditional upon the knowledge that E is true, satisfying the consistency requirements C1 and C2.

The formula (2.1) shows what the ingredients of the inductive recipe are, according to this view:

(a) the background or initial information, measured by P(H) and P(E);
(b) the new information provided by the observation of the occurrence of E, measured by the likelihood P(E|H).

Unfolding P(E) as we have done in (2.1) is important in order to show that the observer should take into account not only the likelihood of the observed phenomenon if the hypothesis in question is true, but its likelihood if the hypothesis is false as well, that is, $P(E|\neg H)$, and the background information about the existence of possible alternative hypotheses to the extent of explaining E, that is, $P(\neg H)$, including the possibility of an error of measurement.

2.2. THE MACHINE

The ingredients listed in the preceding paragraph must be stored in a probabilistic knowledge base. Usually, fixed likelihoods are assumed and ratios are used to measure the relative support the evidence provides to a particular hypothesis confronted to its alternatives, as in Pearl (1985) and (1986). It is easy to see from (2.1) that it holds:

$$P(H|E)/P(\neg H|E)=[P(E|H)P(H)]/[P(E|\neg H)P(\neg H)]. \quad (2.2)$$

Formulas like (2.2) are suitable to measure the "uncertainty factor" assigned to a rule like (1.1): the likelihoods and the initial probability P(H), or some satisfactory approximation to them, are to be elicited from the expert knowledge.

An important remark is appropriate here: it is often said that probabilistic inference requires too much from the expert who is asked to assign a point valued measure of degree of belief. I would like to point out that the qualitative features of the probabilistic analysis of the problem at hand are maybe more important than the numbers involved: to set the problem in a frame suitable for the mechanical implementation of Bayes theorem, we need to carefully judge, as far as we can, the background information that we possibly have about the relevant hypothesis, the observable events possibly correlated to that hypothesis, and the possible alternative explanations of those observable events. All this stuff is good scientific methodology, independently from the trustworthiness of the numbers we get at the end.

Let's consider now, as an example, how to apply formula (2.1) to the rule (1.1). We shall assume, for short, that "X" and "Y" mean, respectively, "light X is ON" and "valve Y is OPEN" (and, conversely, "$\neg$X" and "$\neg$Y" mean "light X is OFF" and "valve Y is CLOSED"). Suppose, furthermore, that X belongs to the *controller* system and Y to the *controlled* system.

Formula (2.2) becomes:

$$P(Y|X)/P(\neg Y|X)=[P(X|Y)P(Y)]/[P(X|\neg Y)P(\neg Y)]. \quad (2.3)$$

Here, the background probability $P(Y)$ might be supplied by a consideration of the information about the working of the controlled system independently from the information provided by the controller, and the information about the reliability of that particular piece of machinery.
The likelihoods $P(X|Y)$ and $P(X|\neg Y)$ will depend upon the reliability of the controller but, possibly, about other factors as well. As a first approximation, suppose we take into account the reliability of the controller only: in doing that, we introduce another variable S, the state of the controller, representing by "S" the hypothesis that it is working well (and by "¬S" the hypothesis it has suffered some breakdown affecting X). Now, the background probability $P(S)$ will be given by the information concerning the reliability of the controller, and we can reasonably assume that:

(A) $P(S)$ is independent from $P(Y)$;

(B) given the simplifying hypothesis that we take into account the reliability of the controller only, if the controller is working well, then we can be certain that the state reported by the controller is the true state of the controlled system, and if the controller is not working well, then we can be certain that the state reported is *not* the true state of the controlled system:

$$P(X|YS)=P(\neg X|\neg YS)=P(X|\neg Y\neg S)=P(\neg X|Y\neg S)=1.$$

Given the assumptions (A) and (B) the underlying probability space, constructed by means of the product rule, will be rather simple:

$$\begin{aligned}&P(XYS)=P(X|YS)P(Y|S)P(S)=P(Y)P(S);\\&P(X\neg Y\neg S)=P(X|\neg Y\neg S)P(\neg Y|\neg S)P(\neg S)=P(\neg Y)P(\neg S);\\&P(\neg XY\neg S)=P(\neg X|Y\neg S)P(Y|\neg S)P(\neg S)=P(Y)P(\neg S);\\&P(\neg X\neg YS)=P(\neg X|\neg YS)P(\neg Y|S)P(S)=P(\neg Y)P(S);\end{aligned} \tag{2.4}$$

Applying formula (2.1) in the right member of (2.3) and substituting in it the values found in (2.4), we'll obtain:

$$P(Y|X)/P(\neg Y|X)=[P(S)P(Y)]/[P(\neg S)P(\neg Y)]. \tag{2.5}$$

In this particular case, we observe that $P(X|Y)=P(\neg X|\neg Y)$, and $P(\neg X|Y)=P(X|\neg Y)$, from which it follows: $P(X|\neg Y)=1-P(X|Y)$. This is due to assumption (B), but it couldn't be the case, in general, if we drop (B).
If, for instance, we refine the analysis by including into the picture the communication channel between the controller and the controlled system and we assume that in case of a communication breakdown the light X will be off, whatever the state of Y and of the controller, then $P(X|Y)\neq P(\neg X|\neg Y)$, and $P(X|\neg Y)\neq 1-P(X|Y)$.
Setting the problem into the frame requested by Bayes theorem allows the expert to apportion the responsibility of the result to the various components of the inference, that is, background knowledge and observations, and to refine the analysis, if it is needed, maintaining the global coherence with those elements left unchanged.
In those cases where the evidence itself is uncertain, as when the very observational data are uncertain, or when the probable consequent of a rule has to be taken as the antecedent of another rule, a generalization of Bayes theorem has been proposed, by some authors like Duda, Hart and Nilsson (1976) on purely practical grounds, and by other authors like Jeffrey (1983) on more theoretical bases. The formula is the following one, where P is the initial probability and P' the final probability:

$$P'(H)=[P(H|E)P'(E)]+[P(H|\neg E)P'(\neg E)]. \tag{2.6}$$

Likelihoods and conditional probabilities are kept fixed, as they express the theoretical knowledge embodied in a model of the system, and the inputs are not given any more by the news that "E is true", but by the news that "E with probability P'(E)".

2.3. PROBLEMS AND CRITICISMS.

An immediate consequence of the additivity rule is that:

$$P(\neg H|E)=1-P(H|E). \tag{2.7}$$

Some people have claimed, as for instance Buchanan and Shortliffe (1984), that evidence for H is not necessarily evidence against $\neg$H, and that sometimes we might prefer to leave the degree of belief in "$\neg$H" given "E" unassigned.
This criticism is based upon a confusion between the *posterior degree of belief* and the *degree of support* provided by the evidence E to the hypothesis H. The latter is given by the new information contained in the observation of E, while the former is given by the combination of this new information with the background information. If we accept the "likelihood principle", which says that all the information provided by an experiment is contained in the likelihood function, then the criticism is immediately met by observing that likelihoods, in general, are not additive, as shown in the preceding pharagraph.
The "likelihood principle" is still waiting for some serious objection.
A serious challenge for A.I. implementations of probability theory is the *independence* problem. Its origins lay in the fact that, in rule based expert systems, there are very often two or more rules with the same consequent:

"if B then A"; "if C then A". (2.8)

Even a simple conditional probability like P(A|BC) is quite complex to calculate, because we have to known the joint probabilities P(ABC) and P(BC). Worse than that, probabilistic dependence, when it exists, hinders the basic advantage of a rule based expert system: its modularity and the sequential character of the computations involved.
In the system PROSPECTOR, Duda, Hart and Nilsson (1976) made the explicit assumption that the antecedents of any rule to which Bayesian updating is to be applied must be probabilistically independent, so that:

$$P(A|BC)=[P(B|A)P(C|A)P(A)]/[P(B)P(C)]. \tag{2.9}$$

This quite restrictive assumption raised a lot of criticism, see for instance Pednault, Zucker and Muresan (1981).
Attempts to meet this problem has lead Pearl to develop *Bayesian networks* in his (1986) and (1987). They are essentially a way of preserving the modularity of rule based systems even when some nodes of the network representing the set of rules are not probabilistically independent. The trick consists in taking advantage of a well known property of Markov chains: when two chains start from the same event, then they are probabilistically independent, given the truth of that particular event. If we represent the network of rules as an oriented graph, then, while many-to-one links in it create dependence, one-to-many links create independence, at par with Markov chains.
In (2.8), "B" and "C" might or might not be independent but, if we are able to introduce two new rules like:

"if E then B"; "if E then C", (2.10)

then B and C become conditionally independent, given E. In such a graph, the underlying joint probabilities are easily obtained by product formulas. If:

$$P(B|CE)=P(B|E);\ P(C|BE)=P(C|E), \tag{2.11}$$

then:

$$P(BCE)=P(B|CE)P(C|E)P(E)$$
$$=P(B|E)P(C|E)P(E). \quad (2.12)$$

By (2.11) and (2.12), (2.9) becomes:

$$P(A|BC)=[P(B|A)P(C|A)P(A)]/[P(B|E)P(C|E)P(E)]. \quad (2.13)$$

Necessary and sufficient conditions for introducing a "dummy" variable E between B and C have been given in Kim (1983), and Kim and Pearl (1983). The shortcomings of this approach are that such a graph, called a *Kim tree*, is, in general, only an approximation of the original network, and that our ability to represent the expert knowledge in this way seems to be restricted by our ingenuity in concocting it as causal knowledge, because "dummy" variables, along with their Markovian analogues, amount to be common causes of the events they connect: on this point, see Shafer (1987).
Another classical criticism of the probabilistic approach is that the expert lacks all the information needed to assign the initial probabilities or, in other words, the background knowledge very often is not adequate to assign a point valued function to our beliefs. The assignment of the joint distribution requires the simultaneous solution of a set of equations of the following form:

$$P[(+)X1...(+)Xn]=P[(+)Xn|(+)Xn\text{-}1...(+)X1]P[(+)Xn\text{-}1|(+)Xn\text{-}2...(+)X1]...$$
$$...P[(+)X2|(+)X1]P[(+)X1], \quad (2.14)$$

where (+) is to be substituted by "¬" or by nothing.
Even if take for granted that the independence issue has been solved in a satisfactory way, many of the members of the product (2.14) will be "unknown", and we have to assign "default" probability values. To perform this task, some people, like Lemmer and Barth (1982), and Cheeseman (1983) and (1984), have make the proposal of considering the problem of the prior probability assignment as a problem of *constrained maximization*, so as to apply to it the usual techniques: the known probabilities are transformed into linear constraints, and a *maximun entropy algorithm* is used, obtaining the initial distribution with the maximum Shannon's entropy compatible with the given constraints.
The standard objection to this kind of procedure is quite straightforward: why a distribution that maximizes entropy is better than any other initial distribution? The objection is justified as far as the proposed algorithm does not follow, of course, from the basic consistency requirements (C1) and (C2), and it must be warranted on different grounds.
The argument given for the maximum entropy method is essentially an argument by symmetry: it is the only method to assign "default" probabilities which yields a solution invariant under the group of mathematical transformations, if we are able to identify it, which map the original problem into an equivalent one. In simpler words, it is fair enough to say that it is a method which yields equivalent solutions for equivalent problems: see Jaynes (1983), Shore and Johnson (1980), Tikochinsky, Tishby and Levine (1984), Erickson and Smith (1988).
It is to be mentioned that the generalization (2.6) of Bayes theorem has been shown to be logically equivalent to a principle of minimization of the cross entropy

$$I(P,P')=P'(E)\log[P'(E)/P(E)+P'(\neg E)\log[P'(E)/P(E)] \quad (2.15)$$

that is to say, belief updating by means of (2.6) picks up exactly that posterior P' which minimizes the "relative distance" from the initial P, as measured by (2.15): see Williams (1980) and Domotor (1985).
In other words, (2.6) yields the minimum change justified by the new information P'(E).

3. Belief Function Systems

3.1. THE RULES

A theory accepting weaker consistency constraints has been proposed by Shafer (1976), developing a pionieristic work by Dempster (1967) and (1968). The theory has quickly found a sympathetic audience in the A.I. community, because it seems to meet some of the above mentioned difficulties of the probabilistic approach: see for example Barnett (1981), Garvey, Lowrance and Fischler (1981), Lu and Stephanou (1984), Shafer (1987) and Smets (1988a).

In this approach, the expert is asked to assess her degrees of belief under the form of a real valued function *m* which satisfies only the consistency requirement (C1): additivity is not required any more.

This function *m*, which is called a *basic probability assignment* (the name is ambiguous, though, because it is not a probability, lacking additivity), satisfies the following conditions:

(i) $m(\emptyset)=0$;

$$\text{(i) } \sum_{A\subseteq\Omega} m(A)=1, \tag{3.1}$$

Condition (i) says that the minimum degree of belief must be assigned to a self-contradictory statement, and condition (ii) says that the sum of the degrees of belief of all the statements belonging to the algebra of statements must be equal to the maximum degree of belief (Ω is the top of this algebra, the tautological statement "$A \vee \neg A$").

Then we can define a *belief function* as follows:

$$\mathrm{Bel}(A)= \sum_{B\subseteq A} m(B). \tag{3.2}$$

From (3.2) and (3.1) we immediately have:

$$\mathrm{Bel}(\Omega)= \sum_{B\subseteq\Omega} m(B)=1, \tag{3.3}$$

that is, belief functions assign the maximum degree of belief to the tautology, so that consistency requirement (C1) is fully satisfied.

The most important piece of the machinery is Dempster's rule, so called after Dempster (1968), which tells us how to combine a belief function Bel, with *m*-values m(A) representing the degree of belief based upon a certain body of evidence, with a belief function Bel', with *m*-values m'(A), representing the degree of belief based upon a second body of evidence, *subjectively independent*, according to the expert's opinion, from the first one. The pooling of these two bodies of evidence yields a belief function Bel*, with *m*-values m*(A). Dempster's rule has the general form:

$$m^*(A)= [\sum_{BC=A} m(B)m'(C)]/K \tag{3.4}$$

with

$$K=1- \sum_{BC=\emptyset} m(B)m'(C). \tag{3.5}$$

The normalization factor 1/K is the proportionality factor needed to satisfy condition (i) in (3.1), and it has its analogous in the factor 1/P(E) in the formula of Bayes theorem.
The updated belief function will have, of course, the form:

$$Bel^*(A)= \sum_{B\subseteq A} m^*(B). \tag{3.6}$$

Dempster's rule is the unique rule that satisfies the consistency requirement:

$$\sum_{(BC)\subseteq\Omega} m(BC)=1, \tag{3.7}$$

which is an immediate extension of condition (ii) in (3.1).

3.2. THE MACHINE

A straightforward way to implement the theory of belief functions in rule based expert systems seems to be to associate to each rule a belief function which is supposed to measure the degree of support the truth of the antecedent of the rule gives to the consequent. This is the way the theory is used in the latest version of the MYCIN system: see Buchanan and Shortliffe (1984), and Gordon and Shortliffe (1985). The antecedent of the rule is interpreted as confirming the conclusion with degree *m*, while the complementary value 1-*m* which, in the original version of MYCIN was left unassigned, according to the idea we have already seen that evidence for "A" is not necessarily evidence for "¬A", is now assigned to the tautology Ω, following Shafer's theory.
The degree of belief assigned to the hypothesis that "valve Y is OPEN" if it is true that "light X is ON", will be expressed by the number:

$$m^*(XY)=x; \tag{3.8}$$

with $m^*(\Omega)=1-x$.
Let's see how we can calculate (3.8) within this framework: we shall make the same assumptions (A) and (B) of paragraph (2.2), introducing the variable "S" again with the same meaning. If we want to take account of the background information about the reliability of Y and of the controller S, we have to combine together these two independent bodies of evidence by means of formula (3.4), so obtaining:

$$\begin{aligned} &m(YS)=m(Y)m(S)\\ &m(\neg Y\neg S)=m(\neg Y)m(\neg S)\\ &m(Y\neg S)=m(Y)m(\neg S)\\ &m(\neg YS)=m(\neg Y)m(S); \end{aligned} \tag{3.9}$$

where m(Y) is the degree of belief in the reliability of Y, and m(S) is the degree of belief in the reliability of S, both of them given the background information. Notice that in (3.9) we have assumed also that both m(Y) and m(S) are additive, given that it is plausible the data we might have about the reliability of the two pieces of machinery are statistical data. Due to the independence hypothesis, (3.9) has exactly the form of the set of equations (2.4), with m(Y)=P(Y) and m(S)=P(S).
The *m* so obtained give our initial degree of belief in "YX" that has to be combined, by means of Dempster's rule again, with the new information provided by

$$m'(X)=1;\ m'(\neg X)=0, \tag{3.10}$$

to get the desired m*(XY). First of all, we compute:

$$
\begin{aligned}
&m^*(XYS)=m(YS)m'(X)=m(YS)\\
&m^*(XY\neg S)=m(Y\neg S)m'(X)=m(Y\neg S)\\
&m^*(X\neg YS)=m(\neg YS)m'(X)=m(\neg YS)\\
&m^*(X\neg Y\neg S)=m(\neg Y\neg S)m'(X)=m(\neg Y\neg S)\\
&m^*(\neg XYS)=m(YS)m'(\neg X)=0\\
&m^*(\neg XY\neg S)=m(Y\neg S)m'(\neg X)=0\\
&m^*(\neg X\neg YS)=m(\neg YS)m'(\neg X)=0\\
&m^*(\neg X\neg Y\neg S)=m(\neg Y\neg S)m'(\neg X)=0.
\end{aligned}
\tag{3.11}
$$

But, given assumption (B), $(XY\neg S)=(X\neg YS)=\emptyset$, therefore, by (3.4):

$$m^*(XY)=m^*(XYS)/\{1-[m^*(XY\neg S)+m^*(X\neg YS)]\}. \tag{3.12}$$

But, by (3.9), we have:

$$m^*(XY)=m(Y)m(S)/(1-\{[m(Y)m(\neg S)]+[m(\neg Y)m(S)]\}). \tag{3.13}$$

Given the additivity of the function *m*, it holds:

$$[m(Y)m(S)]+[m(\neg Y)m(\neg S)]=1-\{[m(Y)m(\neg S)]+[m(\neg Y)m(S)]\}. \tag{3.14}$$

Therefore, under the assumption of additivity plus (A) and (B), (3.12) yields the same result that Bayes theorem, the form of which in this particular example is:

$$P(Y|X)=[P(S)P(Y)]/[P(S)P(Y)+P(\neg S)P(\neg Y)]. \tag{3.15}$$

We have seen that Dempster's rule is another way of combining together the basic ingredients of the inductive recipe which Bayes theorem makes use of as well, and a way which produces equivalent results when the degrees of belief are additive, that is, when the consistency requirement (C2) is satisfied and belief functions turn out to be probability functions. We combine together background information and likelihoods, disguised in the normalization factor. The real difference is when we use not additive functions, and when we drop the assumption of the independence of the bodies of evidence, in our example when we drop assumption (A). More about this point will be said in the next paragraph.
It is left to be said here that Dempster's rule applies immediately when the evidence is incertain, the case for formula (2.6) in probabilistic inference: exactly the same machinery does work if, instead of $m'(X)=1$, the news is received that $m'(X)<1$.
The theoretical relationships between (2.6) and Dempster's rule are discussed more at length in Garbolino (1987).

3.3. PROBLEMS AND CRITICISMS

The most heated discussions have been centered about the issue of non-additivity and the refusal, from hard-core supporters of the theory, of (C2) as a necessary condition of consistency for inference under uncertainty. The most serious attempt to criticize de Finetti's, Savage's, and Lindley's arguments for additivity is that one by Shafer (1986). Here I shall raise some maybe more practical points.
There was a widespread opinion, when belief function theory entered the A.I. arena, that it was easier to be implemented than probability theory because it was apted to fit a less precise state of knowledge than probability with its request for the computation of a joint probability distribution, which raises so many troubles, as we have seen above. I think that this opinion was based upon a misunderstanding due to the mistake of sweeping conceptual issues under the carpet of the mathematical machinery.
The mathematical machinery of belief function theory is, of course, simpler than that one of probability theory because the former has less constraints to satisfy. But this is just a

mathematical fact, and the interpretation of the mathematical tools of a theory is another story. Using belief function theory we don't have to be worried about those complex joint distributions simply because we don't have the product rule. But the real point, of course, is: do we really need to be worried about joint distributions? In other words: do we need to take account of both background knowledge and likelihoods?

Putting things this way, we see that there is a very basic point at stake: is our inductive recipe good or not? And if it is not, why? It is possible, of course, that the recipe is wrong, but critics of Bayes theorem have not given so far a satisfactory answer to this question, maybe misled by the mathematical fact pointed out before. Less mathematical constraints mean an higher degree of freedom in framing the knowledge base: using probability theory, we have to use the additivity rule and the product rule and, insofar as Bayes theorem is a logical consequence of them, the expert is obliged to make up her mind about background information and likelihoods. There is no other way to work out a quantitative inference. Using belief function theory, we can do that as well, as we have seen, but it is possible to drop out one of the ingredients and the machinery still works.

Suppose that in our example the expert takes into account only the reliability of the controller, say x. Then the initial belief would be, with the same assumptions that before:

$$m[(XYS)v(\neg X\neg YS)]=x; \quad m[(\neg XY\neg S)v(X\neg Y\neg S)]=1-x; \tag{3.16}$$

and , after having received the news that the light X is on, the updated belief would be:

$$m^*(XYS)=x$$
$$m^*(X\neg Y\neg S)=1-x. \tag{3.17}$$

It has been claimed, as for instance in Smets (1988b), that (3.17) is a valid result when the expert has no background knowledge at all about the reliability of Y. It is easy to check that Bayes theorem yields the same result if, lacking any relevant knowledge about the reliability of Y, we put P(Y)=0.5, that is, Laplace's uniform prior distribution (which we get as well as a particular case of the maximum entropy principle when the only constraint we know is the number of alternative and mutually exclusive hypotheses). Indeed, under the usual assumptions :

$$P(Y|X)=[P(S)(0.5)]/(0.5)=P(S). \tag{3.18}$$

Smets is right that it is a valid result, but we can see in (3.18) the reason why it is so: if we have no background knowledge, all the relevant information is contained in the likelihood.

Now the funny thing is that people, who claim that P(Y)=0.5 is a totally unwarranted assignment, introducing in the problem information which we don't have, eventually find (3.17) a satisfactory solution.

This is what I mean by "sweeping problems under the carpet of mathematics". There is a lot of difference between saying that, when the expert is sure she has no background knowledge whatever about the problem at hand, she has to do the best she can in this situation, and claiming that background information is not to be considered in making inferences under uncertainty.

But if it is to be considered, then analyzing the problem within the framework of belief function theory is not less complex than doing it within the probabilistic framework: we have to take account of the same factors. Moreover, if we drop assumption (A), that is, stochastic independence, in probability theory we can manage the problem by means of the product rule: if A and B are not independent, we still know that it must hold:

$$P(AB)=P(A|B)P(B)=P(B|A)P(A), \tag{3.19}$$

with $P(A|B)\neq P(A)$ and $P(B|A)\neq P(B)$.

But we cannot apply any more Dempster's rule to calculate m*(AB) from m(A) and m'(B), because the very condition of its applicability fails.
This raises the point of the *commutativity* of Dempster's rule: it has been said that it is more suitable for the combination of several uncertain bodies of evidence, because it is always commutative while, on the other hand, the probabilistic rule (2.6) is commutative only under more restrictive conditions: see Diaconis and Zabell (1982). But this generalized commutativity is a consequence of the independence condition: when the latter fails, the former does too.
With respect to the other "independence" problem we have when there are two rules with a common consequent, belief function meet the same troubles that probability does. If we represent two rules with the same consequent by means of two numbers m(A) and m'(A), we can combine them by means of Dempster's rule only if their antecedents are independent. The problem has been addressed recently by Shenoy and Shafer (1986), and by Mellouli, Shenoy, and Shafer (1987). Their method is an extension of the above mentioned method of *Kim trees*: instead of having a single "dummy" variable we have a partition which renders conditionally independent all the partitions which have it as their common root. This tree of partitions is called a qualitative Markov tree and has the property that the disconnected branches which result from the removal of a partition are conditionally independent given that partition. Belief functions based upon the disconnected partitions can be thus combined according to Dempster's rule.

4. Possibility Theory

4.1 THE RULES

In possibility theory we have a further weakening of the logical constraints imposed to degrees of belief. In particular, the constraint that the degree of belief of a self-contradictory statement must be the minimum is preserved:

$$\Pi(\emptyset)=0, \tag{4.1}$$

and simple additivity, as in probability theory, and sub-additivity, as in belief function theory, is substituted by the *maximum* operation:

$$\Pi(A \vee B)=\max\{\Pi(A),\Pi(B)\}. \tag{4.2}$$

A *possibility* function Π is obtained fron the knowledge of the occurrence of an event B by letting $\Pi(A)=1$ if $(AB)\neq\emptyset$, and $\Pi(A)=0$ otherwise: $\Pi(A)=1$ means that A is *possible* given that B has occurred.
The interesting case, of course, occurs when $\Pi(A)$ is neither 0 nor 1 but an intermediate value between them. This occurs when either the description of the event B is vague or imprecise, or the event A is vague, or both. In this case, A and B are represented by *fuzzy sets* and the definition of possibility measures is:

$$\Pi(A)= \max_{\omega\in A} \mu(\omega) \tag{4.3}$$

where μ is the membership function of the fuzzy set B: a fuzzy membership function takes its values in the interval of real numbers, that is, $\mu\in[0,1]$, while a classical membership function f is a two-valued function, that is, $f\in\{0,1\}$. (4.3) measures the degree to which the event A is *compatible* with the event B, when they are vaguely described.

We can notice that possibility theory is a weakening of belief function theory by observing that, in the latter, if we know for certain an event B, then it is possible to define Bel(A)=1 if B A, and Bel(A)=0 otherwise.
In possibility theory we have the dual of the possibility measure Π, the *necessity* function N, defined as:

$$N(A)=1-\Pi(\neg A), \tag{4.4}$$

satisfying the *minimum* operation:

$$N(AB)=\min\{N(A),N(B)\}. \tag{4.5}$$

The function N(A) measure the degree of "necessity" of A. Possibility and necessity functions are a kind of quantified fuzzy modal operators corresponding to the two "possibility" and "necessity" operators in classical (non fuzzy) modal logic.

4.2. THE MACHINE

Possibility theory is a powerful tool for reasoning under uncertainty when the rules express vague or ill-defined knowledge. It allows to equip an inferential engine with logical rules, rules of fuzzy logic, even if it is not possible for the expert to provide crisp knowledge: for an overview of the basket of the mathematical tools supplied by possibility theory see Dubois and Prade (1988a) and (1988b); Zadeh (1983) and (1986). The possibility function Π, and its dual N, measure the uncertainty about the truth of the consequent of a rule arising from the vagueness of the relationship between the antecedent and the consequent itself.
The rule (1.1) as it is and we have used it so far, it is not a suitable instance of a possibilistic updating rule, because statistical knowledge or uncertainty arising from random "noises" have nothing to do with vagueness. Possibility measures have the following property:

$$\max\{\Pi(A),\Pi(\neg A)\}=1, \tag{4.6}$$

which makes them unsuitable for representing randomness since it will hold in general that $\Pi(A)+\Pi(\neg A)>1$, except for events which occurrence is known for sure, for them are equivalent to the tautological event and, from (4.2) and (4.6), we have that $\Pi(A \vee \neg A)=1$.
An example of a vague rule is:

"if the deuterium boils then a fusion is occurring in the palladium bar". (4.7)

What characterizes a vague rule is the presence of predicates of natural language which are tipically vague. The statement (4.7) can be paraphrased as:

"if the deuterium enjoys the property of being in a state of boiling,
then a fusion is occurring in the palladium bar". (4.8)

What is our probability that a process of fusion is occurring, if we have observed that the deuterium "boils"? The obvious answer is: no meaningful probability can be assigned to such an hypothesis without knowing exactly what it is meant by "boils". Why do we want to eliminate the "vagueness" of the term "boils"? The answer is interesting for it will show the demarcation criterium between probabilistic (and belief function) inferences and possibilistic (and fuzzy) inferences.
Let's consider again our two inductive ingredients: we can conceive of the prior probability of a nuclear fusion under such and such conditions (actually very low, I think); what about the likelihood P(deuterium boiling| fusion)? No likelihood can be reasonably given, not a qualitative one either, if the vagueness is not eliminated by the observational

report. To begin with, the measurement of the quantity of energy produced in the system will be necessary, for the energetic balance is relevant in estimating the likelihood.
With vague rules, one of the ingredients of the inductive recipe is wanting: likelihoods. The reason is very simple: no vague statements are acceptable as reports of "scientific" experiments or observations.
This does not imply, of course, that there is no room for vague rules and for a logic of vague inference: they are important if either the expert is not able to dispose of vague predicates or it is useful or appropriate to represent knowledge in ordinary language andthe rule embodies common sense knowledge. A typical example of vague rules are rules of the form:

"if X is (more or less) P then Y is (more or less) Q", (4.9)

but the uncertainty attributed in this kind of rules is *semantical* uncertainty, that is, uncertainty about the *meaning* of the statements.
Part of the meaning of a predicate is its *extension*: the set of objects which enjoy the property expressed by the predicate. When a predicate Q is vague, then this set is a fuzzy set: we are uncertain whether a particular object on the borderline belongs or not to the set, that is, if it enjoys the property Q. This is a different problem from the problem of calculating my degree of belief in the hypothesis that the object enjoys the property Q, given that I have observed that it enjoys the property P (and R, S, ...). In performing the latter inferential task when natural language predicates are involved, we usually fix their meanings and we keep them constant. This is a kind of idealization, of course, but it is necessary because only in this way we can investigate the theoretical relationships between the predicates P and Q (and R,S,...) the knowledge of which allows us to estimate the likelihood of observing P if the object enjoys Q.

4.3. PROBLEMS AND CRITICISMS

From the point of view put forward in the previous paragraph, the issue of non-additivity is not relevant for criticizing possibility theory. If our aim is to produce a logic of vagueness, it is quite obvious that weaker consistency requirements must hold, and a certain measure of "inconsistency" from the point of view of a "crisp" logic not only must be allowed, but it is compulsory, otherwise we were not doing "vague" inferences.
A more controversial point regards the plausibility of a "vague" inductive inference. The point of view stated above, doubtless, is skeptical about this plausibility: here we shall discuss this point with some more details.
Dubois and Prade (1988c) have proposed a "vague" generalization of the product rule for the dual measures of possibility and necessity:

$$\Pi(AB)=\Pi(A|B)\bullet\Pi(B)$$
$$N(AB)=N(A|B)\bullet N(B). \tag{4.10}$$

These two dual forms are not compatible, though, because $N(A|B)$ is undefined as soon as $N(B)=0$, what we obtain from (4.4) as soon as $\Pi(\neg B)=1$ (remember that we can have both $\Pi(B)=1$ and $\Pi(\neg B)=1$, the case of total uncertainty).
Therefore, we consider only $\Pi(AB)$. The two main solutions, for the operation •, compatible with the mathematical properties of Π are the *minimum* operation and the product. Let's be:

$$\Pi(AB)=\min\{\Pi(A|B),\Pi(B)\}. \tag{4.11}$$

Then , given to the monotonicity of Π, according to which it must hold $\Pi(AB)\leq\Pi(B)$, either

$$\Pi(AB)=\Pi(A|B), \text{ if } \Pi(AB)<\Pi(B) \tag{4.12}$$

or

$$\Pi(A|B)\in[\Pi(B),1], \text{ if } \Pi(AB)=\Pi(B), \tag{4.13}$$

that is, $\Pi(A|B)$ can assume any value between $\Pi(B)$ and 1, because either $\Pi(A|B)$ is the minimum, in which case it must be equal to $\Pi(B)$, or $\Pi(A|B)>\Pi(B)$ and the maximum is 1.
But (4.12) and (4.13) lead to what seems to be an intuitively unsatisfactory results, with rispect to the given characterization of inductive inference. Let's see what they are.
We have also:

$$\Pi(AB)=\Pi(B|A)\bullet\Pi(A). \tag{4.14}$$

Therefore, by (4.11), either

$$\Pi(AB)=\Pi(B|A), \text{ if } \Pi(AB)<\Pi(A), \tag{4.15}$$

or

$$\Pi(A|B)\in[\Pi(A),1], \text{ if } \Pi(AB)=\Pi(A). \tag{4.16}$$

In the latter case we have, by (4.13), $\Pi(A)=\Pi(B)$, the initial plausibilities of A and B are the same, so that it turns out to be a particular case. In the general case, where both $\Pi(AB)<\Pi(B)$ and $\Pi(AB)<\Pi(A)$, we obtain, by (4.15) and (4.12):

$$\Pi(A|B)=\Pi(B|A). \tag{4.17}$$

This property of symmetry might be a reasonable property, indeed, for the degree of uncertainty about the semantical relatedness of two vague statements A and B, supposed than one of them is known to be true. On the other hand, (4.17) is absurd as a property of inductive inference, for the latter is surely not symmetric in this sense.
The reader will notice that the result (4.17) is basically due, *via* (4.11), to the non additivity of Π. This fact points out to the *liaisons dangereuses* which exist between additivity and the product rule on one side, and inductive inference on the other, *liaisons* at the roots of Cox's argument (1961).
If then we consider the product as the second solution for the operation •, we get:

$$\Pi(A|B)=[\Pi(B|A)\Pi(A)]/\Pi(B). \tag{4.18}$$

But this formula is again unsatisfactory, on intuitive grounds, if Π is a possibility function.
Suppose we are in the following situation: we have a theoretical model in which A is an explanation of B in the strong sense that the truth of A implies the truth of B, given certain background conditions we can postulate to be satisfied in our observational setup; on the other hand, B can have other explanations as well, but we are totally ignorant what they might be. This in an idealized situation but common enough in scientific inference.
Such a situation of total ignorance would be shaped in possibility theory by the assignments:

$$\Pi(B|\neg A)=\Pi(\neg B|\neg A)=1. \tag{4.19}$$

Moreover we have, according to the conditions of the problem:

$$N(B|A)=1;\ N(\neg B|A)=0, \tag{4.20}$$

from which we get:

$$\Pi(B|A)=1-N(\neg B|A)=1; \tag{4.21}$$

Therefore, substituting in (4.18) the values given by (4.19) and (4.21), we obtain:

$$\Pi(A|B)=\Pi(A)/\Pi(B);\quad \Pi(\neg A|B)=\Pi(\neg A)/\Pi(B), \tag{4.22}$$

so that the posterior degrees of belief are simply proportional to the initial degrees of belief:

$$\Pi(A|B)/\Pi(\neg A|B)=\Pi(A)/\Pi(\neg A). \tag{4.23}$$

The observation of B has not changed the *relative* "plausibility" of "A" and "$\neg$A" and this fact seems rather counterintuitive. The usual way out from the shortcomings of Π is to appeal to the dual measure N: in this case the information change due to the observation of B would be measured by N(A|B) and N($\neg$A|B). Indeed, by (4.20) we have:

$$N(A|B)=[N(B|A)N(A)]/N(B)=N(A)/N(B);\quad N(\neg A|B)=[N(\neg B|A)N(A)]/N(B)=0. \tag{4.24}$$

Unfortunately, in this case the answer runs directly into a contradiction, due to the fact that the dual formulas of conditioning are not compatible.
Let Π^* and N^* be the updated degrees of possibility and necessity after the first step of the inferential process. Then:

$$\Pi^*(A)=\Pi(A|B);\ \Pi^*(\neg A)=\Pi(\neg A|B); \tag{4.25}$$

and

$$N^*(A)=N(A|B);N^*(\neg A)=N(\neg A|B). \tag{4.26}$$

Therefore, by (4.24) and (4.26) we have:

$$N^*(\neg A)=0. \tag{4.27}$$

Now, by definition of necessity functions, it must hold:

$$N^*(\neg A)=1-\Pi^*(A). \tag{4.28}$$

Thus, by (4.28) and (4.25) we have also:

$$N^*(\neg A)=1-\Pi(A|B). \tag{4.29}$$

Hence, from (4.29), (4.27), and (4.22), we immediately obtain that, unless it is the case that $\Pi(A)=\Pi(B)$,

$$N^*(\neg A)\neq N^*(\neg A). \tag{4.30}$$

A contradiction.
If Π is an upper probability function, a similar result obtains: in general $\Pi^*(A)\neq\Pi(A|B)$, see Kyburg (1987).

5. An Overall Appraisal

5.1. DIFFERENT KINDS OF UNCERTAINTY.

In the preceding paragraphs it has been argued that there exist two different kinds of uncertainty: the first one related to the quantity of information we have about the *truth* of a certain statement, and the second one related to the quantity of information we have about the *meaning* of a certain statement.

Therefore, there are two different measures of uncertainty: the first one measures the degree of belief about the truth of a certain statement; the second one measures the degree of belief about the meaning of a certain statement. Probabilities and belief functions belong to the first kind of measures; possibility and necessity to the second kind.

I think that there are good arguments to support this point of view, and some of them have been given in this paper. To make clear this distinction is the more important insofar as there is a lot of conceptual confusion and misunderstandings right now in the field of A.I. Confusion and misunderstandings for which a valid apology can be found in the scientific youthness of the field and its impressively fast growth in the last decades. Growth that has already given remarkable offsprings, on the other hand: I put the development of a large bag of logical tools to deal with the vagueness of natural languages among these noticeable and useful offsprings. My skepticism concerns the validity of the attempts to extend them to the process of inductive inference and learning from new data.

In probability theory and in belief function theory it is possible to implement a step-by-step process of knowledge acquisition because it is always possible to take the posterior degree of belief P(A|B) and Bel*(A) obtained at the inductive step n as the "prior" or initial degree of belief for the next inductive step $n+1$. The arguments of paragraph (4.3) seem to cast some doubts about the plausibility of doing the same for possibility theory. If this is true, the reason is that the latter represents a different kind of uncertainty. Of course, new data can increase or decrease the fuzziness of a given statement: my point is simply that *this* kind of learning is not learning by conditionalization.

When facing the task of building a knowledge base, vagueness of descriptions and of natural language rules can be an important feature which is not possible to get rid of. Another example is information retrieval with multi-valued and incomplete described criteria: then Π and N can provide an interval-valued degree of "fitness" of a given object to those criteria.

But if the expert system is supposed to perform sequential inferences about the degree of truth of an hypothesis given partial information, then it is better to supply the inference engine with the rules of probability theory, or belief function theory. I dare say that for doing this kind of inference we need sharpening the expert knowledge and the rules as far as we can, going from fuzziness to crispness.

I have already said something about the choice between probability and belief function theory. Some more remarks are proper here. Dempster's rule is perfectly compatible with a Bayesian inference engine: we have seen that when the basic assignments m are additive and the variables are independent, Dempster's rule and Bayes theorem are equivalent. The former can be advantageously used in combining together several pieces of information about the same hypothesis, coming from independent sources, so that they can be supposed to be probabilistically independent.

About the issue of additivity, I would like to say that stronger mathematical constraints oblige us to develop a deeper analysis of the problem at hand; in doing that, we decrease, even though we cannot completely eliminate, the possibility that some logical contradictions are hidden in the logical consequences of what we *prima facie* believe. Of course, there is a price to be paid in terms of computational resources and time. Unlèss we think that human experts are always well-calibrated in their judgments under uncertainty, the real trade-off is not between "psychological realism" and "logical idealism", but between *consistency* and *efficiency*. Unfortunately, we don't have a theory to trace this

indifference curve, to measure how much of consistency can be paid for a little more efficiency, and conversely.
And it does not look probable that in the near future we shall have it.
Reading the A.I. literature, you sometimes can get the feeling that the philosophical stone has been discovered: all the troubles of probability theory overcome by some brand-new *adhockery* and this is almost done cost-free.
It is not sufficient that the machine does work, outputting some numbers when you feed it with other numbers, for being a good inference engine. In the next paragraph I shall give a simple example to make the point.

5.2. MECHANICAL RULES CANNOT HELP IF YOU GOT THE WRONG PROBLEM

Consider the following case; there are 100 taxi-cabs in town; 1 is green and 99 are yellow. A pedestrian has been killed by a car. An eye-witness says that it was a green cab. The eye-witness is 80% reliable in the sense that there are eight chances out of ten that he saw what he says. What is your degree of belief that it was indeed the green cab?
A Bayesian may assign a prior distribution P(G)=0.01 and P(Y)=0.99. At first sight this prior might look quite arbitrary, because there could be reasons behind the accident which have nothing to do with the colour of the cab, like that someone of the taxi drivers is very uncareful, or that someone of them was drunk that night. But it looks less meaningless if we argue this way: being a dangerous driver or being drunk is uncorrelated with the colour of the cab but, if we are totally ignorant about the propensity of each individual driver to drive dangerously or to drink, then we have to assign to each individual driver the same chances of being an uncareful driver or of being drunk that night. Now, it happens that one out of one hundred of these guys is driving a green cab.
Thus, P(G)=0.01 and P(Y)=0.99, and

$$\begin{aligned} P(G|\text{reported } G) &= [P(\text{rep.}G|G)P(G)]/[P(\text{rep.}G|G)P(G)+P(\text{rep.}G|Y)P(Y)] \\ &= (0.8\text{x}0.01)/[(0.8\text{x}0.01)+(0.2\text{x}0.99)] \\ &= 0.038. \end{aligned} \tag{5.1}$$

A Bayesian prior might be also P(G)=P(Y)=0.5, and in this case we shall obtain:

$$P(G|\text{rep.}G)=0.8. \tag{5.2}$$

The difference in the priors makes room for a very big difference for the final result, but explicitly assessing them gives a chance of criticizing them, eventually changing them. We can criticize the argument leading to the prior behind (5.1), or we can criticize the prior behind (5.2) saying, for example, that assigning P(G)=0.5 means to have some reasons to believe that the individual driving the green cab has much more propensity than any other individual to drive dangerously or to get drunk, so that we cannot claim to be in a state of total ignorance.
Now suppose we want to use belief functions: in this case as well we have the choice among several options. For example, we might use additive functions *m*. Combination by means of Dempster's rule with m(G)=0.01, m(Y)=0.99, m'(G)=0.8, and m'(Y)=0.2, then yields:

$$\begin{aligned} &m(G)m'(G)=0.008; \\ &m(G)m'(Y)=0.002; \\ &m(Y)m'(G)=0.792; \\ &m(Y)m'(Y)=0.198. \end{aligned} \tag{5.3}$$

But (GY)=Ø. Thus, renormalizing we obtain:

$$\mathrm{Bel^*(G)}=0.008/(1-0.794)$$
$$=0.038. \tag{5.4}$$

Then let's represent the problem with the same numerical data but with a not additive function for the reliability of the witness, on the grounds that his evidence supports"G" but it is not evidence for "¬G", a MYCIN-like argument: m(G)=0.01, m(Y)=0.99, m'(G)=0.8, and m'(Ω)=0.2. Then:

$$\begin{aligned} &m(G)m'(G)=0.008;\\ &m(G)m'(\Omega)=0.002;\\ &m(Y)m'(G)=0.792;\\ &m(Y)m'(\Omega)=0.198. \end{aligned} \tag{5.5}$$

Given that (GΩ)=G and (YΩ)=Y, we obtain:

$$\begin{aligned} &\mathrm{Bel^*(G)}=0.01/(1-0.792)\\ &\quad=0.048;\\ &\mathrm{Bel^*(Y)}=0.198/(1-0.792)\\ &\quad=0.952. \end{aligned} \tag{5.6}$$

Now, suppose we accept Shafer's suggestion that total ignorance is to be represented by the vacuous belief function defined as:

$$\mathrm{Bel(A)}=1 \text{ if } A=\Omega; \text{ otherwise } \mathrm{Bel(A)}=0. \tag{5.7}$$

Then our prior is m(G)=0, and m(Ω)=1, and, if m'(G)=0.8 and m'(Y)=0.2, we have:

$$\begin{aligned} &m(G)m'(G)=0;\\ &m(G)m'(Y)=0;\\ &m(\Omega)m'(G)=0.8;\\ &m(\Omega)m'(Y)=0.2, \end{aligned} \tag{5.8}$$

and:

$$\begin{aligned} &\mathrm{Bel^*(G)}=0.8.\\ &\mathrm{Bel^*(Y)}=0.2. \end{aligned} \tag{5.9}$$

And if we take *m'* not additive? In that case the result would be:

$$\begin{aligned} &m(G)m'(G)=0;\\ &m(G)m'(\Omega)=0;\\ &m(\Omega)m'(G)=0.8;\\ &m(\Omega)m'(\Omega)=0.2, \end{aligned} \tag{5.10}$$

and:

$$\begin{aligned} &\mathrm{Bel^*(G)}=0.8.\\ &\mathrm{Bel^*(Y)}=0. \end{aligned} \tag{5.11}$$

Many different numerical results for many different frameworks of the same problem (and we don't have exhausted all of them). For each of these frameworks a combination rule produces some numbers: it works, the machine is running but which framework is the right one?

6. References

BARNETT, J.A. (1981) 'Computational methods for a mathematical theory of evidence', Proc. of the 7th IJCAI, 868-875.

BUCHANAN,B.C. and SHORTLIFFE, E.H. (eds.) (1984) Rule-Based Expert Systems: The MYCIN Experiments of the Stanford Heuristic Programming Project , Addison--Wesley, Reading MA.

CHEESEMAN, P. (1983) 'A method of computing generalized Bayesian probability values for expert systems', Proc. 8th IJCAI, 198-202.

CHEESEMAN, P. (1984) 'Learning of expert systems from data', IEEE Proc. of Worshop on Principles of Knowledge-Based Systems, 112-115.

CHEESEMAN, P. (1988) 'An inquiry into computer understanding', Computational Intelligence 4, 58-66.

COX, R.T. (1961) The Algebra of Probable Inference, The Johns Hopkins Press, Baltimora.

DE FINETTI, B. (1974) Probability Theory, Wiley, New York.

DEMPSTER, A. (1967) 'Upper and lower probabilities induced by a multiple-valued mapping', Annals of Mathematical Statistics 38, 325-339.

DEMPSTER, A. (1968) 'A generalization of Bayesian inference', Journal of the Royal Statistical Society Ser.B 30, 205-247.

DIACONIS, P. and ZABELL, S. (1982) 'Updating subjective probability', Journal of the American Statistical Association 77, 822-830.

DOMOTOR, Z. (1985) 'Probability kinematics, conditionals, and entropy principles', Synthese 63, 75-114.

DUBOIS, D. and PRADE, H. (1988a) Possibility Theory: An Approach to Computerized Processing of Uncertainty, Plenum Press, New York.

DUBOIS, D. and PRADE, H. (1988b) 'An introduction to possibilistic and fuzzy logics' in P.SMETS,H.E.MAMDANI,D.DUBOIS,H.PRADE (eds.), Non-Standard Logics for Automated Reasoning, Academic Press, London.

DUBOIS, D. and PRADE, H. (1988c) 'Conditioning in possibility and evidence theories. A logical viewpoint', in B.BOUCHON, L.SAITTA, R.R.YAGER (eds.), Uncertainty and Intelligent Systems, Lecture Notes in Computer Sciences 313, Springer, Berlin, 401-408.

DUDA, R., HART, P.E. and NILSSON, J. (1976) 'Subjective bayesian methods for rule-based inference systems, Proc. AFIPS Nat. Comp. Conf. 47, 1075-1082.

ERICKSON, G.J. and RAY SMITH, C. (eds.) (1988) Maximum Entropy and Bayesian Methods in Science and Engineering. Vol. 1: Foundations, Kluwer Academic Publishers, Dordrecht.

GARBOLINO, P. (1987) 'A comparison of some rules for probabilistic reasoning', International Journal of Man-Machine Studies 27, 709-716.

GARVEY, T.D., LOWRANCE, J.D. and FISCHLER, M.A. (1981) 'An inference technique for integrating knowledge from disparate sources', Proc. 7th IJCAI, 319-325.

GORDON, J. and SHORTLIFFE, E.H. (1985) 'A method for managing evidential reasoning in a hierarchical hypotheses space', Artificial Intelligence 26, 323-357.

JAYNES, E.T. (1983) Papers on Probability, Statistics, and Statistical Physics, Reidel, Dordrecht.

JEFFREY, R.C. (1983) The Logic of Decision, Chicago University Press.

KEMENY, J.G. (1955) 'Fair bets and inductive probability', Journal of Symbolic Logic 20, 263-273.

KIM, J.H. (1983) 'CONVINCE: a conversational inference consolidation engine', Ph.D. Dissertation, Dept. of Computer Science, Univ. of California, Los Angeles.

KIM, J.H. and PEARL, J. (1983) 'A computational model for causal and diagnostic reasoning in inference systems', Proc. 8th IJCAI, 190-193.

KYBURG, H.E. (1987) 'Bayesian and non-Bayesian evidential updating', Artificial Intelligence 31, 271-294.

LEMMER, J.F. and BARTH, S.W. (1982) 'Efficient minimum information updating for Bayesian inferencing in expert systems', Proc. AAAI Conf., 424-427.

LINDLEY, D.V. (1982) 'Scoring rules and the inevitability of probability', International Statistical Review 50, 1-26.

LU, S.Y. and STEPHANOU, H.E. (1984) 'A set-theoretic framework for the processing of uncertain knowledge', Proc. AAAI Conf., 216-221.

MELLOULI, K., SHENOY, P.P. and SHAFER, G. (1987) 'Propagating belief functions in qualitative Markov trees', International Journal of Approximate Reasoning 1, 349-400.

PEARL, J. (1985) 'How to do with probabilities what people say you can't', Proc. 2nd Conf. on AI Applications: the engineering of knowledge-based expert systems, IEEE Press, Washington, 6-12.

PEARL, J. (1986) 'Fusion, propagation, and structuring in belief networks', Artificial Intelligence 29, 241-288.

PEARL, J. (1987) 'Distributed revision of composite beliefs', Artificial Intelligence 33, 173-215.

PEDNAULT, E.P.D., ZUCKER, S.W. and MURESEN, L.V. (1981) 'On the independence assumption underlying subjective bayesian updating', Artificial Intelligence 16, 213-22.

RAMSEY, F.P. (1931) 'Truth and probability' in F.RAMSEY, The Foundations of Mathematics and Other Logical Essays, Routledge & Kegan Paul, London.

SAVAGE, L.J. (1954) The Foundations of Statistics, Wiley, New York.

SHAFER, G. (1976) A Mathematical Theory of Evidence, Princeton University Press, Princeton.

SHAFER, G. (1986) 'Savage revisited', Statistical Science 1, 463-501.

SHAFER, G. (1987) 'Probability judgment in artificial intelligence and expert systems', Statistical Science 2, 3-44.

SHENOY, P.P. and SHAFER, G. (1986) 'Propagating belief function with local computations', IEEE Expert 1, 43-52.

SHIMONY, A. (1955) 'Coherence and the axioms of confirmation', Journal of Symbolic Logic 20, 1-28.

SHORE, J.E. and JOHNSON, R.W. (1980) 'Axiomatic derivation of the principle of maximum cross-entropy', IEEE Transactions on Information Theory IT-26, 26--37.

SMETS, P. (1988a) 'Belief functions' in P.SMETS, E.H.MAMDANI, D.DUBOIS, H.PRADE (eds.), Non-Standard Logics for Automated Reasoning, Academic Press, London.

SMETS, P. (1988b) 'The transferable belief model: comparison with Bayesian models', unpublished paper, IRIDIA, Universite Libre de Bruxelles.

TIKOCHINSKY, Y., TISHBY, N.Z. and LEVINE, R.D. (1984) 'Consistent inference of probabilities for reproducible experiments', Physical Review Letters 52, 1357--1360.

WILLIAMS, P.M. (1980) 'Bayesian conditionalization and the principle of minimum information', British Journal for the Philosophy of Science 31, 131-144.

YAGER, R.R. (1980) 'A foundation for a theory of possibility', Journal of Cybernetics 10, 177-204.

ZADEH, L.A. (1978) 'Fuzzy sets as a basis for a theory of possibility', Fuzzy Sets and Systems 1, 3-28.

ZADEH, L.A. (1983) 'The role of fuzzy logic in the management of uncertainty in expert systems', Fuzzy Sets and Systems 11, 199-227.

ZADEH, L.A. (1986) 'Is probability theory sufficient for dealing with uncertainty in AI? in L.N. KANAL, J.F.LEMMER (eds.), Uncertainty in Artificial Intelligence, North Holland, Amsterdam.

RELIABILITY PARAMETER ESTIMATION BY COMBINATION OF DATA FROM VARIOUS SOURCES

A.G. COLOMBO
Commission of the European Communities
Joint Research Centre - Ispra Establishment
21020 Ispra (Va) - Italy

ABSTRACT. The paper deals with the estimation of reliability parameters by combination of data from various data sources. The problem is discussed in a Bayesian framework and with reference to the estimation of failure rates. First the combination of expert opinions is considered. The well known Wash-1400 and IEEE-Std 500 approaches are recalled briefly. Then the additive error model and the multiplicative error model are discussed in some detail. The Bayesian parametric approach is illustrated as the reference method to combine expert opinions and experimental data. Finally the combination of computed parameters is discussed in both cases: homogeneous data and non-homobeneous data. As a conclusion, some comments on the various methods are given.

1. INTRODUCTION

To assess the reliability of a system, one needs to assign numerical values to the parameters of the system-model. The parameters usually considered in a reliability analysis are failure and repair rates (generally assumed to be time-independent) and failure probabilities on demand (implicitly assumed to be independent of the number of demands). The same model can be used to estimate a failure rate, or a repair rate, or a failure probability on demand /1,8/. Hence, in the following we only discuss the estimation of a constant failure rate. We discuss the problem in a Bayesian framework, we thus assume that the parameter to be estimated is a random variable, and consider the following data-sources: expert opinions, experimental data and already computed parameters.

In section 2 we present methods of combining expert opinions. Section 3 illustrates the Bayesian parametric approach to combine expert opinions and experimental data. Section 4 reports two different methods of combining computed parameters. Finally, in section 5, we comment on the various methods.

A. G. Colombo and A. Saiz de Bustamante (eds.), Systems Reliability Assessment, 273–283.

2. COMBINATION OF EXPERT OPINIONS

2.1 Two Reference Methods

Two well-known studies that use expert opinions extensively for the estimation of reliability parameters are the Wash-1400 /2/ and the IEEE Std-500 /3/ studies.

Wash-1400 considers components of nuclear power plants. It implicitly assumes a Bayesian approach, and a lognormal distribution is assumed as prior for the failure rate of the components. The parameters of each prior are determined from point estimates of failure rates requested from various organizations. The method is the following:

i) For each component, a range in which its failure rate is most likely to lie was defined subjectively, on the basis of the information collected;

ii) The end points of the range were assumed to be the 5-th percentile and 95-th percentile of the lognormal prior.

The distributions thus defined can be used directly in a reliability analysis, or combined with experimental data as discussed in section 3.

IEEE Std-500 considers electrical and electronic components. First, the failure rate estimates of some 200 experts were collected according to a Delphi procedure. Then, for each component, a group estimate was computed taking the geometric average, i.e.

$$\lambda_j = \left[\prod_{i=1}^{N} \lambda_i \right]^{1/N}$$

where:

λ_j failure rate of the j-th component
λ_i estimate of the i-th expert (i = 1,2,...,N)
N number of experts.

This simple approach is justified as follows: "In the analysis and pooling of the failure rate estimates, care must be taken to ensure that the analytical techniques are no more sophisticated than the pool of estimates to which they are applied." Thus "It was decided that the simple procedure of unweighted geometric averaging would be preferred, since it was observed that the distribution of estimates was better represented by failure rate exponent averages than by mantissa averages."

2.2 Combination of Expert Opinions via Bayes Theorem

In this approach /4/, expert opinions are treated as evidence to be incorporated in the decision maker's body of knowledge in a coherent manner using Bayes theorem.

Denote:

X parameter to be estimated (assumed to be a random variable (r.v.));

$\pi_o(x)$ prior distribution of X. It expresses the decision maker's prior state of knowledge about X (prior to receiving the opinions of the experts);

S* set of experts' opinions about X;

$L(S^*|x)$ likelihood of the evidence S*, given that the value of the parameter to be estimated is x (to be defined by the decision maker);

$\pi(x|S^*)$ posterior distribution of X. It expresses the decision maker's posterior state of knowledge about X, given that he received the set of expert's opinions S*:

K normalization factor.

Bayes theorem gives:

$$\pi(x|S^*) = \pi_o(x)L(S^*|x)/K. \qquad (2.1)$$

The problem is thus reduced to the assessment of π_o and L. In /4/ two models are proposed: additive error and multiplicative error. These models are discussed in this section considering a single expert. Ref./4/ also discusses the case of more than one expert.

In the case of a single expert, Bayes theorem can be written as follows:

$$\pi(x|x^*) = \pi_o(x)L(x^*|x)/K \qquad (2.2)$$

where x* is the expert estimate and $L(x^*|x)$ is the decision maker's model of the expert's ability to make a quantitative judgement about the parameter to be estimated. The shape and functional form of the likelihood may differ from one decision-maker to another modelling the same expert. Generally, the form of the likelihood is also different for a given expert's assessment of different quantities. The decision maker must also define the shape and the functional form of the prior distribution $\pi_o(x)$.

In sections 2.2.1 and 2.2.2 we discuss the two models: additive error and multiplicative error, assuming that the parameter to be estimated is the failure rate of an item. As above, we will use capital letters to denote r.v.'s and lower case letters to denote specific values (constants).

2.2.1 Additive error. The decision-maker treats the information obtained by the expert as a r.v., sum of two terms:

$$\Lambda_1 = \lambda_1 + R_1 \qquad (2.3)$$

where λ_1 is a "central value" of the distribution of the failure rate Λ and R_1 is the error term (a r.v.).

Taking the mean value of both sides of (2.3) we obtain:

$$E(\Lambda_1) = E(\lambda_1) + E(R_1) = \lambda_1 + b_1. \qquad (2.4)$$

This means that the decision maker expects the expert's judgement to be biased by the additive quantity b_1. Under the assumption, usually made in the theory of errors, that the error term R_1 is normally distributed, and denoting by σ_1^2 the variance of the distribution, the likelihood function is:

$$L(\lambda_1^*|\lambda) = \frac{1}{\sqrt{2\pi}\,\sigma_1} \exp\left\{-\frac{1}{2}\left[\frac{\lambda_1^*-(\lambda_1+b_1)}{\sigma_1}\right]^2\right\}. \tag{2.5}$$

Equation (2.5) has the following meaning: the decision maker's probability that the expert will provide the estimate λ_1^* in the small interval $(\lambda_1^*, \lambda_1^* + \Delta\lambda_1^*)$ is $L(\lambda_1^*|\lambda)\Delta\lambda_1^*$.

To apply Bayes theorem, one needs the prior distribution $\pi_o(\lambda)$. To keep the mathematics simple, ref./4/ assumes that $\pi_o(\lambda)$ is also normally distributed. As parameters of the prior let us assume mean = λ_o and variance = σ_o^2. As the prior and likelihood are normally distributed, the posterior distribution is also normal. Its mean and variance are:

$$\lambda_p = \frac{\sigma_1^2}{\sigma_o^2+\sigma_1^2}\lambda_o + \frac{\sigma_o^2}{\sigma_o^2+\sigma_1^2}\lambda_1 \tag{2.6}$$

$$\sigma_p^2 = \left(\frac{1}{\sigma_o^2} + \frac{1}{\sigma_1^2}\right)^{-1} \tag{2.7}$$

respectively.

Let:

$$\frac{\sigma_1^2}{\sigma_o^2+\sigma_1^2} = w_o \quad ; \quad \frac{\sigma_o^2}{\sigma_o^2+\sigma_1^2} = w_1 \tag{2.8}$$

Eq.(2.6) can be written as:

$$\lambda_p = w_o\lambda_o + w_1(\lambda_1^* - b_1). \tag{2.9}$$

Equation (2.9) shows that the decision maker's posterior mean is a weighted average of his prior expected value and the expert estimate "corrected" by subtracting the amount of bias b_1. The weights w_o and w_1 are proportional to the variances of the likelihood and prior, respectively. They represent the decision maker's relative confidence in the two estimates. Some comments follow:

i) A perfect expert is modelled by assigning no bias ($b_1 = 0$) and $\sigma_1 \to 0$. In this case the likelihood is a delta function concentrated around λ_1^*, and the posterior mean is entirely dominated by the expert's opinion, i.e. $\lambda_p \to \lambda_1^*$, $\sigma_p \to 0$;

ii) An expert believed to be "ignorant" is represented by $\sigma_1 \to \infty$ (flat likelihood). In this case the posterior distribution is dominated by the prior, i.e. $\lambda_p \to \lambda_o$, $\sigma_p^2 \to \sigma_o^2$;

iii) If the expert estimate is equal to the decision maker's prior expected value, and no bias is assumed, then $\lambda_p = \lambda_o = \lambda_1$. What the decision maker gains is more confidence in his estimate.

2.2.2 Multiplicative error. In this case the expert's estimate is treated by the decision maker as a r.v., product of two terms:

$$\Lambda_1 = \lambda_1 R_1 \tag{2.10}$$

where, as in section 2.2.1, λ_1 is a "central value" of the distribution of λ and R_1 is a r.v. accounting for the error term.

Taking logarithms of (2.10) we obtain:

$$\ln \Lambda_1 = \ln \lambda_1 + \ln R_1 \tag{2.11}$$

$$E(\ln \Lambda_1) = E(\ln \lambda_1) + E(\ln R_1) = \ln \lambda_1 + \ln b_1. \tag{2.12}$$

Equations (2.11) and (2.12) are of the same form as (2.3) and (2.4), respectively.

If the assumption that $\ln R_1$ is normally distributed is justified, and the distribution has variance σ_1^2, then one can write:

$$L(\lambda_1^*|\lambda) = \frac{1}{\sqrt{2\pi}\,\sigma_1\lambda_1^*} \exp\left\{-\frac{1}{2}\left[\frac{\ln\lambda_1^* - (\ln\lambda_1 + \ln b_1)}{\sigma_1}\right]^2\right\}. \tag{2.13}$$

A lognormal distribution for expert estimates is discussed and justified in refs./5,6/.

We note that the median, mean and variance of R_1 are b_1, $b_1 \exp\{\sigma^2/2\}$ and $b_1^2 \exp\{\sigma^2\}[\exp\{\sigma^2\}-1]$, respectively.

Assuming that the prior is also lognormally distributed, with parameters $\ln \lambda_o$ and σ_o, i.e.:

$$\pi_o(\lambda) = \frac{1}{\sqrt{2\pi}\,\sigma_o\lambda} \exp\left\{-\frac{1}{2}\left[\frac{\ln\lambda - \ln\lambda_o}{\sigma_o}\right]^2\right\} \tag{2.14}$$

it follows that the posterior distribution is lognormal too:

$$\pi(\lambda|\lambda_1^*) = \frac{1}{\sqrt{2\pi}\,\sigma_p\lambda} \exp\left\{-\frac{1}{2}\left[\frac{\ln\lambda - \mu}{\sigma_p}\right]^2\right\} \tag{2.15}$$

where $\mu = w_o \ln \lambda_o + w_1(\ln \lambda_1^* - \ln b_1)$ (2.16)

and σ_p, w_1, w_2 are as in (2.7) and (2.8).

The median of the posterior distribution is:

$$\lambda_M = e^{\mu} = \lambda_o^{w_o} \left(\frac{\lambda_1^*}{b_1}\right)^{w_1} \tag{2.17}$$

i.e. it is a weighted product of the decision maker's prior estimate, λ_o, and the expert estimate, λ_1^*, which is adjusted by dividing it by the bias factor. The weights are as in (2.8).

3. COMBINATION OF EXPERT OPINIONS AND EXPERIMENTAL DATA

3.1 The Bayesian Parametric Approach

A typical way to estimate a failure rate is by combination of expert opinions and experimental data by Bayes theorem /7,8/. The expert opinion is explicited by a prior distribution. The experimental data define the likelihood function. The classical estimation methods: maximum likelihood and confidence interval methods, can be seen as a special case of Bayesian estimation.

Bayes theorem was introduced in section 2, formula (2.1). Let us rewrite it as:

$$\pi(\lambda|Z) = \pi_o(\lambda)L(Z|\lambda)/K \tag{3.1}$$

where Z is the experimental sample and K, **L**, π_o, π have the same meaning as in (2.1).

The prior distribution π_o must be defined on the basis of the prior information available (prior to performing the experiment). Ref./9/ shows that some simple and reasonable hypotheses, defined as "ground-hypotheses", imply a beta distribution as prior for failure probabilities on demand and failure rates. As likelihood for the failure process we consider a binomial distribution which describes a constant failure rate process with sufficient accuracy (see ref./8/).

As is well known, if a beta distribution is assumed as prior and a binomial distribution is assumed as likelihood, then the posterior distribution is also beta distributed. We discuss here some of the main points of this approach. For more details see refs./8,9/. Let us denote the parameters of the prior distribution by a and b, and the parameters of the likelihood function by N and r (N = operation time; r = number of failures).

The parameters a and b can be assigned in various ways, e.g.:

- give them directly (a can be seen as the number of failures in a hypothetical sample in which the number of demands is a+b (see ref./10/, p.463);
- give the mean μ and the standard deviation σ of the distribution;
- give two percentiles of the distribution, such as the 50-th (M) and the 95-th (U);
- give the values s and t defined as follows: s = probability that the item will fail during the first hour of operation; t = factor by which s must be multiplied if the item should really fail during

the first hour of operation (see ref./9/).
The parameters of the posterior distribution are a+r and b+N-r.

Assuming other distributions as prior, e.g. a lognormal or a loguniform, the posterior distribution is not defined analytically, it must be computed numerically.

Given the posterior distribution of λ, as point estimate of λ one can assume the mean m or the median M of the posterior distribution. They are defined as follows:

$$m = \int_0^\infty \lambda \pi(\lambda|Z) d\lambda \tag{3.2}$$

$$\int_0^M \pi(\lambda|Z) d\lambda = ½. \tag{3.3}$$

A symmetrical interval estimate of λ is defined by the lower bound λ_ℓ and upper bound λ_u defined by:

$$\int_0^{\lambda_\ell} \pi(\lambda|Z(d\lambda = (1-\alpha)/2 \tag{3.4}$$

$$\int_{\lambda_u}^\infty \pi(\lambda|Z) d\lambda = (1+\alpha)/2 \tag{3.5}$$

respectively, where α is the probability level. Normally, α = 0.90 or α = 0.95 is assumed.

4. COMBINATION OF COMPUTED PARAMETERS

In sections 2 and 3 we implicitly assume that the data to be combined are "homogeneous". In fact, to combine expert opinions (section 2), the decision maker asks the experts for data about a well defined item, working in specified operational conditions. When one defines a prior distribution for a given item (section 3), one refers to an item having the same characteristics and operational conditions as the item for which experimental data are available.

When combining computed parameters, the situation may be different. In fact, the computed parameters available can refer to items that can be considered "similar" to that of interest, but not "identical". The operational conditions can also be considered "similar" but not "identical". Finally, the "quality" of the various data-sources can be judged to be different. In these situations one has available "non-homogeneous" data.

Ref./11/ proposes and discusses one method of combining homogeneous reliability parameters and two methods of combining non-homogeneous reliability parameters. In this section we report the method proposed in ref./11/ to combine homogeneous parameters and the weighted artihmetic averaging method of combining non-homogeneous data. In the latter method we introduce some refinements suggested by the experience we have gained in recent years in reliability data treatment.

For a given item, we consider a "central" estimate and a "pessimistic" estimate of the failure rate λ, supplied by various data-sources. An "optimistic" estimate of λ is usually also supplied. We do not use this information because we feel that it is neither reliable nor useful. In the case where the information of the various data-sources cannot be assumed homogeneous, we assume that we can assign a weight to the pair of values supplied by each data source. The weight takes into account possible (minor) differences in the characteristics of the items considered by the various data-sources, differences in the operational conditions, and the "quality" we assign to each data source.

Denote by:

$S_1, S_2, \ldots, S_i, \ldots, S_n$ the n data sources

$\lambda_{i,M}$ central estimate of λ supplied by data source S_i

$\lambda_{i,U}$ pessimistic estimate of λ supplied by data source S_i

w_i weight of data-source S_i (when assigned).

We take the $(\lambda_{i,M})$s as estimates of the 50-th percentile (median) of the distribution of λ, and the $(\lambda_{i,U})$s as estimates of the 95-th percentile of the distribution of λ. Furthermore, we denote:

λ_M 50-th percentile (median) of the distribution of λ

λ_U 95-th percentile of the distribution of λ

μ_λ mean value of the distribution of λ

σ_λ standard deviation of the distribution of λ

a,b parameters of the distribution of λ, assumed to be beta.

4.1 Homogeneous Data

In the case of homogeneous data a binomial failure process /8/ and a beta prior /9/ are assumed. Each pair of values $(\lambda_{i,M}, \lambda_{i,U})$ defines a beta distribution with parameters (a_i, b_i). Each pair (a_i, b_i) determines a hypothetical sample of $r_i = a_i$ failures in operation time $T_i = a_i + b_i$ (see ref./10/, p.463, and ref./12/). One can thus combine the data by simply pooling the samples, thus obtaining:

$$r = \sum_{i=1}^{n} r_i, \qquad T = \sum_{i=1}^{n} T_i \tag{4.1}$$

and then:

$$a = r, \qquad b = T - a. \tag{4.2}$$

which are the parameters of the posterior distribution of λ. The 50-th and 95-th percentiles of the distribution are obtained by integrating the posterior density f(a,b). The mean μ_λ and the standard deviation σ_λ of the distribution of λ are given by:

$$\mu_\lambda = \frac{a}{a+b} \quad \text{and} \quad \sigma_\lambda = \left(\frac{ab}{(a+b)^2(a+b+1)}\right)^{\frac{1}{2}} \tag{4.3}$$

respectively.

4.2 Non-homogeneous Data

The distribution of λ is assumed to be beta, as in the case of homogeneous data. The shape of the distribution is defined by λ_M and λ_U, which are computed by a weighted arithmetic averaging of the data-source percentiles:

$$\lambda_M = \sum_{i=1}^{n} \lambda_{i,M} w_i \Big/ \sum_{i=1}^{n} w_i \tag{4.4}$$

$$\lambda_U = \sum_{i=1}^{n} \lambda_{i,U} w_i \Big/ \sum_{i=1}^{n} w_i \ . \tag{4.5}$$

The parameters a,b, the mean μ_λ and the standard deviation σ_λ of the distribution of λ are computed as in section 4.1, starting from λ_M and λ_U .

Each weight w_i can be split into three terms: a_i, b_i, c_i, that account for: (minor) differences in the item characteristics, differences in the item oprational conditions, and differences in the data-source quality, respectively.

Care must be taken in pooling data. If the numerical value of a percentile $\lambda_{i,M}$ or $\lambda_{i,U}$ is very different from the others (n-1), one must investigate the cause and, possibly, consider it as an "outlier", then exclude it from the averaging.

5. SOME COMMENTS

Obviously, the method to be used to combine data in a specific situation depends on the data available. Nevertheless, we would stress the fact that, if in a special case the use of more than one method is envisaged, then the Bayesian parametric approach that combines prior information and experimental data must be preferred. It is the most typical and convenient way of making an inference.

Concerning the various methods, we can make the following comments. The combination of expert opinions requires considerable information, and no datum is taken directly from experiments: all figures derive from an interpretation of the reality made by the experts and the decision maker. A decision maker is often reluctant to weight the experts. Many experts judge the method too sophisticated. Nevertheless, this method has the advantage of being well formalized and easy to understand and use.

The combination of a prior information and experimental data via Bayes theorem is, in our opinion, the best way of making an inference.

As we said before, when a choice is possible, this method must be preferred. It is coherent in the sense that if one makes successive updates, then the final result does not depend on the specific sequence of updates considered.

To combine computed parameters, the binomial-beta model is well supported from a theoretical point of view; nevertheless, it must be critically considered as with an increasing number of data sources the distribution of λ may become very tight. The weighted arithmetic averaging method is a very simple method that allows us to combine non-homogeneous data. Nevertheless, as was remarked in section 4, attention must be paid to mixing data that are very different. If the item characteristics, operational conditions and quality of the data sources are not identical, but can be considered as very similar, then one can use this last method and assign the same weight to all data. This method is not coherent in the sense discussed before; nevertheless, it reduces the variability of λ less drastically than the binomial-beta model.

As a final remark, we would stress the fact that we have discussed the estimation of constant parameters only. This is mainly for two reasons: (i) the reliability analysis of a system is generally based on constant parameters; (ii) very few methods have been developed to estimate non-constant reliability parameters by data from different sources. As a method of combining expert opinions and experimental data to estimate time-dependent failure rate we mention the Bayesian nonparametric approach /13,14/. This method allows us to combine coherently a prior failure distribution defined by an expert and experimental failure data. The same method can be used to estimate time-dependent repair rates and non-constant failure probabilities on demand.

REFERENCES

1. Colombo A.G., 'A single model to estimate failure rate and failure probability on demand', in preparation.
2. U.S. Nuclear Regulatory Commission, 'Reactor Safety Study', Wash-1400 (NUREG-75/014), 1975.
3. Institute of Electrical and Electronics Engineers, 'IEEE Guide to the Collection and Presentation of Electrical, Electronic and Sensing Components Reliability Data for Nuclear Power Generating Stations', IEEE Std.500, 1977.
4. Mosleh A. and Apostolakis G., 'Models for the use of expert opinions', paper presentd at the Workshop on Low-Probability/High-Consequence Risk Analysis, Arlin, Virginia, June 1982.
5. Dalkey N.C., 'An Experimental Study of Group Opinion', RM-5888-PR, The RAND Corporation, Santa Monica, California, 1969.
6. Martino J.P., 'Lognormality of Delphi estimates', Technological Forecasting, 1, 355, 1970.
7. Martz H.F. and Waller R.A., 'An Exploratory Comparison of Methods for Combining Failure-Rate Data from Different Data Sources', LA-

7556, Los Alamos Scientific Laboratory, 1977.
8. Colombo A.G. and Jaarsma R.J., 'Bayesian estimation of constant failure rate and unavailability', IEEE Transactions on Reliability, R-31 (1982) 84-86.
9. Colombo A.G. and Costantini D., 'Ground-hypotehses for beta distribution as Bayesian prior', IEEE Transactions on Reliability, R-29 (1980) 17-21.
10. Hayes H.L. and Winkler R.L., 'Statistics, Probability, Inference and Decision'. Vol.1, New York, Holt, Rinehart and Winston, Inc., 1970.
11. Colombo A.G. and Jaarsma R.J., 'Combination of reliability parameters from different data sources', Proc. 4th EuReDatA Conf., Venice, March 1983.
12. Colombo A.G. and Jaarsma R.J., 'Approximation of the parameters of the gamma distribution by two percentiles'. Rivista di Statistica Applicata, 15 (1982) 154-160.
13. Ferguson T.S., 'A Bayesian analysis of some non-parametric problems', Ann. Statist., 1 (1973) 209-230.
14. Colombo A.G. and Jaarsma R.J., 'Bayes non-parametric estimation of time-dependent failure rate', IEEE Transactions on Reliability, R-34 (1985) 109-112.

UNCERTAINTY MODELLING IN SYSTEMS RELIABILITY ANALYSIS

C. GUEDES SOARES
Technical University of Lisbon
Instituto Superior Técnico
Av. Rovisco Pais
1096 Lisboa, Portugal

ABSTRACT: An overview is presented of the methods available to quantify uncertainties and to combine them in a systems reliability analysis. The different sources of uncertainty are identified and a discussion is provided to illustrate their different nature. The methods of reliability analysis are then described showing how they can incorporate parameter and model uncertainty. Finally Bayesian methods are presented illustrating how to update probabilistic statements with new information and how to incorporate parameter and model uncertainty in the reliability predictions.

1. INTRODUCTION

The study of systems reliability can be required in a number of different circumstances which in turn dictate the particular features of the system model to be developped.

An important field of application is the probabilistic risk assessment in different industries, with particular relevance to nuclear engineering where the consequences of failure are often largest. Risk assessments can also be the basis for the design of structures and equipment which are intended to be operational even in the occurrence of some hazard situations. Furthermore, system reliability studies are of interest in the planning and organization of maintenance schemes and on their integration in the operational life of production systems.

The main features of the reliability analysis are, in all cases, the identification of the modes of failure and of the parameters that influence failure, the modelling of their interrelations and the representation of the chain of events leading to the different types of failure.

Most parameters that are included in such probabilistic models are variables subjected to a random variability which is inherent to the physics of the process. The characterization of each variable is made on the basis of a statistical analysis of data which provides estimates of the parameters that govern the probability distribution of the variables. These estimates have obviously some confidence bounds which can also be interpreted as uncertainty statements.

In constructing the computational model a choice exists between obtaining an approximate solution to the exact problem or an exact solution to the approximate problem. Most often both model and solution methods are not exact, implying some degree of uncertainty about their results.

The modelling of the relationships between variables is partially a subjective process in which one is often faced with the dilema of being too detailed or too crude in the level of representation. In the first case too many variables can be introduced in the analysis and some

A. G. Colombo and A. Saiz de Bustamante (eds.), Systems Reliability Assessment, 285–303.

of them as well as their relationship may be less well known in statistical terms. Furthermore, they may originate a complicated computational model which can bring numerical inaccuracies. On the other hand, including a smaller number of variables the accuracy of the model cannot be expected to be as good as when more variables are accounted for.

In representing the chain of events leading to failure i.e. the failure paths, use is made of established techniques applied by experienced persons which guarantee the quality of the model. However, in all cases it is conceivable that some unidentified paths may not be included in the model whose predictions will thus not always be correct.

This brief discussion has indicated that in addition to the variability of basic events, one deals with the uncertainty in the parameters estimated to describe those events and in the mathematical models that are constructed to represent the system. The recent trend in reliability analysis is to build mathematical models that combine the randomness of the variables with these two different sources of uncertainty and that reflect these uncertainties in the final estimates of system reliability. This work will present an overview of the methods used currently in those studies.

2. UNCERTAINTIES IN SYSTEM RELIABILITY ANALYSIS

The most commonly used methods of quantification of systems reliability are the analytical method using the full probability distributions of the variables or only second moment statistical information, the Fault-Tree Analysis and the Monte-Carlo method which can be applied to a detailed description of the system [1-4].

In performing risk assessments, the probability of failure predicted by a system reliability analysis is only one part of the study since the risk is the product of this probability by the consequences of the failure [5,6]. In the assessment of the consequences of failure another source of uncertainty is introduced.

The quantification of the uncertainty of the variables can be made by estimating the parameters of the probabilistic models adopted to describe the data or by estimating which is the proportion of possible outcomes of a variable which can be expected to fall in a given interval. The most commonly used description of the first type is the standard deviation or even better, the coefficient of variation, while confidence intervals and fractiles are used in the second case.

Estimating the parameters of a distribution is a basic problem of statistics but differences exist between the classical and the Bayesian viewpoint. In the first case the parameters, although unknown, have a fixed value while in the latter approach they are considered a random variable. The Bayesian viewpoint, which is discussed in section 4, is especially important in the analysis of rare events which by their nature always constitute very small samples to be analysed.

Often it is not possible determine failure rates from the analysis of failure data and it is necessary to construct mathematical models that represent the physical or mechanical relation between variables. For example the ultimate strength of a column will depend on its dimensions and on the material properties. The uncertainty resulting from these models is of a different nature than in the choice of the probability model or distribution that describes a variable. as will be discussed in more detail in sections 2.3 and 2.4.

2.1. Uncertainties in Component Failure Data

The initiating events of a system failure, which are random occurrences, can be internal or external to the system. In the first case one can have failures equipments, of controlling systems or of components which can be mechanical, electrical or structural. In the second case the external causes can be environmental factors such as earthquakes, ocean waves or wind or they can be accidents such as collisions or explosions.

In general, equipments are characterized by their failure rates which are derived from statistical analysis of failures. However, data on failure rates can be very misleading [3]. Data on equipment failure is only started to be collected after a reporting system has been developed and implemented. Often companies do not put enough human resources to report and record all detected failures. Manufacturers are continuously improving the design of the components which makes data from earlier designs obsolete. Furthermore, maintenance and testing procedures will also change during the life of the system, afecting its failure rate. Failure data on an equipment is often correlated with the environment in which it is operating, a feature that is not normally incorporated in the analysis [7].

When using component failure data, judgements must be made as to whether particular components in a system are similar enough to the ones for which data exists, which is another potential source of data misusage.

In addition to equipment failure data, one should also consider the variability with which humans perform their tasks under varying conditions and stress levels and the difficulty of collecting this type of data [8].

2.2. Measurement Uncertainties

Many of the variables used in engineering models are evaluated through measurements. However, measurements may be subjected to random and to systematic errors which need to be quantified [9]. Measurements can be described by their accuracy and precision which can only be assessed in repeatable and reproduceable measurements.

The repeatability of measurements is determined from the closeness of the agreement between the results of successive measurements of the same quantity subjected to the same conditions as regards the method of measurement, the observer, the measuring instrument, the location, the conditions of use, and in addition the measurement must be made over a short period of time.The reproducibility of measurements is defined as the closeness of the agreement between the measurements of the same parameter, when they are carried out under changing conditions, as regards the factors mentioned previously. Both concepts can be quantified by the dispersion of the results.

The precision of a measurement is the closeness of the agreement between the results of applying the experimental procedure several times under the same prescribed conditions. The accuracy is the closeness of the agreement between the results and the true value of the measured parameter. Thus, while imprecision is quantified by the dispersion of the results about the sample mean, inaccuracy is reflected by the shift of the sample mean from the true mean.The quantification of the bias and uncertainty of a measuring process is a prerequisite to the correct derivation of the natural variability of the quantities that are being measured.

The uncertainty of different variables is generally quantified by the analysis of samples of the variable outcomes. Common descriptors in a sample of N outcomes are the sample mean $\overline{X}$ and variance $\hat{s}_X^2$:

$$\overline{X} = \sum_{i=1}^{N} X_i / N \tag{1}$$

$$\hat{s}_X^2 = \sum_{i=1}^{N} (X_i - m_X)^2 / (N-1) \tag{2}$$

which are estimators of the mean m_X and variance $s_X{}^2$ of the random variable. The coefficient of variation V_X is the non dimensional form of quantifying the uncertainty:

$$V_X = s_X/m_X \tag{3}$$

The uncertainty in the estimates of the population parameters depends on the size of the sample. For example, the variance of the sample mean is related to the variance of the variable by:

$$V[m_X] = V[X]\ /N \tag{4}$$

and it is seen to decrease with increasing number of observations.

Measurement uncertainties can be represented by an uncertain factor B that multiplies the correct value of the variable X to yield the measured value $\hat{X}$:

$$\hat{X} = B\ \ X \tag{5}$$

If the measurement error and the measured variable are statistically independent, the mean and the uncertainty of the variables are given by:

$$\overline{X} = \overline{B}\ m_X \tag{6}$$

$$\hat{V}_X^2 = V_B^2 + V_X^2 \tag{7}$$

where m_X is the true mean of X, $\overline{X}$ is the mean of the measurements, $\hat{V}_X^2$ is the coefficient of variation of the measured values, V_B and V_X are the coefficient of variation of the measurement error and of the variable itself. If the mean of the variable is known, we can define the bias of the measurements from the expected value of the variables i.e.:

$$B = \overline{X}/\ m_X \tag{8}$$

In this case the variance of B is given by:

$$V[B] = \frac{V[\overline{X}]}{m_X^2} = \frac{V[X]}{N\ m_X^2} \tag{9}$$

where the variance of the mean is related with the variance of the variable by eqn. 4. This expression shows clearly that the measurement uncertainty will decrease with increasing number of observations.

An interesting case worth considering is the indirect observation in which case the value of $\hat{X}$ is estimated from measuring $\hat{Y}$:

$$\hat{X} = a + b\hat{Y} + e \tag{10}$$

where a and b are regression coefficients and e is a normally distributed random variable with zero mean and standard deviation s_e, which quantifies the dispersion relative to the regression line. Considering a and b and s_e as constants, it results in:

$$\overline{X} = a+b\,\overline{Y} \tag{11}$$

$$V\,[X] = b^2\, s_y^2 - s_e^2 \tag{12}$$

Reference 10 provides an example of an analysis of experimental results in which the measurement uncertainty is separated from the randomness of the parameters and various modelling situations are examined in detail.

2.3. Probabilistic Model Uncertainty

Many physical phenomena and system behaviour are modelled as random variables which follow a specified probability distribution. The variable is completely described by the knowledge of the type of distribution and of its parameters.

The choice of the type of distribution or of the probabilistic model is made on the basis of considerations about the nature of the process under study and eventually of formal statistical tests of fit of the theoretical model to the measured data [11, 12].

Whenever there is enough measured data one can use the tests of fit to yield definite statements about the adequacy of the probabilistic models to describe the existing data. However, in many situations one is faced with little data or one is interested in the tails of the distribution to which data is always insufficient. In these cases the choice of the distribution must rely more heavily on judgement and often it is not clear that one model is preferable to another. This situation is described by saying that there is an uncertainty concerning the correct probabilistic model. The methodology to incorporate probabilistic model uncertainty in the analysis is discussed in sections 3 and 4.

2.3.1 *Parameter Uncertainty*

Even in the cases in which the type of probabilistic model is perfectly well defined, it only becomes fully described when its parameters are quantified. There are several methods available to estimate the probability distribution parameters or to fit distributions to data which are methods that depend on the type of distribution. However, general approaches are the methods of least squares, of moments and of maximum likelihood, in addition to the empirical approach of using probability paper to plots the data [11-15].

Graphical methods are inaccurate but have many advantages at the early stages of data analysis. The method of least squares fits a line to a series of points whose coordinates have been previously transformed according to the type of distribution under consideration. In the method of moments, the theoretical moments of the distribution are equated to the sample moments. The method of maximum likelihood, which is the most correct one from a statistical point of view is based on the likelihood function which gives the probability of obtaining the given sample if the estimator equals the sample value.

Independently of the method used, it will yield estimates of the parameters which have confidence intervals or an uncertainty associated with them. This uncertainty is called parameter uncertainty and again sections 3 and 4 will indicate how to include such information in the system reliability analysis.

2.3.2. *Uncertainties in Fault-Tree Analysis*

One difficulty in fault-tree analysis is the completeness of the model, since one cannot be sure to have identified all the paths leading to the top event. However, this problem is more important in accident studies than in the reliability analysis of well defined electrical, mechanical or structural systems [3,4].

Another source of uncertainty is the dependency between components which will yield common cause failures [14]. This type of failures can also be induced by unexpected human errors which are often not modelled [8].

Fault-trees are based on binary states in which the component can be in a working or in a failed state. In general, real components exhibit continuously degrading characteristics, which implies that they can be modelled as functioning even when having a performance much less than normal [15]. The modelling and quantification of the effect of repairable components is a further source of uncertainty [8].

In the evaluation of the fault-trees very often a complete analysis is not performed when dealing with very large systems. Only the most important minimal cut sets are included, which makes the truncation effect dependent on the ordering criterium and even on the capacity of the computer code or of the computer used [18,19].

2.4. Physical Model Uncertainty

While in low reliability components or even in sub-systems their failure rate can be derived from a statistical analysis of failures, that is not possible in the high reliability range. In these situations, the component reliability is determined from computational models that represent the effect of different parameters on the failure of the component. The predictions of these models which represent the interaction between the different physical or mechanical variables, are subjected to an uncertainty that depends on their level of complexity.

Whenever the basic events are physical processes they can be represented by probabilistic models whose output has obviously some variability. Examples are the modelling of earthquake intensities [20], of the height of ocean waves [21,22] of wave induced forces [23] or of the strength of a machine part or of a structural component [10].

These probabilistic models of physical processes depend on basic physical variables, most of which are random in nature. Thus the description of the processes must be based on a model of the relationship between those variables and on a statistical analysis of the measurable parameters. The statistical analysis is always performed on a sample of the population, yielding estimates which have well defined confidence intervals.

Certain physical quantities keep their properties during measurement and after it. In these cases a series of measurements will reflect the random variability of the quantity, in addition to the measurement uncertainties. However, when the variables of interest are related with the behaviour of the components, the measurements may imply changing their initial characteristics. Examples are destructive testing of materials to assess their strength or the assessment of the strength of materials subjected to fire. In these cases and in others in which testing is not even possible, like the ultimate strength of a nuclear plant, it becomes necessary to construct mathematical models that represent the interaction between various measureable physical variables and the quantity of interest, the strength in the referred examples.

Various theoretical models are possible of development, some more accurate than others. This implies that when one uses one theoretical model to predict a specific quantity, it has an uncertainty associated with it [24,25]. This uncertainty is different for the various computational models that represent the same variable as can be seen for example in [26].

The main difference between probabilistic model uncertainty and physical model uncertainty is that the first concerns the different probability distributions than can describe the variability of a variable while the second is related with the mathematical model that represents the relations between various physical variables and the one of interest.

A common way of representing this model uncertainty B is to multiply it by the model prediction $\hat{X}$ to yield its correct value X [27,10] in an equivalent way as represented in eqn. 5 for the measurement uncertainty. Ditlevsen [28] has even shown that a general representation of model uncertainty that is compatible with first order second moment reliability methods is:

$$X = A + B\,\hat{X} \tag{13}$$

where A and B are constants.

Noting that these expressions are equivalent to eqn. 10 used for measurement uncertainties, the physical model uncertainty can be expressed by eqns. 11 and 12. Its incorporation in an uncertainty analysis involves exactly the same methods as for measurement uncertainty which was treated in the previous section and later in section 4.

Having the full mathematical expression that gives the model prediction X in terms of n variables X_i (i=1,2,...,n) one can say that:

$$X = \hat{g}(X_1, X_2,..., X_n) \tag{14}$$

The model predictions are corrected by Bg to yield the improved value of X:

$$g\,(\underline{X}) = B_g \cdot \hat{g}(\underline{X}) \tag{15}$$

Each variable can itself be subjected to model or to measurement errors:

$$X_i = B_i\,\hat{X}_i \tag{16}$$

leading to

$$X = B_g \cdot \hat{g}(B_1 X_1,\; B_2\, X_2,\, ...,\, B_n\, \hat{X}_n) \tag{17}$$

In the scope of first order second moment methods this function can be expanded in a Taylor series, so that the mean and variance of X is given by:

$$m_X = \overline{B}_g \cdot g(\overline{B}_1\, \overline{X}_1, \overline{B}_2\, \overline{X}_2, ..., \overline{B}_n\, \overline{X}_n) \tag{18}$$

$$s_X^2 = s_B^2 + \sum_i \sum_j \left(\frac{\partial g}{\partial x_i}\right)_m \left(\frac{\partial g}{\partial x_j}\right)_m \; s_i\, s_j\, r_{ij} \tag{19}$$

where the variance of the correction factor is s_B^2 and the double summation gives the variance of the function g in terms of a second moment aproximation, which will be considered in detail in section 3.3. The parenthesis enclosing the partial derivations $(\ )_m$ indicate that they are evaluated at the mean values of the variables, and s_i is the standard deviation associated with variable X_i.

The correlation coefficient r_{ij} between variables X_i and X_j will often be estimated from the data:

$$\hat{r}_{ij} = \sum_k \; [(X_{ik} - \overline{X}_i)\; (X_{jk} - \overline{X}_j)] \,/\, (N-1)\, s_i\, s_j \tag{20}$$

However in many cases the limiting situations of perfect correlation (r_{ij}= 1) or of statistical independence (r_{ij}= 0) yield results that are adequate for pratical purposes.

3. METHODS FOR UNCERTAINTY ANALYSIS

Consider the general case of a system output Y which depends in a deterministic way on various random variables or uncertain quantities X_i and on parameters a_i which are in general functions of X_i and are also uncertain:

$$Y = g(X_1, X_2, ..., X_n, a_1, a_2, ..., a_m) \tag{21}$$

The objective of the uncertainty analysis is the quantification of the uncertainty of Y as a result of the uncertainty in X_i and a_i. The function $g(\underline{X},\underline{a})$ is most general, representing the physical response of engineering systems or components or the reliability of an engineering system, quantified by any of the available methods. By incorporating the parameters a_i in the formulation of g(X) one can deal in a unified way with parameter, measurement and model uncertainty [29-31].

The following sections will deal with different techniques of quantifying the uncertain quantities. Often after predictions are made one is interested in knowing the sensitivity of the predictions to the uncertainty of each variable, which can be determined by methods of sentivity analysis. However, an uncertainty analysis should only be conducted after a preliminary sensitivity analysis is performed. This is required so as to make sure that no unnecessary variable is being included in the computational model. Thus a discussion of sensitivity analysis is presented before the methods of uncertainty analysis.

3.1. Methods for Sensitivity Analysis

These techniques are intended to identify the variables that have significant influence on the result, which is an important information for uncertainty analysis since only the most influential ones are worth including in a computational model. This allows the models to be reduced to the smaller size compatible with the desired accuracy [32].

In addition to being a prerequesite to the probabilistic modelling analysis can be also a posterior study in a reliability analysis, by indicating which variables should be affected to increase the reliability of the system.

The most simple method is the subjective one in which experts choose some variables and discard others based on their judgement. A more quantitative procedure is differential sensitivity analysis in which one calculates the partial derivatives of the output with the respect to each of the input variables. The sensitivity coefficient is then:

$$a_i = \partial Y/\partial X_i \tag{22}$$

where $\partial/\partial x_i$ represents a partial derivative with respect to X_i. An estimate of the sensitivity coefficient can be obtained from the ratio of the change of the output by the change of the input variable [33]:

$$a_i = \Delta Y/\Delta X_i \tag{23}$$

as was done for example in [23].

This estimate is reasonable for linear and almost linear relationships. If the relationship is not linear the sensitivity coefficient gives an estimate of the slope of the function $g(\underline{X})$ along the X_i axis in the neighbourhood of the point considered.

An extension of the differential sensitivity method is to estimate sensitivity coefficients in the one-at-a-time design where each input is evaluated at its mean and then at its mean plus or minus some multiple of its standard deviation. This requires (2k+1) runs for k variables and has the benefits that these results can be saved and used with a fractional factorial experiment to estimate a response surface, as discussed in section 3.5.

Measures of the effect that X_i has on the output are the difference between the output values at the extreme values of X_i or the standard deviation of the output resulting from those evaluations.

Another method uses partial rank correlation coefficients between each variable and the output, which can be the basis of ordering the variables by their influence on the output [34,35]. It is based on substituting the values of the sampled values by their rank and calculating the correlation coefficient between the ranks and the output.

For large problems a useful method is the use of a Latin Hypercube Sampling scheme to obtain a response surface which substitutes the original response model. This response surface, which is easy to compute, can be used as a basis of calculating sensitivities, either by any method discussed previously or directly by sampling the original variables and analysing the results in the system output [36]. The details of sampling and of construction of response surfaces are dealt with in sections 3.4 and 3.5. In adopting this last procedure care must be taken that the final results may depend strongly on the assumptions made about the type of probabilistic distribution of the original variables, i.e. the probabilistic model uncertainty may be large.

Finally the adjoint method provides a rigorous mathematical approach to sensitivity analysis and yields the sensitivity coefficients for any value of the X_i and not just around their mean value, which does not require any assumption about linearity [37]. This method is the most appealing from a theoretical point of view, but even for moderate size problems it is impractical due to the difficulty in obtaining the adjoint equations [33]. For moderate size problems the one-at-a-time method and the rank order correlation have been recommended in [33].

3.2. Full Distribution Methods

The uncertainty of Y is quantified by integrating the joint distribution of X_i and a_i over the appropriate domain of the variables [38]. When the variables are statistically independent the joint distribution is the product of the distributions of the individual variables:

$$f(Z_1, Z_2, ..., Z_n) = f_1(Z_1) ... f_n(Z_n) \qquad (24)$$

where the generic variable Z_i represents either a random variable X_i or a parameter a_i.

Whenever variables are correlated, the joint distribution can be constructed from the conditional distributions:

$$f(Z_1, Z_2) = f(Z_1 | Z_2)\, f(Z_2) \qquad (25)$$

This procedure must be applied sequentially to all the appropriate variables, to obtain the joint distribution of the n variables i.e.:

$$f(Z_1, Z_2, ..., Z_n) = f(Z_1, Z_2, ..., Z_{n-1} | Z_n) . f(Z_n) \qquad (26)$$

The evaluation of the integral of multidimensional probability distributions is extremely difficult whenever the number of variables is large and it is only feasible for special cases [39]. One approximate method of performing such calculation which was developed in the framework of structural reliability theory [40] consists in transforming the input variables into a set of uncorrelated normal variables which can then be integrated.

Another approximate method which became popular among nuclear engineers is the direct application of the distribution functions of the variables. The probability distribution of the random variables are discretized into histograms which are combined two at a time to yield the histogram of the combined variable. Successive applications of the method will yield the final result [41]. The efficiency of the method can be improved by using equal probability width intervals [42]. The error involved in the discretization can be reduced by increasing the number of intervals in which the distribution is discretized.

3.3. Second Moment Methods

Often the information that is of interest is only the first two statistical moments of the variables. For probability distributions that have two parameters such as the normal, this provides the complete description of the variable.

These methods were known as the theory of error in the scope of measurement theory, and they are the basis of the present structural reliability theory. They are based on expanding the functional relationship defining the failure condition in a multivariable Taylor series about a point Z':

$$Y = g\,(Z_1^*, Z_2^*, \dots, Z_n^*) + \sum_i \left(\frac{\partial g}{\partial Z_i}\right)_{\underline{Z}^*} (Z_i - Z_i^*) + \\ + \sum_i \sum_j \left(\frac{\partial^2 g}{\partial Z_i \partial Z_j}\right)_{\underline{Z}^*} (Z_i - Z_i^*)(Z_j - Z_j^*) \qquad (27)$$

This multidimensional point $Z^* = (Z_1^*, Z_2^*, \dots, Z_n^*)$ can be either the mean value [43] or other more appropriately chosen point [44].

The mean value E[.] and standard deviation D[.], of this linearized function is easily calculated and a reliability index is defined as [43]:

$$\beta = E[Y]/D[Y] = 1/V_y \qquad (28)$$

On the assumption of normally distributed variables the associated probability of failure is

$$P_f = \Phi\,(-\beta) \qquad (29)$$

where N is the standardized normal distribution.

Further developments of the theory have shown that beta is the minimum distance from the design point to the failure surface [44,45]. Furthermore, the reliability index has been generalized to be an integrated value over the failure surface instead of only a point estimate [45].

The theory has been developed to handle multivariable and multifailure cases [46] becoming the basis of system reliability analysis [47,48], which has been developed as an association of these concepts for component reliability assessments and fault-tree methodology for system predictions [49,50].

These methods can also deal with time varying situations as explained in [51] for the general case and in [52,53] for a particular case.These methods have been developed mostly in the field of structural engineering and most of their practical applications were in the development and calibration of design codes [54,55].

3.4. Monte-Carlo Methods

The basic version of this method is based on a direct sampling which consists in generating the values of the random variables from their original distribution and evaluating the system function (eqn. 21) for each set of sampled values. In this way the probability distribution can be generated and synthetized by a histogram or by statistical moments, such as the mean and variance [56-58].

The main advantage of this method is its general applicability, which is very important whenever the functional relation is very complicated [59]. However, to obtain reliable information for these cases with this method it may be necessary to conduct a relatively large number of runs becoming a very costly procedure to adopt.

Furthermore, the Monte Carlo approach tells very little about the contribution of each variable to the output uncertainty and about the sensitivity of the results to the assumed input distributions.

The basic Monte Carlo method can be improved by variance reduction techniques which minimize the variability of the output by utilizing additional information that may be available about the simulated process. However, because the randomness of the generation process is being altered, care must be taken not to bias the results [60-61].

Random sampling chooses the values of the input variables X_i by sampling directly from their probability distributions $F(X_i)$. Importance sampling will choose more points in the region where the probability density function is largest [62].

In stratified sampling the sample space of X is partitioned into N disjoint strata S_i. In the sampling procedure, values of X are drawn from each strata and if more samples are taken from the important subregions this method should be equally efficient as importance sampling.

The Latin Hypercube sampling, which has become popular, divides the range of each variable X_i into N strata of equal marginal probability 1/N and then samples once from each stratum of each variable, combining them randomly. One advantage of this method occurs when the output Y is dominated by only few components of X. This method ensures that each of those components is represented in a fully stratified manner no matter which components turn out to be important.

3.5. Response Surface Methods

These methods were developed to overcome the computational disadvantage of Monte-Carlo methods. The basic feature of the method consists in replacing the system function (eqn. 21) by an analytical expression

$$Y = R(Z_1, Z_2, ..., Z_n) \quad (30)$$

which is called a response surface. In general, a statistical design is used to choose the values of the input variables at which the output is calculated. The function R is then determined by using a least squares fitting procedure to adjust the coefficients of a set of simple functions to those points. Once the response surface has been fitted to the design function, the uncertainty of Y can be determined by any of the three previously mentioned methods [33, 63, 64].

The main advantage of the response surface approach is the economy of calculations resulting from avoiding the time consuming evaluations of g(Z). However, it is difficult to estimate the accuracy with which the response surface approximates the original function, which is a disadvantage of the method. Because this is the most promising method to solve large scale problems, a more detailed description of the procedure will be given here.

The first step in this procedure, which is common to any uncertainty analysis is the screening so as to reduce the number of variables to be considered by discarding the ones that do not influence the uncertainty of the output. The choice of the variables is made on the basis of a sensitivity analysis which has been discussed in section 3.1. Once the most important subset of input variables has been chosen, a response surface model may be fitted to the output of the response function.

Sometimes a systematic approach is taken calculating the output with a complete factorial design which leads to NP evaluations of the response functions, where P is the number of variables in the subset and N is the number of values that each variable experiences. The important limitation of this approach is that the number of the points of the design grows exponentially with the number of variables.

To solve this problem a fractional factorial design can be used, retaining a fraction (1/NK) of the points of a complete factorial design, which leads to a total number of runs of N^{P-K}. However, some information will be lost because not all possible combinations are being considered [33].

Another possible method is the central composite design which allows more than two values per variable but keeps the total number of evaluations to a reasonable level.

Often the response surfaces that are chosen to fit the output data are of second order, of the type [33]:

$$Y_k = A_o + \sum_i A_i Z_{ik} + \sum_i \sum_j A_{ij} Z_{ik} Z_{jk} \tag{31}$$

where Y_k is the kth response of the model, Z_{ik} is the kth simulated value of the ith input variable X_i, i.e. $Z_{ik} = (X_{ik} - m_i)/s_i$ where m_i and s_i are the mean and standard deviation of X_i. The regression coefficients A_i and A_{ij} are determined from a least squares fit which minimizes the error which in the kth run is given by:

$$r_k = \hat{Y}_k - Y_k \tag{32}$$

A quantitative measure of the goodness of fit is provided by the multiple correlation coefficient:

$$R = [\sum (\hat{Y}_i - \overline{Y})^2 / \sum (Y_i - \overline{Y})^2]^{1/2} \tag{33}$$

After determining the regression coefficients that define the response surface the final step of the method is to determine the uncertainty of the response as results from the uncertainty of the parameters and variables. This can be done by any of the methods discussed in the previous sections.

4. COMBINATION AND UPDATING OF UNCERTAINTIES

The methods of reliability analysis have been originally developed to combine random variables in order to assess the reliability of a system in terms of the reliability of its components. However, recent developments have indicated that uncertain quantities should also be included in the analysis, raising the important question about how to combine the randomness of variables with the uncertainty of parameters [65].

In section 2, uncertainties of different types have been identified due to measurements, statistical parameter estimation and physical models. It was indicated that all of them could be represented by a new random variable B (eqn. 16) with appropriate mean and standard deviation. In this way uncertain quantities are treated as random variables and can be combined with the true random variables using any of the methods discussed in section 3.

The interesting problem that still needs to be considered is how to use outcomes of random variables to update existing information about them in terms of type of probability distribution and value of the parameters in the distribution. This updating can be made with the use of Bayesian methods, whose relevance to reliability based design has been indicated in [66]. An introductory treatment can be found in [67] while [68] and [69] presents more advanced topics.

Bayesian analysis is a probabilistic framework by which the uncertainty in any design variable and the existing knowledge about the variable can be accounted together with the risk involved in a decision process or in accepting an estimated value of the variable.

In making inferences from historical data one analyses the generating process so as to choose a probabilistic model that describes its variability. The recorded data is then used to estimate the parameters of the distribution and then inferences can be made about the occurrence of future events. Consider that the variable Q is described by a probability density function $f_Q(q|i)$ conditional on the uncertain parameters i. The prior probability distribution of the parameters is based on some initial information I_o i.e. $f(\underline{\tilde{\theta}} \mid I_o)$.

The sample of observations $\underline{Q}$ is used to update the distribution of i, providing the posterior distribution of the parameters:

$$f''(\underline{\tilde{\theta}}) = f''(\underline{\tilde{\theta}} \mid \underline{Q}, I_o) = f(\underline{Q} \mid \underline{\tilde{\theta}}) \cdot f'(\underline{\tilde{\theta}} \mid I_o)/K \tag{34}$$

where K is a normalising constant. Note that the conditional distribution $f(\underline{Q}|\underline{\tilde{\theta}})$ is the sample likelihood function.

$$f(\underline{Q} \mid \underline{\tilde{\theta}}) = f(Q_1 \mid \underline{\tilde{\theta}}) . f(Q_2|\underline{\tilde{\theta}}) ... f(Q_n|\underline{\tilde{\theta}}) \tag{35}$$

Now the Bayesian distribution of the variable Q is obtained by integrating the postulated model by the posterior distribution over all values of i:

$$f_Q(q) = \int_{\underline{\tilde{\theta}}} f_Q(q|\underline{\tilde{\theta}})\, f''(\underline{\tilde{\theta}})\, d\underline{\tilde{\theta}} \tag{36}$$

Inferences can now be made from the updated Bayesian distribution.

It is easy to show that using point estimates for the uncertain parameters underestimates the variance in the original variable Q. Consider that a distribution has the mean value $\tilde{\theta}$ as the only uncertain parameter and that it is described by the mean $\tilde{\theta}$ and standard deviation s_Q i.e. it can be represented as $f(q|\tilde{\theta}, s_Q)$. Consider that the mean value of Q is described by a distribution $f(\tilde{\theta})$ with a mean $m_{\tilde{\theta}}$ and a variance $s^2_{\tilde{\theta}}$.

For a quadratic loss function the Bayes estimator of $\tilde{\theta}$ is its mean. Thus the mean of the updated Bayesian distribution is the mean of $m_{\tilde{\theta}}$:

$$\tilde{m}_Q = \int_{\tilde{\theta}} m_Q\, f(\tilde{\theta})\, d\tilde{\theta} = m_{\tilde{\theta}} \tag{37}$$

since m_Q is the mean of $f_Q(q|m_{\tilde{\theta}}, s^2_Q)$ i.e. it is $m_{\tilde{\theta}}$.

However, the variance of the updated distribution is:

$$\tilde{s}_Q^2 = s_Q^2 + \int \tilde{\theta}\ (m_Q\text{-}\tilde{m}_Q)\ f(\tilde{\theta})\ d\tilde{\theta} \tag{38}$$

where the second term represents the increase in variance due to the uncertainty in $\tilde{\theta}$.

The same type of reasoning can be made for probabilistic model uncertainty. If a set of data could be reasonably described by various probability distributions, the use of only one would underestimate the uncertainty of the process.

Consider the set of probabilistic models that may describe the data:

$$f_1(q \mid \underline{A}_1),\ f_2(q \mid \underline{A}_2)\ ...,\ f_n(q \mid \underline{A}_n) \tag{39}$$

which are conditional on the unknown parameters A_i. The model parameter set $\underline{\tilde{\theta}}$ describes the probability of each model being correct and must satisfy the condition

$$\sum_i \tilde{\theta}_i = 1 \tag{40}$$

The composite model is given by:

$$\tilde{f}(q \mid \underline{A},\tilde{\theta}) = \tilde{\theta}_1\ f_1(q|\underline{A}_1) + ... + \tilde{\theta}_n\ f_n(q|\underline{A}_n) \tag{41}$$

The Bayesian distribution for the composite model is given by:

$$\tilde{f}(q) = p''(\tilde{\theta}_1)\quad f_1(q) + ... + p''(\tilde{\theta}_n)\ f_n(q) \tag{42}$$

where $\tilde{f}_i(q)$ is the Bayesian distribution of model i and $p''(\tilde{\theta}_i)$ is the posterior probability of the model.

An application of Bayesian analysis that is of interest indicates how the descriptive statistics estimated from two samples can be combined to yield an updated estimate of minimum variance. It can be shown that the updated mean m is obtained from the mean of each set m_1 and m_2 weighted by the inverse of their variances $V[m_1]$ and $V[m_2]$ [70,71]:

$$m = \frac{V[m_2]\ m_1 + V[m_1]\ m_2}{V[m_1] + V[m_2]} \tag{43}$$

The updated variance of the mean is:

$$V[m] = \frac{V[m_1]\ V[m_2]}{V[m_1] + V[m_2]} \tag{44}$$

In general, the first data set will represent the prior knowledge about the variable while the second set is the new information that became available.

The Bayesian approach can also be adopted to update the failure rates estimated from experimental data [72-74].

5. CONCLUSIONS

This work has presented an overview of the state of the art concerning uncertainty modelling related with systems reliability analysis. It was shown that variability and uncertainty can both be modelled by random variables and thus they can be handled by the same computational techniques.

The basic methods of uncertainty analysis have been presented, including analytical approaches, simulation and response surface methods which represents a sort of compromise between the complexity both approaches.

Finally, Bayesian methods have been discussed, showing their possibilities for including prior information about variables and to deal with parameter and probabilistic model uncertainty.

6. ACKNOWLEDGEMENT

This work has been conducted in the scope of the research activities developed with the support of INIC, the National Institute for Scientific Research through CEMUL, the Centre for Mechanics and Materials of the Technical University of Lisbon.

7. REFERENCES

1. R.E. Barlow and H.E. Lambert (Eds), Reliability and Fault Tree Analysis, Society for Industrial and Applied Mathematics (SIAM), 1975.
2. E.J. Henley and H. Kumamoto, Reliability Engineering and Risk Assessment, Prentice-Hall, Englewood Cliffs, 1981.
3. P.S. Jackson, R.W. Hockenbury and M.L. Yeater, "Uncertainty Analysis of System Reliability and Availability Assessment", Nuclear Engineering and Design, Vol. 68,1981, pp. 5-29.
4. P. Kafka and H. Polke, "Treatment of Uncertainties in Reliability Models", Nuclear Engineering and Design, Vol. 93, 1986, pp. 203-214.
5. S. Kaplan and B.J. Garrick, "On the Quantitative Definition of Risk", Risk Analysis, Vol. 1, 1981, pp. 11-27.
6. G.W. Parry and P.W. Winter, Characterization and Evaluation of Uncertainty in Probabilistic Risk Analysis", Nuclear Safety, Vol. 22, 1981, pp. 28-42.
7. A.G. Colombo and P. Ihm, "A Quasi-independence Model to Estimate Failure Rates", Reliability Engineering and System Safety, Vol. 21, 1988, pp. 309-318.
8. I.A. Watson, "Human Factors in Reliability and Risk Assessment", Reliability Engineering, A. Amendola and A. Saiz de Bustamante, (Eds) Kluwer Acad. Pub., Dordrecht, 1988, pp. 257-300.
9. W. Zijp, "The Treatment of Measurement Uncertainties", Ispra Course on Data Uncertainties, Sensitivities, Consistency and Adjustment, Ispra, April 1986.
10. C. Guedes Soares, "Uncertainty Modelling in Plate Buckling", Structural Safety, Vol. 5, 1988, pp. 17-34.
11. N.R. Mann, R.E. Schafer and N.D. Singpurwalla, Methods for Statistical Analysis of Reliability and Life Data, J. Wiley & Sons, New York, 1974.
12. K.V. Bury, Statistical Models in Applied Science, J. Wiley & Sons, New York, 1975.
13. A.Z. Keller, "Estimation of Parameters of Distribution", Reliability Engineering, A. Amendola and A. Saiz de Bustamante (Eds), Kluwer Acad. Pub., Dordrecht, 1988, pp. 27-47.
14. G. Apostolakis and P. Moieni, "The Foundations of Models of Dependence in Probabilistic Safety Assessment", Reliability Engineering, Vol. 18, 1987, pp. 177-195.
15. Z.W. Birnbaum, J.D. Esary and A.W. Marshall, "A Stochastic Characterization of Wear-Out for Components and Systems", Annals of Mathematical Statistics, Vol. 37, 1967, pp. 816-825.
16. J.M. Kontoleon, "A General Approach for Determining Reliability Measures of Repairable Systems", Reliability Engineering, Vol. 2, 1981, pp. 57-63.
17. W.E. Vesely, "Minimal Cut Set Failure and Repair", Reliability Engineering, Vol. 5, 1983, pp.

65-72.
18. J.B. Fussel, "A Formal Methodology for Fault Tree Construction", Nuclear Science and Engineering, Vol. 52, 1973, pp. 421-432.
19. A. Poucet, "Fault Tree and Event Tree Techniques", Reliability Engineering, A. Amendola and A. Saiz de Bustamante (Eds), Kluwer Acad. Pub., Dordrecht, 1988, pp. 129-169.
20. E. Rosenblueth, "Probabilistic Design to Resist Earthquakes", J. Eng. Mech. Div., ASCE, Vol. 90, 1964.
21. J.A. Battjes, "Long-Term Wave Height Distribution at Seven Stations Around the British Isles", Deutsche Hydrographische Zeitschrift, Vol. 25, 1972, 179-189.
22. C. Guedes Soares, "Bayesian Prediction of Design Wave Heights", Reliability and Optimization of Structural Systems'88, P. Thoft-Christensen (Ed.), Springer-Verlag, 1989, pp. 311-323.
23. C. Guedes Soares and T. Moan, "On the Uncertainties Related to the Extreme Environmental Loading on a Cylindrical Pile", Reliability Theory and its Application in Structural and Soil Mechanics, P. Thoft-Christensen (Ed), Martinus Nijhoff Pub., The Hague, 1983, pp. 351-364.
24. A. Molesh, "Model Uncertainty and its Application in Probabilistic Risk Assessments", Proc. 8th Int. Conf. on Structural Mechanics in Reactor Technology, Brussels, 1985.
25. N. Siu and G. Apostolakis, "On the Quantification of Modelling Uncertainties", Proc. 8th Int Conf. on Structural Mechanics in Reactor Technology, Vol. M2, 1985, pp. 375-378.
26. C. Guedes Soares and T.H. Soreide, "Behaviour and Design of Stiffened Plates under Predominantly Compressive Loads", International Shipbuilding Progress, Vol. 30, 1983, pp. 13-27.
27. A.H-S. Ang and C.A. Cornell, "Reliability Bases of Structural Safety and Design", J. Structural Div. ASCE, Vol. 100, 1975, pp. 1755-1769.
28. O. Ditlevsen, "Model Uncertainty in Structural Reliability", Structural Safety, Vol. 1, 1982, pp 73-86.
29. D.C. Cox and P. Baybutt, "Methods for Uncertainty Analysis: A Comparative Survey", Risk Analysis, Vol. 1, 1981, pp. 251-258.
30. R.L. Iman and J.C. Helton, "An Investigation of Uncertainty and Sensitivity Analysis Techniques for Computer Models", Risk Analysis, Vol. 8, 1988, pp. 71-90.
31. A.G. Colombo, "Uncertainty in System Reliability Analysis", Reliability Modelling and Applications, A.G. Colombo and A.Z. Keller (Eds), Kluwer Acad. Pub., Dordrecht, 1987, pp. 71-81.
32. H. Rief, "Statistical Sensitivity and Uncertainty Analysis", Ispra Course on Data Uncertainties Sensitivities, Consistency and Adjustement, Ispra, April 1986.
33. D.J. Downing, R.H. Gadner and F.O. Hoffman, "An Examination of Response-Surface Methodologies for Uncertainty Analysis in Assessment Models", Technometrics, Vol. 27, 1985 pp. 151-163.
34. W.J. Conover and R.L. Iman, "On some Alternative Procedures Using Ranks for the Analysis of Experimental Designs", Communications in Statistics, Vol. A5, 1976, pp. 1349-1368.
35. R.L. Iman, J.C. Helton and J.E. Campbell, "An Approach to Sensitivity Analysis of Computer Models: Part I-Introduction, Input Variable Selection and Preliminary Variable Assessment", J. Quality Technology, Vol. 3, 1981, pp. 174-183.
36. R.L. Iman, J.C. Helton and J.E. Campbell, "An Approach to Sensitivity Analysis of Computer Models: Part II-Ranking of Input Variables, Response Surface Validation, Distribution Effect and Technique Synopsis", J. Quality Technology, Vol. 13, 1981, pp.232-240.
37. D.G. Cacuci, "Sensitivity Theory for Non-Linear Systems", J. Math. Phys., Vol. 22, 1981, pp. 2794-2812.
38. A.M. Freudenthal, J.N.Garretts and M. Shinozuka, The Analysis of Structural Safety, J. Struct. Div., ASCE, Vol. 92, 1966, pp. 235-246.
39. K. Breitung, "Asymptotic Approximations for Multinormal Integrals", J. Engrg Mech. Div. ASCE, Vol. 110, 1984, pp. 357-366.

40. M. Hohenbichler and R. Rackwitz, "Non-normal Dependent Vectors in Structural Reliability", J. Engrg Mech. Div., ASCE, Vol. 107, 1981, pp. 1227-1238.
41. S. Kaplan, "On the Method of Discrete Probability Distributions in Risk and Reliability Calculations - Application to Seismic Risk Assessment", Risk Analysis, Vol. 1, 1981, pp. 189-196.
42. A.G. Colombo and R.J. Jaarsma, "A Powerful Numerical Method to Combine Random Variables", IEEE Trans. on Reliability, Vol. R29, 1980.
43. C.A. Cornell, "Structural Safety Specifications based on Second Moment Reliability Analysis", Final Report, Symposium on Concepts of Safety of Structures and Methods of Design, IABSE, Zurich, 1969, pp. 235-246.
44. A.M. Hasofer and N.C. Lind, "An Exact and Invariant First-Order Reliability Format", J. Engrg Mech., ASCE Vol. 7, 1979, pp. 435-451.
45. O. Ditlevsen, "Generalised Second Moment Reliability Index", J. Structural Mechanics, Vol. 7, 1979, pp. 435 451.
46. O. Ditlevsen, "Narrow Reliability Bounds for Structural Systems", J. Structural Mechanics, Vol. 7, 1979, pp. 453-472.
47. M. Hohenbichler and R. Rackwitz, "First Order Concepts in System Reliability", Structural Safety, Vol. 1, 1983, pp. 127-188.
48. O. Ditlevsen and P. Bjerager, "Methods of Structural System Reliability", Structural Safety, Vol. 3, 1986, pp. 195-229.
49. P. Bjerager, A. Karamchandani and C.A. Cornell, "Fault-Tree Analysis in Structural System Reliability", Proc. 5th Int. Conf. on Applications of Statistics and Probability in Soil and Structural Engineering, ICASP 5, Vancouver, 1987.
50. Y. Murotsu et al, "Probabilistic Collapse Analysis of Offshore Structure", J. Offshore Mechanics and Artic Engineering, ASME, Vol. 109, 1987, pp. 270-277.
51. R. Rackwitz, "Reliability Analysis of Structural Components and Systems", Reliability Theory and its Application in Structural and Soil Mechanics, P. Thoft-Christensen (Ed.), Martinus Nijhoff Pub., 1983, pp. 173-214.
52. C. Guedes Soares, "Reliability of Marine Structures", Reliability Engineering, A. Amendola and A. Saiz de Bustamante (Eds), Kluwer Acad. Pub., Dordrecht, 1988, pp. 513-559.
53. C. Guedes Soares and L.D. Ivanov, "Time Dependent Reliability of the Primary Ship Structure", Reliability Engineering and System Safety, Vol 26 , 1989, pp. 59-71.
54. J. Ferry Borges, "Implementation of Probabilistic Safety Concepts in International Codes", Proc. 2nd Int. Conf. on Struct. Safety and Reliability of Engrg Structures, (ICOSSAR)", Munchen, 1977, pp. 121-133.
55. C. Guedes Soares and T. Moan, "Uncertainty Analysis and Code Calibration of the Primary Load Effects in Ship Structures", Structural Safety and Reliability, I. Konishi, A. H-S Ang and M. Shinozuka (Eds) ,IASSAR, 1985, Vol. III, pp. 501-512.
56. J. M. Hammersley and C. D. Hadscomb, Monte Carlo Methods, Methuen Pub, 1964.
57. R. Y. Rubinstein, Simulation and the Monte Carlo Method, John Wiley & Sons,1981.
58. M. Perlado,"Monte Carlo Simulation", Advanced Systems Reliability Simulation, A. G. Colombo and A. Saiz de Bustamante (Eds.), Kluwer Acad Pub, Dordrecht, 1989.
59. M. Shinozuka, "Basic Analysis of Structural Safety", J. Struct. Div. ASCE, Vol. 109, 1983, pp. 721-740.
60. A.C. Lucia, "Treatment of Uncertainties", Advanced Seminar on Structural Reliability, ISPRA, May 11-15 1987.
61. M.D. McKay, R.J. Beckman and W.J. Conover, "A Comparison of Three Methods for Selecting Values of Input Variables in the Analysis of Output from a Computer Code", Technometrics, Vol. 21, 1979, pp. 239-245.
62. G.I. Schueller, C.G. Bucher, U. Bourgund and W. Ouypornprasert, "On Efficient Computational Schemes to Calculate Structural Failure Probabilities", Stochastic Structural Mechanics, Y.K.

Lin and G.I. Schueller (Eds), Springer Verlag, 1987, pp. 388-407.
3. P.H. Meyers, Response Surface Methodology, Allyn and Bacon Inc., Boston, 1971.
4. L. Olivi (Ed.), Response Surface Methodology, CEC Joint Research Centre, Ispra 1984.
5. A. G. Colombo,"Reliability Parameter Estimation by Combination of Data from Various Sources" A. G. Colombo and A. Saiz de Bustamante (Eds.), Kluwer Acad. Pub, Dordrecht 1989.
6. C.A. Cornell, "Bayesian Statistical Decision Theory and Reliability-Based Design", Proc. Int. Conf. on Structural Safety and Reliability, A.M. Freudenthal (Ed.), Pergamon Press, 1972, pp. 47- 68.
7. J.R. Benjamin and C.A. Cornell, Probability, Statistics and Decision for Civil Engineers, McGraw-Hill Book Co., New York, 1970.
8. G.E.P. Box and G.C. Tiao, Bayesian Inference in Statistical Analysis, Addinson-Wesley, Reading, MA, 1973.
9. H. Martz and R.A. Waller, Bayesian Reliability Analysis, J. Wiley & Sons, 1982.
0. W.H. Tang and A.H-S. Ang, "Modelling Analysis and Updating of Uncertainties", ASCE Nat. Struct. Engrg. Meeting, 1973, Paper 2016.
1. W.H. Tang, "Bayesian Frequency Analysis", J. Hidraulics Div. ASCE, Vol. 106, 1980, pp. 1203-1218.
2. S. Kaplan, "On a Two-Stage Bayesian Procedure for Determinig Failure Rates from Experimental Data", IEEE Trans. on Power Apparatus and Systems, Vol. PAS-102, 1983, pp. 195-202.
3. A.G. Colombo and R.J. Jaarsma, "Bayesian Estimation of Constant Failure Rate and Unavailability", IEEE Trans. on Reliability, Vol. R-29, 1980, pp. 17-21.
4. A.G. Colombo, D. Constantini, R.J. Jaarsma, "Bayes Nonparametric Estimation of Time Dependent Failure Rate", IEEE Trans. on Reliability, Vol. R-34, 1985, pp. 109-112.

8. NOTATION

X_i - ith random variable from a sequence of independent and identically distributed variables
$\hat{X}$ - Measured value of X
$\overline{X}$ - Sample mean value of X
m_X - Estimated mean value of the population of X
$\hat{s}_X$ - Sample standard deviation of X
$\hat{s}^2_X$ - Sample variance of X
σ_X - Standard deviation of X
V_X - Coefficient of variation of X
$\hat{V}_X$ - Coefficient of variation of measured value of $\hat{X}$
N - Number of outcomes in a sample
E [X]- Expected value of X
V [X]- Variance of X
D [X]- Standard deviation of X
r_{ij} - Correlation coefficient between X_i and X_j
B - Uncertain correction factor
a, b - Regression coefficients
e - Normally distributed, zero mean random variable
$\hat{g}(\underline{X})$ - Prediction equation, function of $\underline{X}$
$g(\underline{X})$ - Corrected prediction equation
$f(\underline{Z})$ - Joint probability density function of variables $\underline{Z}$
$f_i(Z_i)$- Probability density function of variable Z_i

β - Reliability index
Q - Observed values of a random variable
I_o - Prior information
$\underline{\theta}$ - Vector of the parameters of a probability density function
$\bar{\underline{\theta}}$ - Bayesian updated vector of random parameters

SUBJECT INDEX

GPSR Compliance

The European Union's (EU) General Product Safety Regulation (GPSR) is a set of rules that requires consumer products to be safe and our obligations to ensure this.

If you have any concerns about our products, you can contact us on ProductSafety@springernature.com

In case Publisher is established outside the EU, the EU authorized representative is:

Springer Nature Customer Service Center GmbH
Europaplatz 3
69115 Heidelberg, Germany

Batch number: 10607222

Printed by Printforce, the Netherlands